Developments in Sustainable Chemical and Bioprocess Technology

Ravindra Pogaku · Awang Bono
Christopher Chu
Editors

Developments in Sustainable Chemical and Bioprocess Technology

Springer

Editors
Ravindra Pogaku
Awang Bono
Christopher Chu
School of Engineering and Information Technology
University Malaysia Sabah
Kota Kinabalu
Malaysia

ISBN 978-1-4614-6207-1 ISBN 978-1-4614-6208-8 (eBook)
DOI 10.1007/978-1-4614-6208-8
Springer New York Heidelberg Dordrecht London

Library of Congress Control Number: 2013950365

Printed on acid-free paper

Springer is part of Springer Science+Business Media (www.springer.com)

Preface

Sustainable Chemical and Bioprocess Technology is a field having numerous applications in the production of chemicals, fuels, materials, pharmaceuticals, and medicines at the genetic level. The field's novelty is in dealing with various topics of environmental engineering, chemical and bioprocesses, separation processes, advanced materials. The objective is to identify specific cases or environmental manipulations that result in improvements in yield and productivities of chemical and biotechnological processes.

To address cutting-edge topics in the key areas of chemical and bioprocess engineering, the content of the chapters of this book is divided into four main parts comprising of 47 chapters. Environmental Engineering (Chaps. 1–16); Chemical and Bioprocesses (Chaps. 17–29); Separation Processes (Chaps. 30–36); Advanced Materials (Chaps. 37–47).

The emphasis is laid on sustainability in chemical and bioprocess technology. The content related to process development, sustainability analysis in chemical and bioprocessing, industry case studies, supply case studies, and outlines of the latest technology of sustainability developments are reviewed. These topics include sustainability security and diversity of supply, costs and process economics, environmental and emissions legislation, technology development, public perceptions, market opportunities, energy conservation, and the role of alternative and renewable energy.

There is potential hub for chemical and biotechnology industries world over with its diverse flora and fauna. By introducing new and innovative methods and technologies in the chemical and bioprocessing of natural resources, one can expect to see a synergistic combination that promises greater sustainability.

Key Features

- Highlights the issues of sustainability development and technology related to environmental engineering, chemical and bioprocesses, separation processes, advanced materials
- Demonstrates engineering in action with numerous examples of chemical and bioprocesses

- Includes methods for identifying key enzymes in complex bioprocesses
- Contains a comprehensive review of chemical and biochemical engineering
- Discusses processes at the micro, macro, molecular, enzyme, and cell levels
- Explains concepts of stoichiometry, kinetics, and thermodynamics of separation processes
- Comprehensive overview of advanced materials
- Minimizes mathematical complexity
- Links the traditional processes to a sustainable environment models

Acknowledgments

We are grateful to Prof. Datuk Dr. Mohd. Harun Abdullah, the Vice Chancellor of University Malaysia Sabah and Associate Prof. Ir. Dr. Rosalam Hj. Sarbatly, the Dean of School of Engineering and Information Technology, University Malaysia Sabah, for their encouragement and extending constant support in realizing our dream to publish a book.

Special thanks are due to Dr. Abu Zahrim Yaser and Ms. Norfirdayu Ibrahim for their assistance in coordinating the book publication.

Last but not least, we thank dedicated reviewers of the chapters for their qualitative input: A. A. H. Kadhum, A. A. A. Raman, A. Z. Yaser, A. M. Shariff, A. N. Anuar, A. Ibrahim, B. H. Hameed, C. G. Joseph, D. Foo, C. S. Sipaut, D. Krishnaiah, H. A. Bakar, I. N. H. M. Amin, I. C. M. Leng, J. Janaun, M. N. A. Seman, M. A. Hussain, M. A. Ahmad, M. S. Kalil, M. S. Salit, M. K. A. Hamid, M. Sundang, N. Aziz, N. A. Hashim, O. A. Hamid, R. F. Mansa, R. Baini, S. M. Anisuzzaman, S. Abang, S. W. Lau, S. Kumaresan, S. A. Rahman, S. M. Saufi, T. Y. Wu, T. K. Wei, T. H. Jin, W. M. H. W. Yussof, Y. Samyudia.

Contents

Part IV Advanced Materials

Part I
Environmental Engineering

Coagulation/Flocculation of Anaerobically Treated Palm Oil Mill Effluent (AnPOME): A Review

A. Z. Yaser, B. Nurmin and S. Rosalam

Abstract Inefficiency of palm oil mill effluent (POME) treatment cause considerable environmental problems including from aesthetic point of view and inhibits the growth of the desirable aquatic biota necessary for surface water self-purification. Due to its low operating cost, anaerobic digestion of palm oil mill effluent (POME) is widely accepted by the managers, yet the colour of treated effluent (AnPOME) turned to dark brown. In this paper, the colourants in the AnPOME and its possible treatments have been reviewed with greater emphasis was put on coagulation/flocculation method. Application of anionic polymer as pre-treatment for integrated system has been shown to be necessary for sustainable AnPOME treatment including its sludge disposal.

Keywords Palm oil mill effluent · Anaerobic digestion · Decolourisation

Introduction

Palm oil mill industry is among major agricultural industry in Malaysia. However, the palm oil mills generate large amounts of highly polluting effluent called palm oil mill effluent (POME). Palm oil mill effluent (POME) is rich in carbohydrates, proteins, nitrogenous compounds, lipids, minerals, cellulose, hemicelluloses and lignin (Hii et al. 2012). Currently, anaerobic digestion of POME is widely accepted by the managers due to its low operating cost. Although there is

A. Z. Yaser (✉) · S. Rosalam
Chemical Engineering Programme, School of Engineering and Information Technology, Universiti Malaysia Sabah, Jalan UMS, 88400 Kota Kinabalu, Sabah, Malaysia
e-mail: zahrim@ums.edu.my

B. Nurmin
Civil Engineering Programme, School of Engineering and Information Technology, Universiti Malaysia Sabah, Jalan UMS, 88400 Kota Kinabalu, Sabah, Malaysia

R. Pogaku et al. (eds.), *Developments in Sustainable Chemical and Bioprocess Technology*, DOI: 10.1007/978-1-4614-6208-8_1,

significant organic matter reduction during anaerobic digestion treatment, the colour of effluent (AnPOME) turned to dark brown (Fig. 1) (Zahrim et al. 2012) and it is also contained bioflocs, anaerobic microorganisms and macrofibrils (Ho and Tan 1989). These days, due to public's increased consciousness, the appearance from the AnPOME becomes a great concern. Several studies [e.g. (Jakobsen et al. 2007; Fathahi 2010)] reported the occurrence of water pollution which is caused by improper treatment of palm oil mill effluent (POME). The colour of the effluent might be contributed by the residual lignin (Poh et al. 2010), tannin, humic acid- and fulvic acid-like substances (Edem 2002; Kongnoo et al. 2012) as well as anaerobic fermentation by product, e.g., melanoidin (Zahrim et al. 2009; Bunrung et al. 2011). The objective of this review is to discuss the presence of colourant in AnPOME and its possible treatment by coagulation/flocculation.

Anaerobic Digestion and Colourants

Generally, anaerobic digestion mechanism consists of several "stages": (1) hydrolysis, (2) fermentation (acidogenesis) and (3) methanogenesis (Metcalf and Eddy 2004). During hydrolysis, two mechanisms could happen: (1) the particulate material [i.e. plant cell debris and less than 50 % of total pollutant level (Wu et al. 2010)] is converted to the soluble compound (Metcalf and Eddy 2004) and/or (2) the complex material (carbohydrate, lipids and protein) is converted to simple compounds (sugar, amino acids, etc.) (Poh and Chong 2009).

Lignin, a main plant component, is a heterogeneous aromatic polymer interspersed with hemicellulose and occurs surrounding microfibrils. Lignin (density: 1.3–1.4 g cm^{-3} and brown in colour) contains P-hydroxy-phenyl, syringyl and guaiacyl units (Mohan and Karthikeyan 1997). The lignin content in POME is

Fig. 1 **a** Raw POME **b** Anaerobically treated POME

around 1700–7890 mg/L (Poh et al. 2010; Hii et al. 2012). During fermentation stage, the amino acids, sugars and fatty acids are degraded to several compounds, i.e., lactate, propionate, acetate, formate. (Metcalf and Eddy 2004). Besides that, natural condensation between sugars (carbonyl groups) and amino acids or proteins (free amino groups) through Maillard reaction could produce another colourant, i.e., melanoidin (Chandra et al. 2008). It has been reported that the wastewater from distilleries and fermentation industries also contain melanoidins (Chandra et al. 2008). Due to its structural complexity, dark colour and offensive odour, it poses serious threat to soil and aquatic ecosystem (Chandra et al. 2008). Presently, there is no report on Maillard reaction products in AnPOME. Finally, during anaerobic digestion, the methanogenic substrates are converted to methane and carbon dioxide (Metcalf and Eddy 2004). Other important colourant that could present in AnPOME might be tannin (Edem 2002). Tannins are complex dark-coloured non-crystalline substances composed of polyhydroxy phenolic (aromatic hydroxyl) compounds, related to catechol, glycosides or pyrogellol, which vary in composition. Tannins extracted from wood, bark and leaves are used extensively in the preservation of animal skins (Mohan and Karthikeyan 1997).

Coagulation/Flocculation for AnPOME Treatment

Over the past years, several studies have been dedicated for treatments of AnPOME. Ho and Tan (1988) carried out AnPOME treatment via combination of dissolved air flotation (DAF)-pressurised activated sludge. Decolourisation of AnPOME by means of conventional activated sludge (CAS) and activated sludge-granular activated carbon (ASGAC) sequencing batch operation was reported by Zahrim et al. (2009). During the period of treatment, colour removal for the CAS and ASGAC was about 7 and 35 %, respectively. An attempt to treat biologically treated palm oil mill effluent using water hyacinth (*Eichhornia crassipes*) was reported by Yeoh (1993). With a hydraulic retention time (HRT) of 5 days, the decolourisation in this system is not significant due to the presence of recalcitrant organic such as lignin. Wah et al. (2002) investigated ultrafiltration (UF) of AnPOME with several pre-treatments, i.e., sand filtration, centrifugation and coagulation. Combination of filtration–UF treatment gave the best overall treatment efficiency, with an overall reduction of 93 % for total nitrogen, suspended solids, turbidity and colour content. Chemical oxidation by using hydrogen peroxide with iron catalyst (Fenton process) for AnPOME treatment was studied by Aris et al. (2008). In this study, they found that the COD and colour removal were around 75–82 % and 92–95 %, respectively. The treated water will end up in acidic condition which is not suitable to be discharged in surface water. Integration Fenton process with biofilm attached on granular activated carbon has been shown effectively improved the decolourisation performance (Rakmi et al. 2004). Adsorption of AnPOME using palm fibre ash as an adsorbent was investigated by Bunrung et al. (2011).

Coagulation of coloured effluents has been used for many years, either as main or pre-treatment, due to its low capital cost. Coagulation is the destabilisation of coloured effluent using coagulant(s), which can be classified into two main categories, i.e., metal coagulants and polymers. Coagulation tends to overcome the factors that promote coloured effluent stability and form agglomerates or flocs. Flocculation in other words is the process of whereby destabilised particles or particles formed as a consequence of destabilisation are induced to come together, make contact and thereby form large(r) agglomerates (Bratby 2006). Advancement in polymer synthesising and purification resulted in the development of vast types of polymers. The best polymer should be selected to ensure the highest performance of coagulation as well as to reduce the chemical cost (Zahrim et al. 2011). Ho and Tan (1989) studied AnPOME treatment using coagulation (aluminium sulphate)/flocculation (cationic polyacrylamide) (CF), DAF and CF-DAF which is believed to be beneficial in reducing treatment time and area. Despite the fact that both methods were able to achieve a 97 % removal of the suspended solids of the digested liquor, the removal of soluble solid is very difficult. The authors stated that the total solid removal for CF, DAF and CF-DAF is 56, 59 and 63 % respectively (Ho and Tan 1989). The effects of synthetic non-ionic and cationic polymers, chitosan and chemical (alum) coagulant on the removal of lignin were investigated by Ganjidoust et al. (1997). It has been found that a non-ionic polymer had poor effect as compared to cationic polymers. Cationic polymers resulted in removal of about 80 % of colour and 30 % of TOC. The percentage colour removed by alum was the same as the one removed by cationic polymers, but the TOC removed was higher by alum (about 40 %). The natural coagulant, chitosan, resulted in the highest removal in both colour and TOC as compared to both synthetic polymers and chemical coagulant. Up to 90 % of colour and 70 % of TOC were observed to be removed by chitosan (Ganjidoust et al. 1997). A study on coagulation/flocculation process for the treatment of molasses wastewater has been studied by Liang et al. (2009). Experimental results indicate that ferric chloride was the most effective among the conventional coagulants, achieving 89 % COD and 98 % colour eliminations; while aluminium sulphate was the least effective, giving COD and colour reductions of 66 and 86 %, respectively. In addition, cationic polymer was shown to be more effective than anionic polymer in enhancing settleability of flocs formed (Liang et al. 2009).

Compared to cationic polymers, anionic polymers were reported to be more biodegradable (Zahrim et al. 2011; Weston et al. 2009). This characteristic is important for converting generated sludge (due to coagulation/flocculation process) into soil conditioner (Robinson et al. 2001; Yaser et al. 2007). Recently, Jami et al. (2012) applied anionic polymer as coagulant aids and compare the use of coagulants ferric chloride and aluminium sulphate to reduce turbidity. The result of the coagulation process showed that ferric chloride gave a better reduction in turbidity at dosage of 100 mg/L, pH of 8 and with polymer dose of 100 mg/L than alum. Previously, Poon and Chu (1999) studied coagulation/flocculation of sewage sludge using anionic polymer. They found that the addition of 30 mg/L of $FeCl_3$ and 0.5 mg/L polymer could provide a reduction in SS, total nitrogen (N) and total phosphorous (P) higher than 80, 70 and 40 %, respectively. Study on tannery wastewater coagulation/flocculation was conducted by Haydar and Aziz (2009). Alum was used as coagulant while anionic polymers as coagulant aid. The combination of alum with anionic polymer resulted in effluent turbidity removal of 99.7 %, TSS removal of 96.3 %, COD removal of 48.3 % and chromium removal of 99.7 %. It was reported in this study that the application of polymer has reduced sludge volume by 60–70 % and cost of chemicals by 50 % for comparable removal efficiencies. Synthetic dye solution treatment has been reported by Mishra and Bajpai (2005) and the authors found that the anionic food-grade polymer is able to reduce 35–71 % colour.

Conclusions

Due to recalcitrance nature of POME, the treated effluent from anaerobic treatment system is always in dark brownish colour. The colour of the effluent is due to the presence of residual lignin, tannin, humic acid- and fulvic acid-like substance. Natural condensation between sugars (carbonyl groups) and amino acids or proteins (free amino groups) through Maillard reaction could produce another type of colourant, i.e., melanoidin. Over the years, several treatment methods such as biological, advanced oxidation, membrane separation, adsorption and coagulation/flocculation have been studied. None of the stand-alone treatments could fully decolourised AnPOME, and hence, integrated system is required. Coagulation/flocculation could be applied as pre-treatment. Since many of the anionic polymers are reported to be biodegradable, application of anionic polymer has potential for sustainable AnPOME flocculant and further study are required.

Acknowledgments The authors would like to thank the Universiti Malaysia Sabah for funding this project under grant SLB0041-TK-2012 and PHD0007-TK-2012. Thanks to Izreen, Nasimah & Hillery for the photos.

References

Aris, A., Siew, O. B., See, K. S., & Ujang, Z. (2008). Tertiary treatment of palm oil mill effluent using fenton oxidation. *Malaysia Journal of Civil Engineering, 20*, 12–25.

Bratby, J. (2006). *Coagulation and flocculation in water and wastewater treatment*. London: IWA Publishing.

Bunrung, S., Prasertsan, S., & Prasertsan, P. (2011). Decolourisation of Biogas Effluent of Palm Oil Mill using Palm Ash. *TIChE International Conference* 2011. Hatyai, Songkhla Thailand.

Chandra, R., Bharagava, R. N., & Rai, V. (2008). Melanoidins as major colourant in sugarcane molasses based distillery effluent and its degradation. *Bioresource Technology, 99*, 4648–4660.

Edem, D. O. (2002). Palm oil: Biochemical, physiological, nutritional, hematological and toxicological aspects: A review. *Plant Foods for Human Nutrition (Formerly Qualitas Plantarum), 57*, 319–341.

Fathahi, T. K. T. (2010). Water quality and sources of pollution of the Sg Kinabatangan basin. *Seminar and Workshop POMET³* Sabah, Malaysia.

Ganjidoust, H., Tatsumi, K., Yamagishi, T., & Gholian, R. N. (1997). Effect of synthetic and natural coagulant on lignin removal from pulp and paper wastewater. *Water Science and Technology, 35*, 291–296.

Haydar, S., & Aziz, J. A. (2009). Coagulation-flocculation studies of tannery wastewater using combination of alum with cationic and anionic polymers. *Journal of Hazardous Materials, 168*, 1035–1040.

Hii, K.-L., Yeap, S.-P., & Mashitah, M. D. (2012). Cellulase production from palm oil mill effluent in Malaysia: Economical and technical perspectives. *Engineering in Life Sciences, 12*, 7–28.

Ho, C. C., & Tan, Y. K. (1988). The treatment of anaerobically digested palm oil mill effluent by pressurised activated sludge. *Journal of Chemistry Technology Biotechnology, 41*, 75–84.

Ho, C. C., & Tan, Y. K. (1989). Comparison of chemical flocculation and dissolved air flotation of anaerobically treated palm oil mill effluent. *Water Research, 23*, 395–400.

Jakobsen, F., Hartstein, N., Frachisse, J., & Golingi, T. (2007). Sabah shoreline management plan (Borneo, Malaysia): Ecosystems and pollution. *Ocean and Coastal Management, 50*, 84–102.

Jami, M. S., Muyibi, S. A., & Oseni, M. I. (2012). Comparative study of the use of coagulants in biologically treated palm oil mill effluent (POME). *Advances in Natural and Applied Sciences, 6*, 646–650.

Kongnoo, A., Suksaroj, T., Intharapat, P., Promtong, T., & Suksaroj, C. (2012). Decolorization and organic removal from palm oil mill effluent by Fenton's process. *Environmental Engineering Science, 29*, 855–859.

Liang, Z., Wang, Y., Zhou, Y., Liu, H., & Wu, Z. (2009). Variables affecting melanoidins removal from molasses wastewater by coagulation/flocculation. *Separation and Purification Technology, 68*, 382–389.

Metcalf & Eddy (2004) *Wastewater engineering—Treatment and reuse.*, New York:McGraw-Hill Companies.

Mishra, A., & Bajpai, M. (2005). Flocculation behaviour of model textile wastewater treated with a food grade polysaccharide. *Journal of Hazardous Materials, 118*, 213–217.

Mohan, S. V., & Karthikeyan, J. (1997). Removal of lignin and tannin colour from aqueous solution by adsorption onto activated charcoal. *Environmental Pollution, 97*, 183–187.

Poh, P. E., & Chong, M. F. (2009). Development of anaerobic digestion methods for palm oil mill effluent (POME) treatment. *Bioresource Technology, 100*, 1–9.

Poh, P. E., Yong, W.-J., & Chong, M. F. (2010). Palm Oil Mill Effluent (POME) characteristic in high crop season and the applicability of high-rate anaerobic bioreactors for the treatment of POME. *Industrial and Engineering Chemistry Research, 49*, 11732–11740.

Poon, C. S., & Chu, C. W. (1999). The use of ferric chloride and anionic polymer in the chemically assisted primary sedimentation process. *Chemosphere, 39*, 1573–1582.

Rakmi, A., Norzaini, M., Abu-Zahrim, Y., & Zalina, H. (2004) Chemibiological treatment of difficult wastewater: Case study on tanning wastewater. In Z. Ujang, & M. Henze, (Eds.) *Environmental Biotechnology: Advancement in Water and wastewater applications in the tropics.* London: IWA Publishing.

Robinson, T., McMullan, G., Marchant, R., & Nigam, P. (2001). Remediation of dyes in textile effluent: a critical review on current treatment technologies with a proposed alternative. *Bioresource Technology, 77*, 247–255.

Wah, W. P., Sulaiman, N. M. N., Nachiappan, M., & Varadaraj, B. (2002). Pre-treatment and membrane ultrafiltration using treated palm oil mill effluent. *Songklanakarin Journal of Science Technology, 24*, 891–898.

Weston, D. P., Lentz, R. D., Cahn, M. D., Ogle, R. S., Rothert, A. K., & Lydy, M. J. (2009). Toxicity of anionic polyacrylamide formulations when used for erosion control in agriculture. *Journal of Environmental Quality, 38*, 238–247.

Wu, T. Y., Mohammad, A. W., Jahim, J. M., & Anuar, N. (2010). Pollution control technologies for the treatment of palm oil mill effluent (POME) through end-of-pipe processes. *Journal of Environmental Management, 91*, 1467–1490.

Yaser, A. Z., Abd Rahman, R., & Kalil, M. S. (2007). Co-composting of palm oil mill sludge-sawdust. *Pakistan Journal of Biological Science, 10*, 4473–4478.

Yeoh, B. G. (1993). Use of water hyacinth (Eichhornia crassipes) in upgrading in small agroindustrial wastewater treatment plants. *Water Science and Technology, 28*, 207–213.

Zahrim, A. Y., Fansuri, M. B., Nurmin, B., & Rosalam, S. (2012). A review on the decolourisation of anaerobically treated palm oil mill effluent (AnPOME). In A. Bono, & C. S. Sipaut, (Eds.), *Proceedings of 26th Symposium of Malaysian Chemical Engineers.* Kota Kinabalu, Sabah, Malaysia.

Zahrim, A. Y., Rachel, F. M., Menaka, S., Su, S. Y., Melvin, F., & Chan, E. S. (2009). Decolourisation of anaerobic Palm Oil Mill effluent via activated sludge-granular activated carbon. *World Applied Sciences Journal, 5*, 126–129.

Zahrim, A. Y., Tizaoui, C., & Hilal, N. (2011). Coagulation with polymers for nanofiltration pre-treatment of highly concentrated dyes: A review. *Desalination, 266*, 1–16.

Dewaterability of Anaerobic Digested Sludge with Cations and Chitosan as Dual Conditioners

S. W. Lau, S. H. Chong, H. M. Ang, T. K. Sen and H. B. Chua

Abstract Anaerobic digested sludge dewatering is very important in sludge management and disposal. In this study, flocculation and dewatering behaviour of anaerobic digested sludge were investigated using metal cations (sodium, magnesium, calcium, ferric and aluminium salts) and chitosan as dual conditioners. The trivalent ferric and aluminium ions were found to be more effective than the monovalent and divalent cations in improving sludge dewaterability as measured by capillary suction time (CST). Post-addition of chitosan into the sludge pre-coagulated with metal cations further enhanced the sludge dewaterability. This study suggests that pre-destabilisation of sludge by coagulation with metal cations followed by chitosan addition as polymeric flocculant may improve the dewaterability of anaerobic digested sludge.

Introduction

Sewage sludge production has increased dramatically during the last decades due to increased urban and industrial water purification activities (Saveyn et al. 2005). It is commonly known that sludge handling is an expensive process which may represent up to 40 % of the capital cost and 50 % of the operation cost of a wastewater treatment plant (Spellman 2003). The high water content and the colloidal and compressible nature of sludge make its handling and processing

S. W. Lau (✉) · H. B. Chua
Department of Chemical Engineering, School of Engineering and Science, Curtin University Sarawak Malaysia, CDT 250, 98009 Miri, Sarawak, Malaysia
e-mail: lau.shiew.wei@curtin.edu.my

S. H. Chong · H. M. Ang · T. K. Sen
Department of Chemical Engineering, Faculty of Science and Engineering, School of Chemical and Petroleum Engineering, Curtin University, Gpo Box U1987, Perth, WA 6845, Australia
e-mail: m.ang@curtin.edu.au

R. Pogaku et al. (eds.), *Developments in Sustainable Chemical and Bioprocess Technology*, DOI: 10.1007/978-1-4614-6208-8_2,

extremely costly in terms of both money and time. Sludge dewatering and anaerobic digestion are two widely used operations in wastewater treatment in recent years. Dewatering is essential to reduce the volume of sludge and is still a bottleneck for sludge treatment. Sewage sludge may be pre-treated or conditioned by biological, chemical and/or physical treatment to enhance dewaterability. Until recently, there has been little published information on dewaterability of digested sludge and the influencing factors of sludge dewaterability. Thermal pre-treatment has been used for sludge dewatering, but it is less popular nowadays due to its high capital cost and a number of operational problems such as odour control and scale formation (Turovskiy and Mathai 2006). Metal salts also play an important role in sludge flocculation, settling and dewatering (Pevere et al. 2007).

With the advances in development of the polymer industry, polymer conditioning has now become the most extensively used, economical and operable practice of all the pre-treatment methods for improved dewatering process (Fu et al. 2009). Cationic polymers are the most widely used polymers for sludge since most sludge solids carry negative charges due to the presence of extracellular polymeric substances (EPS) (Chitikela and Dentel 1998; Mikkelsen and Keiding 2002). Recently, there has been considerable interest in using alternative substances such as surfactants, acids, enzymes and natural polymers to replace or to supplement polymers in sludge conditioning. This is due to high polymer costs, high moisture content in the resulting sludge cake, and concern over biodegradation and toxicity of polymers since the maximum allowable concentration is 1 ppm in potable water (Ayol 2005; Chen et al. 2001; Chitikela and Dentel 1998; Fu et al. 2009; Rose and John 2010).

Although there are many reported works on the thermal pre-treatment, polymer dosing and role of metal salts on dewaterability of sludge, little research has covered the use of the mono-, di- and trivalent metal salts in anaerobic digested sludge dewatering. Moreover, the effect of dual conditioners such as the combination of metal cations with biopolymers on the sludge dewaterability is addressed here for the first time. It is commonly known that the sludge particle has a net negative charge and forms a double layer with the counter-ion charge in the solution. The double layer results in an osmotic repulsion of adjacent particles and keeps a relatively stable system. Consequently, the double layer inhibits aggregation and causes poor settling and dewatering properties. After the addition of metal cations into the sludge system, the negative charge of sludge is expected to be neutralised by the positive charge carried by the metal cations on the surface of sludge. It thus appears plausible that this type of conditioner could supplement the more expensive polymer and neutralise a portion of the required charge. Presumably, some cationic organic polymer would still be needed to bridge the particles and make the sludge sufficiently dewaterable. The natural biopolymer, chitosan, is of considerable interest since it is nontoxic and environmentally friendly. The objective of this research is to investigate the relationship between various metal cations and chitosan, together with their dual-conditioning mechanism in charge neutralisation as well as bridging for enhanced dewaterability of anaerobic digested sludge.

Methodology

Materials

The anaerobic digested sludge used in the current study was obtained from Beenyup Wastewater Treatment Plant of Water Corporation in Perth. Sludge samples were stored at 4 °C and brought to room temperature before use. The commercial polymer, EMA 8845 MBL (SNF, Australia) used in this study, was also a generous gift from Beenyup Wastewater Treatment Plant. It was supplied as a cationic water-soluble polymer after dilution to an average concentration of 0.65 % w/v. The natural biopolymers, low molecular weight chitosan (50,000–190,000 Da) and medium molecular weight chitosan (190,000–310,000 Da) were purchased from Sigma-Aldrich and prepared as 0.5 % w/v stock solutions at pH 2. Cations used in this study were from the laboratory-grade metal salts and prepared as 1.0 % w/v stock solutions.

Analytical Methods and Characterisation of Sludge

The main characteristics of the anaerobic digested sludge were analysed using several methods. Total solids (TS) content and specific gravity (SG) were determined according to standard methods (APHA 2005). Capillary suction time (CST) test was conducted using a CST apparatus from Triton Electronics Ltd., UK. Total dissolved solids (TDS), electrical conductivity (EC), pH and temperature were measured with a bench meter. The zeta potentials of the sludge supernatant were measured with a Malvern Zetasizer Nano ZS (Malvern Instruments Ltd., UK), while particle size analysis was conducted using a Malvern Mastersizer 2000 (Malvern Instruments Ltd., UK). Image analysis was also performed on the sludge samples using a Nikon SMZ800 stereomicroscope. The properties of the sludge used in this study are as follows: pH 6.9–7.1; SG 1.000–1.005; TS 1.4–1.6 %; TDS 0.28–0.29 %; EC 5–6 mS/cm; zeta potential −12 to −13 mV; particle size range 2.5–1,500 μm; mean particle size 100 μm; and CST 140–160 s.

Conditioning and Dewatering Experiments

Sludge samples (100 mL) were conditioned with either chitosan or cation. The experiment was repeated with a number of different conditioner dosages, in a 250-mL beaker. The mixture was rapidly stirred at 500 rpm for 1 min, followed by a slow mixing at 200 rpm for 5 min. The CST test was conducted on the conditioned sludge. 15 mL of the conditioned sludge sample was centrifuged at 2,500 rpm for

15 min, and the volume of centrate collected was used to determine the percentage of water removal from the total sample volume.

For the study of dual conditioning, the sludge was pre-conditioned with cation as described, followed by addition of chitosan with a similar mixing procedure. The CST test was again conducted on the dual-conditioned sludge, and the percentage of CST reduction was calculated by comparison with the CST of the unconditioned sludge. All experiments were conducted at room temperature (24 ± 1 °C).

Results and Discussion

Effect of Polymer Type and Dosage on Sludge Flocculation and Dewatering

Figure 1a illustrates the effect of polymer type and dosage on the CST. With increasing polymer dosage, the CST decreased rapidly and eventually reached a plateau for both the low molecular weight chitosan and medium molecular weight chitosan. Medium molecular weight chitosan was more effective than low molecular weight chitosan in enhancing the sludge dewaterability, but the percentage of water removal was within 3 % variation (Fig. 1b). EMA 8845 showed a gradual decrease in the CST in response to increasing polymer dosage and reached a similar CST as the medium molecular weight chitosan at 19.2 g/kg. Overdosing EMA 8845 (>20 g/kg) led to super-flocculation causing the CST readings that were less meaningful. After overdosing, it was possible to pour only liquor in the CST funnel, rather than sludge solids (CST Equipment Manual 1998).

It is well known that a higher molecular weight polymer may allow its cationic sites to bind to more particles, thus bridging the solid particles in the suspension system (Dentel 2010). Larger flocs are then formed and will be easier to separate from the treated suspension. Low to medium molecular weight chitosan may be more efficient for fine sludge particles at even very low dosages since the polymeric flocculants diffuse more readily to bond the particles forming small flocs which facilitate the sludge dewatering. In a similar observation, Hogg (1999) explained that an enhanced preference for adsorption on larger particles and the smaller number of molecules available made high molecular weight polymers especially ineffective for stable dispersions of very fine particles. Though very large and resilient flocs were formed by EMA 8845 at higher dosages, Fig. 1b shows that the percentage of water removal after centrifugation recorded a minimum at 19.2 g/kg. This suggests that there is a significant amount of water trapped in the flocs and the applied centrifugal force could not separate even the free water out of the flocs. However, at the dosage of 32.0 g/kg, the percentage of water removal is also high. This is probably due to the less resilient super-flocs formed as a result of overdosing.

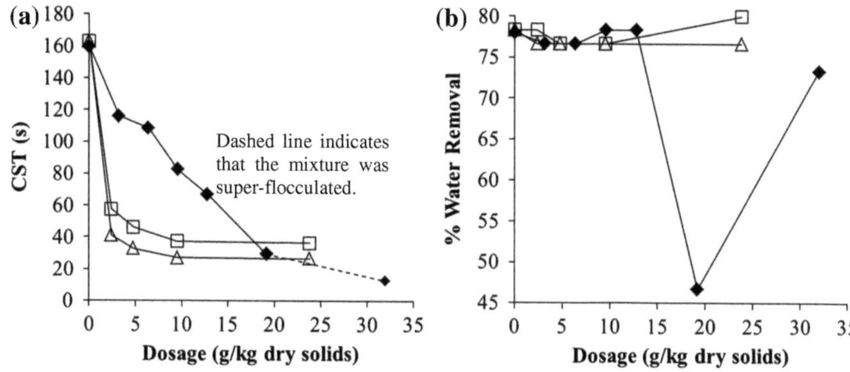

Fig. 1 Effect of polymer type and dosage on sludge dewatering performance measured by **a** CST and **b** % water removal [(*filled diamond*) EMA 8845, (*empty square*) low MW chitosan, (*empty triangle*) medium MW chitosan]

Effect of Cation Type on the Dual Conditioning of Sludge with Chitosan

As shown in Fig. 2, the trivalent ferric and aluminium ions were more effective than the monovalent and divalent cations in improving the sludge dewaterability in terms of CST reduction. When conditioned with ferric and aluminium salts alone, the CST values reduced from 142.7 s for an unconditioned sludge to 83.7 and 80.0 s, respectively, corresponding to 41.3 % and 43.9 % reduction. The increasing order of sludge dewaterability for each metal cation was noted as follows: sodium < calcium < magnesium < ferric < aluminium. According to Sanin et al. (2006), potassium (and other monovalent ions) does not improve the floc structure as much as the divalent cations such as calcium and magnesium. This may be explained by their inability to form linkages between the EPS within the sludge matrix and floc surfaces due to their monovalent nature. In another dual-conditioning study with a different suspension system, Ozkan and Yekeler (2004) found that magnesium was more effective on the coagulation of celestite suspension than calcium at neutral pH because magnesium has a higher ionic potential (ionic charge/radius). This is in line with the findings of this research but is contrary to the findings of Sanin et al. (2006) where magnesium behaviour was found to be intermediate between that of potassium and calcium ions. Aluminium and ferric salts are widely used as coagulants in wastewater treatment. They form cationic hydrolysis products that are strongly adsorbed on negatively charged particles and can give effective destabilisation (Renault et al. 2009). Cations may have significant effects on dewaterability by changing the amount of sludge bound water, due to neutralisation of the negative surface charges and bridging of the floc components, leading to build up of a floc network. However, this work found that

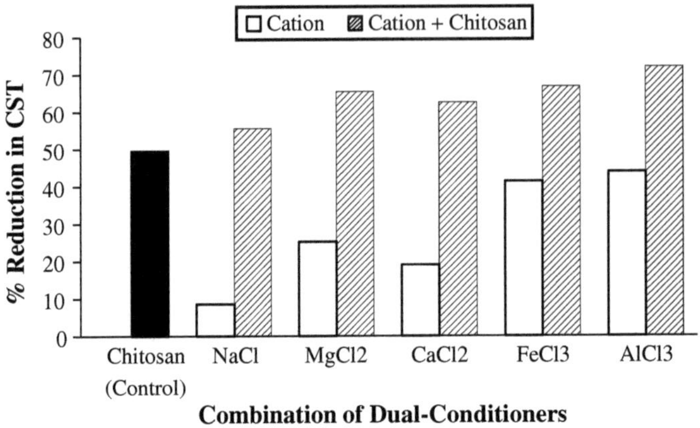

Fig. 2 Reduction in CST at various combinations of dual conditioners (20 g/kg cation +2.5 g/kg chitosan)

none of the metal cations used alone can reduce the CST as much as low molecular weight chitosan used alone.

In order to examine the dual-conditioning effect on the sludge dewaterability, low molecular weight chitosan was selected to feature the charge interactions as proposed by Hogg (1999) where charge interactions are believed to be more important for low molecular weight than for high molecular weight polymers. It can be seen from Fig. 2 that dual conditioning of sludge using metal cations and chitosan improved sludge dewaterability variably depending on the cation type. The combination of aluminium ion and chitosan optimised the sludge dewaterability, leading to a 72 % CST reduction. Microscopic observation indicated that the sludge dual-conditioned with aluminium ion and chitosan forms larger and denser flocs with less trapped water compared to sludge conditioned with chitosan alone.

Conclusions

Ferric and aluminium ions were more effective than the monovalent and divalent cations in improving the sludge dewaterability in terms of CST reduction. They formed cationic hydrolysis products that were strongly adsorbed on the negatively charged particles and bridged the floc components. Post-addition of chitosan further enhanced the dewaterability of the sludge by polymer bridging permitting floc growth to large sizes. This study suggests that the dual-conditioning mechanism works by pre-destabilisation of sludge through coagulation with cations, followed by flocculation induced by chitosan addition as polymeric flocculant. The addition of metal cations can reduce the polymer demand for sludge conditioning which is an advantage in urban and industrial water purification treatment.

Acknowledgments The authors would like to acknowledge Physics Department at Curtin Perth for the image analysis. Dominique Dodge-Wan proofreads this chapter.

References

APHA (2005). *Standard methods for the examination of water and wastewater.* 21st edn. Washington, DC, American Public Health Association, American Water Works Association, and Water Environment Federation.

Ayol, A. (2005). Enzymatic treatment effects on dewaterability of anaerobically digested biosolids-I: Performance evaluations. *Process Biochemistry, 40*(7), 2427–2434.

Chen, Y. G., Yang, H. Z., & Gu, G. W. (2001). Effect of acid and surfactant treatment on activated sludge dewatering and settling. *Water Research, 35*(11), 2615–2620.

Chitikela, S., & Dentel, S. K. (1998). Dual-chemical conditioning and dewatering of anaerobically digested biosolids: Laboratory evaluations. *Water Environment Research, 70*(5), 1062–1069.

CST Equipment Manual (1998). Essex: Triton Electronics Ltd.

Dentel, S. K. (2010). Chemical conditioning for solid-liquid separation processes. *Drying Technology, 28*(7), 843–849.

Fu, J., Jiang, B., & Cai, W. (2009). Effect of synthetic cationic surfactants on dewaterability and settleability of activated sludge. *International Journal of Environment and Pollution, 37*(1), 113–131.

Hogg, R. (1999). The role of polymer adsorption kinetics in flocculation. *Colloids and Surfaces A: Physicochemical and Engineering Aspects, 146*, 253–263.

Mikkelsen, L. H., & Keiding, K. (2002). Physico-chemical characteristics of full scale sewage sludges with implications to dewatering. *Water Research, 36*, 2451–2462.

Ozkan, A., & Yekeler, M. (2004). Coagulation and flocculation characteristics of celestite with different inorganic salts and polymers. *Chemical Engineering and Processing: Process Intensification, 43*(7), 873–879.

Pevere, A., Guibaud, G., van Hullebusch, E. D., Boughzala, W., & Lens, P. N. L. (2007). Effect of Na^+ and Ca^{2+} on the aggregation properties of sieved anaerobic granular sludge. *Colloids and Surfaces A: Physicochemical and Engineering Aspects, 306*, 142–149.

Renault, F., Sancey, B., Badot, P. M., & Crini, G. (2009). Chitosan for coagulation/flocculation processes—an eco-friendly approach. *European Polymer Journal, 45*(5), 1337–1348.

Rose, G. R. & ST. John, M. R. (2010). Encyclopedia of polymer science and technology. (2nd edn., Vol 7. pp. 211–233). Nalco Chemical Company.

Sanin, F. D., Vatansever, A., Turtin, I., Kara, F. & Durmaz, B. (2006). Operational conditions of activated sludge: Influence on flocculation and dewaterability. Drying Technology *24*(10), 1297–1306.

Saveyn, H., Meersseman, S., Thas, O., & van der Meeren, P. (2005). Influence of polyelectrolyte characteristics on pressure-driven activated sludge dewatering. *Colloids and Surfaces A: Physicochemical and Engineering Aspects, 262*, 40–51.

Spellman, F. R. (2003). *Handbook of water and wastewater treatment plant operations*. London: Lewis Publishers.

Turovskiy, I. S. & Mathai, P. K. (2006). *Wastewater sludge processing*. New Jersey: Wiley.

Optimization Study on Bioethanol Production from the Fermentation of Oil Palm Trunk Sap as Agricultural Waste

A. H. Norhazimah and C. K. M. Faizal

Abstract The utilization of oil palm trunk sap as an alternative agricultural waste resource for bioethanol production was investigated. The effects of temperature (25–40 °C), initial pH (3–7), agitation rate (110–250 rpm), and percentage inoculums (5–15 % v/v) on bioethanol yield were identified using modified face-centered central composite design (modified CCD) of response surface methodology (RSM). All the data were statistically validated by analysis of variance (ANOVA). The results showed that the temperature was the most significant factor that influenced the bioethanol yield, followed by the interaction effect of temperature and initial pH, quadratic effect of initial pH, and interaction effects of initial pH and agitation. Using modelling, the maximum bioethanol yield of 0.4926 g g^{-1} was predicted to be achieved at temperature 31.73 °C, initial pH 5.5, and agitation rate 110 rpm. Using these conditions, the final bioethanol yield achieved was in good agreement with the model prediction.

Introduction

The oil palm is the major commodity crops in Malaysia and Indonesia. Replantation of oil palm trees every 25–30 years generates large quantities of oil palm trunk (OPT) waste. The outer region of the trunk is rich with vascular bundles, which might be suitable for manufacturing building material, while the inner region is rich with high content of parenchymas in tissue structure (Tomimura 1992). These parenchymas are not suitable to be used as building materials. In order to be utilized for wood-based products, liquid content (sap) in the OPT has to be eliminated. This sap waste can be regarded as an important source for bioethanol production since it

A. H. Norhazimah (✉) · C. K. M. Faizal
Biofuel Research Group, Faculty of Chemical and Natural Resources Engineering, Universiti Malaysia Pahang, 26300 Gambang, Kuantan, Pahang, Malaysia
e-mail: aimc_86@yahoo.com

R. Pogaku et al. (eds.), *Developments in Sustainable Chemical and Bioprocess Technology*, DOI: 10.1007/978-1-4614-6208-8_3,

contains a lot of readily fermentable sugars and only minimum pretreatment is needed before fermentation. No research has been done to optimize the bioethanol production from the OPT sap using statistical design of experiment (DOE). DOE is a powerful tool that allows the knowledge gain about the process being studied through the iterative process with a minimum number of experiments (Altekar et al. 2006). The conventional method of one-factor-at-a-time (OFAT) is often used to optimize a process condition; however, it frequently fails to locate optimal conditions for the process due to its failure to consider the combined effect of the multiple factors involved (Kalil et al. 2000; Coninck et al. 2004). Optimization by applying response surface methodology (RSM) is quicker and can obtain more accurate information about factor effects including magnitude and direction compared to OFAT approach. Statistical analysis of data generated from the experiment clearly establishes the relationship between response (measured parameter of interest) and factors (process variables) being studied where factors may have individual effects on the response or may have interaction effect. RSM is used when a factorial experiment reveals the presence of nonlinear behavior by checking the significance of curvature. Experimental region that contains the optimal solution tends to have the response that is nonlinearly correlated with variables (Yang and El-Haik 2009). In bioethanol research area, RSM has been applied for bioethanol production from sago starch, cassava mash, and carob pod (Bandaru et al. 2006; Yingling et al. 2010; Turhan et al. 2010; Vaheed et al. 2010). As far as the researchers are concerned, to date, there is no discussion of employing face-centered CCD for bioethanol production from OPT sap. The main objective of this paper is to evaluate the effects of temperature, initial pH, agitation, and inoculum size on the production of bioethanol from OPT sap, and later to optimize the selected variables.

Methodology

Biomass Preparation

Two OPT trunks (25 years old) were obtained from the local plantation area in Pahang, Malaysia. They were squeezed to extract their sap after removing 10 cm of the hard outer layer. Then, the extracted sap was sterilized at 121 °C for 15 min prior to use.

Culture Preparation and Fermentation

Sake yeast strain, *Saccharomyces cerevisiae* Kyokai No. 7 bought from American Type of Culture Collection (ATCC 26422) was selected to ferment OPT sap. Inoculums were grown in YPD medium at 30 °C and 150 rpm until a concentration

Table 1 Actual and coded levels of variables considered in the design of experiments

Variable	Coded symbol	Variation levels		
		(−1)	(0)	(+1)
Temperature (°C)	A	25.0	32.5	40.0
Initial pH	B	3	5	7
Agitation (rpm)	C	110	180	250
Inoculums (% v/v)	D	5	10	15

of the range 0.22–0.24 g/l was achieved within 12–18 h. The bioethanol fermentations were carried out in a 500-ml Erlenmeyer flask (working volume 250 ml). Samples at 48 h were collected and analyzed. The four selected process variables were temperature, initial pH, agitation rate, and percentage inoculums. Levels of the variables are shown in Table 1.

These variables were studied at three levels with four-factor of two-level full factorial design (2^4), 6 replicate central points, and 8 star points ($\alpha = 1.0$). Thus, a total of 30 runs were generated using the Design Expert Version 7.1.6 software (shown in Table 2). The response variable measured was bioethanol yield ($Y_{P/S}$), where the maximum concentration of bioethanol is divided by the total concentration of sugar. The response values were fitted by second-order model in order to associate the response variable to the independent variables. After optimizing the conditions using RSM, the model was validated by confirmations runs on the optimized condition performed in 500-ml shake flask and 2-liter bioreactor.

Results and Discussion

ANOVA and Model Development

The goodness of fit for the model and significance of each regression coefficient was estimated using regression analysis and ANOVA. Each factor with $p < 0.05$ was considered as potentially significant. Smaller *p*-value indicated that the corresponding coefficient was more significant than bigger *p*-value. Based on the results, the model equation adequately describes the response since the *p*-value for the model is very low (less than 0.0001), implying that the model is significant. In this case, the main factors of temperature and agitation (A and C), interaction of temperature, initial pH, and agitation (AB, AC, and BC), the quadratic effect of temperature and initial pH (A^2 and B^2) are significant model terms at the 95 % confidence level. Among the significant variables, the temperature had the largest effect on bioethanol yield. The ranking is as follows: $A > AB > B^2 > BC > A^2 > C > AC$. Although factor B was insignificant, this factor could not be excluded from the model in order to retain the hierarchical as B interaction with

Table 2 Face-centered CCD design layout and experimental results

Standard run number	Coded values of variables				Percentage yield, $Y_{P/S}$ (g g^{-1})		
	A	B	C	D	Observed	Predicted	Residual
1	25.0	3	110	5	–	–	–
2	40.0	3	110	5	0.063	0.083	−0.020
3	25.0	7	110	5	0.280	0.290	−0.010
4	40.0	7	110	5	0.400	0.390	+0.010
5	25.0	3	250	5	–	–	–
6	40.0	3	250	5	0.000	0.006	−0.006
7	25.0	7	250	5	–	–	–
8	40.0	7	250	5	0.087	0.093	−0.006
9	25.0	3	110	15	0.380	0.340	+0.040
10	40.0	3	110	15	0.022	0.083	−0.061
11	25.0	7	110	15	0.300	0.290	+0.001
12	40.0	7	110	15	0.380	0.390	−0.010
13	25.0	3	250	15	0.410	0.440	−0.030
14	40.0	3	250	15	0.000	0.006	−0.006
15	25.0	7	250	15	0.240	0.180	+0.060
16	40.0	7	250	15	0.034	0.093	−0.059
17	32.5	5	180	10	0.350	0.440	−0.090
18	32.5	5	180	10	0.440	0.440	+0.000
19	32.5	5	180	10	0.400	0.440	−0.040
20	32.5	5	180	10	0.420	0.440	−0.020
21	25.0	5	180	10	0.390	0.430	−0.040
22	40.0	5	180	10	0.400	0.250	+0.150
23	32.5	3	180	10	0.410	0.310	+0.010
24	32.5	7	180	10	0.350	0.330	+0.020
25	32.5	5	110	10	0.460	0.480	−0.020
26	32.5	5	250	10	0.380	0.390	−0.010
27	32.5	5	180	5	0.420	0.440	−0.020
28	32.5	5	180	15	0.400	0.440	−0.040
29	32.5	5	180	10	0.410	0.440	−0.030
30	32.5	5	180	10	0.550	0.440	+0.110

other variables is significant. Table 3 shows the ANOVA results for the response surface quadratic model.

The "Lack of Fit" F-value of 0.83 implies the lack of fit is not significant relative to the pure error. The "Lack of Fit" measures the failure of the model to represent data in the experimental domain at points, which are not included in the regression. Therefore, nonsignificant lack of fit is good. The predicted R^2 of 0.6904 is reasonable with the adjusted R^2 of 0.8476, which illustrates a high correlation between the observed values and the predicted values. Adequate precision measures the signal to noise ratio where a ratio greater than 4 is desirable. Adequate precision of 13.058 indicates an adequate signal, which means this model can be used to navigate the design space. Backward deletion of terms was applied to

Table 3 Analysis of variance (ANOVA) for response surface reduced quadratic model

Source	Sum of squares	DF	Mean square	F-value	p-value	Prob>.F
Model	0.610	8	0.076	16.24	<0.0001	Significant
A—temperature	0.100	1	0.100	8.84	<0.0001	Significant
B—initial pH	0.001	1	0.001	0.002	0.5540	Not significant
C—agitation	0.030	1	0.030	4.82	0.0132	Significant
AB	0.090	1	0.090	17.59	0.0002	Significant
AC	0.025	1	0.025	3.76	0.0233	Significant
BC	0.037	1	0.037	4.60	0.0071	Significant
A2	0.031	1	0.031	3.96	0.0121	Significant
B2	0.042	1	0.042	5.88	0.0045	Significant
Residual	0.072	18	0.004			
Lack of fit	0.049	13	0.004	0.83	0.6396	Not significant
Pure error	0.023	5	0.005			
Standard deviation	0.063		R-squared			0.8945
Mean	0.310		Adjusted R-squared			0.8476
Adequate precision	13.058		Predicted R-squared			0.6904

eliminate the statistically nonsignificant terms. The following model describes the bioethanol yield in terms of code variables (Eq. 1):

$$\begin{aligned}\text{Bioethanol yield} &= 0.440 - 0.087 \times A + 0.010 \times B - 0.047 \times C + 0.087 \times A \times B \\ &\quad - 0.045 \times A \times C - 0.054 \times B \times C - 0.096 \times A^2 - 0.110 \times B^2\end{aligned} \tag{1}$$

Overall Graph-Based Discussion and Model Validation

Figure 1 shows the contour plot of bioethanol yield for the interaction between temperature and initial pH, and between temperature and agitation. Based on Fig. 1, the temperatures above 36 °C are not suitable for higher bioethanol yield. The higher temperature above the optimum temperature range will decrease the growth rate (Shuler and Kargi 2002). The bioethanol yield and fermentation rate are generally low at elevated temperatures as some of the bioethanol may be entrained with the intense release of carbon dioxide (Ribereau-Gayon et al. 2006). Fermentation efficiency of *S. cerevisiae* is usually very low at high temperatures due to the increased fluidity of membranes to which yeast responds by changing its fatty acid composition (Breisha 2010). Medium temperature (26–36 °C) and low agitation turn out higher yield of bioethanol than other temperature and agitation settings.

Agitation could reduce the diffusion barriers by improving mass transfer characteristic, thus resulting in improved mixing inside the fermentation broth. It helps to maintain the concentration gradient between the interior and exterior of

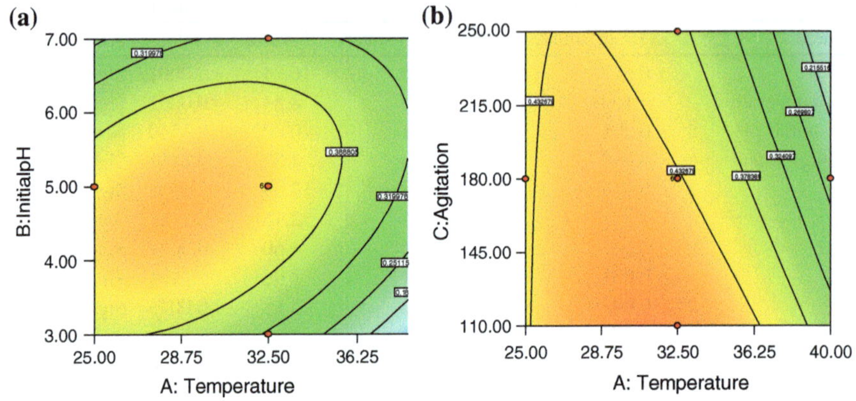

Fig. 1 Bioethanol yield for (**a**) the interaction of temperature and initial pH; (**b**) interaction of temperature and agitation shown by the contour plot at pH 5 and 10 % inoculation

the cell, and maintain satisfactory supply of sugars to the cells (Rodmui et al. 2008). Even though the initial pH factor does not have a significant impact on bioethanol yield, initial pH should be around 3.5–6.0 to obtain a better bioethanol yield. Outside this range, the bioethanol yield will be lower. The best pH is 5.5, which is consistent with the finding by Turhan et al. (2010) on carob pods; they obtained maximum bioethanol production at pH 5.5 in carob raw material. The pH variations of the medium result in changes in the activity of the enzyme and hence change the reaction rate (Shuler and Kargi 2002). Percentage of inoculums, however, does not have any significant effect on bioethanol yield. Numerical optimization methods were carried out using the optimization function embedded in Design Expert 7.1.6 software.

To achieve maximum desirability, temperature, initial pH, and agitation were set within the range, whereas bioethanol yield was set at a maximum theoretical yield over glucose (0.511 g/g). The software generated 17 solutions including the best suggestion conditions, i.e., temperature 31.7 °C and initial pH 5.5. A confirmation experiment under the optimized conditions was performed in 500-ml shake flask and 2-liter bioreactor to verify the goodness of the statistically predicted model. The bioethanol yield obtained reached 90 % of the value predicted by the software, hence confirming that the model was effectively built with good validity.

Conclusions

Temperature has the most pronounced influence on bioethanol yield in the fermentation of OPT sap. Other factors of initial pH and inoculums have a small influence on bioethanol production. The optimum bioethanol yield of 0.49 g bioethanol per g sugar can be achieved at 31.7 °C and 110 rpm.

References

Altekar, M., Homon, C. A., Kashem, M. A., Mason, R. M., Patnaude, L. A., Yingling, J., et al. (2006). Assay optimization: A statistical design of experiments approach. *Journal of the Association for Laboratory Automation, 11*(1), 33–34.

Bandaru, V. V. R., Somalanka, S. R., Mendu, D. R., Madicherla, N. R., & Chityala, A. (2006). Optimization of fermentation conditions for the production of ethanol from sago starch by co-immobilized amyloglucosidase and cells of *Zymomonas mobilis* using response surface methodology. *Enzyme and Microbial Technology, 38*(1–2), 209–214.

Breisha, G. Z. (2010). Production of 16 % ethanol from 35 % sucrose. *Biomass and Bioenergy, 34*(8), 1243–1249.

Coninck, J. D., Leclercq, B., Exbrayat, J. M., & Duyme, F. (2004). Factorial designs: An efficient approach to choosing the main factors influencing growth and hydrolase production by *Tetrahymena thermophila. Journal of Industrial Microbiology and Biotechnology, 31*(5), 204–208.

Kalil, S. J., Maugeri, F., & Rodrigues, M. I. (2000). Responses surfaces analysis and simulation as a tool for bioprocess design and optimization. *Process Biochemistry, 35*(6), 539–550.

Ribereau-Gayon, P., Dubourdieu, D., Doneche, B., & Lonvaud, A. (2006). *Handbook of enology, volume 1: The microbiology of wine and vinifications* (p. 100). England: Wiley.

Rodmui, A., Kongkiattikajorn, J. & Dandusitapun, Y. (2008). Optimization of agitation conditions for maximum ethanol production by coculture. *Kasetsart Journal (Nature Science), 42*, 285–293.

Shuler, M. L. & Kargi, F., (2002). bioprocess engineering basic concepts. (Vol. 75. 2nd edn. pp. 160–166). Prentice Hall International Series, U.S.

Tomimura, Y. (1992). Chemical characteristics and utilization of oil palm trunks. *JARQ, 25*, 283–288.

Turhan, I., Bialka, K. L., Demirci, A., & Karhan, M. (2010). Ethanol production from carob extract by using *Saccharomyces cerevisiae. Bioresource Technology, 101*(14), 5290–5296.

Vaheed, H., Shojaosadati, S. A., & Galip, H. (2010). Evaluation and optimization of ethanol production from carob pod extract by *Zymomonas mobilis* using response surface methodology. *Journal of Industrial Microbiology and Biotechnology, 38*(1), 101–111.

Yang, K., & El-Haik, B. S. (2009). *Design for six sigma a roadmap for product development* (2nd ed.). United States of America: McGraw-Hills Companies.

Yingling, B., Zongcheng, Y., Honglin, W., & Li, C. (2010). Optimization of bioethanol production during simultaneous saccharification and fermentation in very high-gravity cassava mash. *Antonie van Leeuwenhoek, 99*(2), 329–339.

Effect of Isolated Mesophilic Bacterial Consortium on the Composting Process of Pressed–Shredded Empty Oil Palm Fruit Bunch

J. C. Lai, H. B. Chua, A. Saptoro and H. M. Ang

Abstract Empty fruit bunch (EFB) is one of the main solid wastes generated from palm oil mills. Incineration and mulching have been employed traditionally to process EFB. This method of waste treatment becomes a great concern since burning of EFB is prohibited by environmental regulation, and moreover, the process is costly and unsustainable. Composting of EFB has the potential to replace the conventional methods used. Composting is a bioconversion process involving microbial activity that transforms biomass wastes into useful organic substrate or fertilizer. In this work, an experimental study was carried out to investigate the composting of pressed–shredded empty fruit bunch (EFB) with mesophilic bacterial consortium as composting microorganisms. The bacterial consortium was isolated from empty fruit bunch compost and cultivated under mesophilic condition at 35 °C. The experiments were conducted under ambient conditions, and the process was carried out for a duration of 40 days. Changes in moisture content, pH and temperature during composting process were evaluated. The composting temperature varied from 26.5 to 28 °C, and the final pH of compost was found to be 6.91. The moisture of compost fluctuated between 60 and 80 % over the period of study. At the end of composting period, an average of 54 % loss in the dry weight for EFB was observed when the isolated mesophilic bacterial consortium was added during the composting process.

J. C. Lai (✉) · H. B. Chua · A. Saptoro
Department of Chemical Engineering, School of Engineering and Science, Curtin University Sarawak Campus, CDT 250, 98009 Miri, Sarawak, Malaysia
e-mail: jessie_870609@yahoo.com.sg

H. M. Ang
Department of Chemical Engineering, School of Chemical and Petroleum Engineering, Curtin University Bentley Campus, GPO Box U1987 Perth, WA 6845, Australia

R. Pogaku et al. (eds.), *Developments in Sustainable Chemical and Bioprocess Technology*, DOI: 10.1007/978-1-4614-6208-8_4,

Introduction

Due to rapid development of oil palm plantations in Malaysia, the environmental impact of solid wastes from the palm oil production is a major concern. Empty fruit bunch (EFB) is the major solid palm oil residues produced in amount of more than 20 % of the fresh fruit weight (Mohammad et al. 2012). Conventionally, incineration and mulching are used to treat EFB. However, incineration caused air pollution and mulching application increased cost of transportation and distribution of EFB in plantation. To overcome the drawback of the two processes, composting was explored as an alternative waste treatment to manage the increasing amount of EFB.

Composting is a microbiological process to degrade organic waste through the actions of enzymes, microorganisms and oxygen under controlled aerobic and thermophilic conditions (Weidong et al. 2007). The composting highly depends on availability of moisture within the composting substrate to sustain microbial activities for decomposition of organic matters. Richard et al. (2002) reported that moisture content is varied for different compost mixture and composting duration, ranging from 50 to 70 %. Even so, unequal moisture distribution can reduce microbial activities as shown in closed composting system studied by Suhaimi and Ong (2001). Consequently, maintaining moisture content is essential to enhance microbial activities.

Composting temperature tends to change in response to microbial activities. The process has four distinct phases of temperature changes, namely mesophilic phase, thermophilic phase, cooling phase and maturation phase (Tuomela et al. 2000). Mesophiles at initial composting would degrade simple carbohydrates and protein into simpler organic matters (Swan et al. 2002). The large quantities of metabolic heat energy generated during decomposition would result in temperature rising beyond 45 °C. At this stage, thermophiles take over and degrade the organic matter further and cause temperature to rise to 70 °C. Due to depletion of food source, the microbial activities tend to decline and the process arrives at the cooling and maturation phases. The mature compost is attained at ambient temperature (Waldron and Nichols 2009; Swan et al. 2002). Similar temperature profiles were reported from studies by Schuchardt et al. (2002), Zahrim and Asis (2010) and Yahya et al. (2010).

Most of the literature studies reported that pH does not vary significantly during the composting. Microbial fermentation of carbohydrates (cellulose in EFB) is known to form humic acid to increase acidity of compost, but ammonification of inorganic nitrogen would neutralize the pH of compost at the end of composting (Thambirajah et al. 1995). In fact, pH of matured compost depends on the nature of substrate which is being decomposed. Kabbashi et al. (2007) obtained matured compost with pH 5.6 during the co-composting of EFB in a bioreactor, while Haroun et al. (2007) found a pH of 6.6 in the composting treatment of tannery sludge. Similar pH trend was also observed by Singh et al. (2011) in the course of vermicomposting of cattle manure.

Little is known about the effect of adding bacterial consortium on the composting process of EFB. Therefore, the aim of this preliminary study is to investigate the changes in moisture content, pH and temperature of EFB compost during the course of composting with the addition of the isolated mesophilic bacterial consortium. The outcome from this study is expected to determine any beneficial effects on the composting process of EFB by adding exogenous bacteria.

Methodology

Experimental Material

Pressed and shredded empty fruit bunch (EFB) and young compost of EFB (after 15 days composting) from Bintulu Lumber Development (BLD) Sdn Bhd at Miri, Sarawak, Malaysia, were used in the experiments. Young compost of EFB was employed as a source for isolation of mesophilic bacterial consortium.

Preparation of Mesophilic Bacterial Consortium

Mesophilic bacterial consortium was isolated from young compost of EFB (after 15 days composting) supplied by BLD Sdn Bhd. To isolate the bacterial consortium, 10 g of young compost was dissolved in 100 mL distilled water. The compost solution was filtered to remove sand and fibre and was kept as inoculum. To cultivate mesophilic bacterial consortium, 5 % inoculum was taken and transferred to culture medium containing the following (per litre of mineral salt solution): 1L mineral salt solution (source of mineral), 0.5 % urea (nitrogen source), 2 % glucose (nutrient), 1 % yeast extract (vitamin) and is placed in water bath at 35 ± 3 °C. Mineral salt solution was prepared as follow: 3 g Na_2SO_4, 5 g $MgSO_4 \cdot 7H_2O$, 0.4 g NaCl, 2.5 g K_2HPO_4, 2.5g H_2KPO_4, 14 mg $FeCl_3$, 110 mg $MnSO_4 \cdot H_2O$, 25 mg $ZnSO_4 \cdot 7H_2O$ and 0.8 mg $CuSO_4$ were dissolved in 1L distilled water. An UV–VIS photo-spectrometer was used to analyse the growth of the mesophilic bacterial consortium. The consortium was harvested at the active growth phase and stored at 4 °C.

Experimental Procedures

The experiments were carried out in 3 baskets with a height of 14 cm, length of 31 cm and width of 31 cm. Approximately 1 kg of EFB in total wet weight was used for each basket. The 3 baskets were correspondingly labelled as Control (PC),

Pile 1 (P1) and Pile 2 (P2), in which P2 was a replicate set of P1. The mesophilic bacterial consortium was mixed with the EFB at a mass percentage of 0 % in PC and 10 % in P1 and P2 based on total wet weight of approximately 1 kg EFB. Moisture content for the three piles was maintained at range of 60–70 % by adding water periodically during the course of the composting. The experiment was conducted at temperature of 30 ± 2 °C. Air was supplied to composting mixture by natural aeration with regular turning. The composting process was run for a duration of 40 days.

Sampling and Analysis

The sampling was taken periodically for forty days. Approximately 2 g of sample was collected from the top, bottom and centre of the heap from each basket, and they were homogeneously mixed for gravimetric analyses. The samples were oven-dried for 24 h at 120 °C for moisture determination. For pH measurement, the samples were dissolved in distilled water with ratio of 1:10 w/v. The average temperature was taken by inserting thermometer at surface and core of the heap of PC, P1 and P2. All analyses were repeated, and data presented in the paper were the mean results. Changes in the colour, texture, size and possible growth of other organisms such as mushroom, fungus and beetle larva were recorded based on physical observation.

Results and Discussion

Changes in Physical Parameters During Composting Process

Figure 1a shows the variation in the moisture content for the three EFB compost piles over the 40-day composting period. The moisture of the composting piles was observed to fluctuate from 60 to 80 % over the period of study. For PC, the

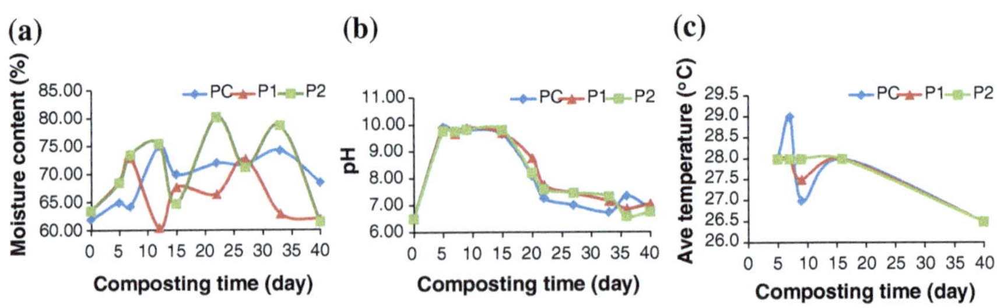

Fig. 1 Changes in moisture (**a**), pH (**b**) and temperature (**c**) in three piles during composting of EFB

Table 1 Physical parameters before and after composting

	Moisture (%)			Total wet weight (g)		Dry weight (g)		Dry weight loss (g)	Dry weight reduction (%)
	Initial	Final	Mean	Initial	Final	Initial	Final		
PC	62.0	68.5	69.2 ± 4.6	1,468	707.0	557.8	222.7	335.1	60.1
P1	63.5	62.0	66.4 ± 4.6	1,538	658.0	561.4	250.0	311.4	55.5
P2	63.5	61.5	70.8 ± 6.7	1,538	693.0	561.4	266.8	294.6	52.5

moisture rose from 62 to 75 % after 12 days of composting, but it became more settled at about 71 % from day 15 onwards. However, for the P1 and P2, the moisture appeared to fluctuate more from as low as 60 % to as high as 80 %. Over the 40-day composting period of study, the mean moisture value of 69.2 ± 4.6 %, 66.4 ± 4.6 % and 70.8 ± 6.7 % was observed for PC, P1 and P2, respectively. Overall, the results showed that moisture of P1 and P2 was less evenly distributed as compared to that of PC although attempts were made to compensate for water loss due to evaporation by periodic addition of water. The fluctuation of moisture may also be attributed to the generation of metabolic water due to the microbial action during the composting process. The fluctuation of the moisture content during the composting was known to impede the decomposition of organic matter (Suhaimi and Ong 2001). Besides, excessive moisture has also been found to reduce the availability of oxygen to the composting microorganisms and cause a slowdown in the composting process (Lu et al. 2001). Following 40 days of composting, it was observed that the PC, P1 and P2 experienced a 60.1, 55.5 and 52.5 % dry weight reduction, respectively, as shown in Table 1. The greater fluctuation of moisture during the composting process might have contributed to the slower rate of composting and corresponding lower weight loss in P1 and P2. Addition of the mesophilic bacterial consortium on EFB composting did not show any significant effect on the rate of EFB decomposition. The finding also suggests that the endogenous microorganisms present in PC were sufficient to initiate the composting process.

Changes in Composting pH

Besides the moisture variation during the composting process, the variation in the pH observed is another indicator of the presence of microbial activity in the composting process. Figure 1b shows the pH changes during the composting of EFB with mesophilic bacterial consortium. The pH for PC, P1 and P2 increased from initial pH 6.51 to pH 9.93, 9.85 and 9.78, respectively, on day 5. The pH increase might be caused by ammonification process, resulting in the release of ammonia and mineralization of organic nitrogen during initial composting process (Gao et al. 2010). With increase in composting time, the pH remained alkaline at above 9. The pH often rises to 8–9 for the fully developed compost (Sundberg et al. 2004). Thus, a pH adjustment was carried out on day 14 to decrease pH below 9 by adding aqueous HCl to the compost. The pH later decreased below pH 8 on day 21. At the end of composting period, the pH of the PC, P1 and P2 settled in the range of 6.8–7.0. Overall, the pH profile for the control and test piles is almost identical throughout the composting period.

Changes in Composting Temperature

Temperature has been recognized as a critical parameter in monitoring a composting process. The variation in temperature is strongly correlated with microbial activity, composting stage and proliferation of different microbial groups (Liu and Price 2011). Figure 1c shows the variation in average temperature during composting process. The temperature in PC increased from 28 to 29 °C on day 7 and then gradually decreased to 27 °C on day 9. However, the temperature in P1 and P2 remained fairly constant with only a slight decrease on day 7 to day 13. The stable temperature profile might be due to condition of high humidity in the three piles. Additionally, small quantity of EFB used in the study might cause heat to be easily dissipated during the composting process. As a result, it could affect microbial activities and slowdown the breakdown of the available organic matters and release of nitrogenous products. In a composting process, thermophiles are essential to further break down simple carbohydrates, which have been degraded initially by mesophiles during mesophilic phase of composting process (Swan et al. 2002).

Conclusion

This study concluded that addition of mesophilic bacterial consortium did not speed up or improve the composting process of EFB. The finding suggests that the existing populations of microorganisms that were present in the EFB compost piles at the beginning of the composting process were able to initiate the biodegradation of EFB. However, less weight loss was observed for the compost piles that experienced larger fluctuation of moisture during the composting process. This observation supports the idea that excessive fluctuation of the moisture content in the compost piles could have negative impact on the composting of EFB. The pH variation due to the microbial activities that was also observed at the initial phase could have an adverse effect on the composting process. Therefore, composting conditions should be operated at some optimal state for effective use of mesophilic bacterial consortium in composting of EFB. Besides, further study can be performed to establish the effect of the pH fluctuation on the rate of the composting process.

Acknowledgments The authors would like to thank Celtex Resources Sdn Bhd and Curtin University, Sarawak, for funding and technical support to this project.

References

Gao, M., Liang, F., Yu, A., Li, B., & Yang, L. (2010). Evaluation of stability and maturity during forced-aeration composting of chicken manure and sawdust at different C/N ratios. *Chemosphere, 78*, 614–619.

Haroun, M., Idris, A., & Syed Omar, S. R. (2007). Characterisation and composting of tannery sludge. *Malaysian Journal of Soil Science, 11*, 71–80.

Kabbashi, N. A., Alam, Z. & Ainuddin, M. (2007). Bio-composting process development by SSF for utilization agro-industrial wastes, 3rd Kuala Lumpur international conference on biomedical engineering 2006. In: F. Ibrahim, N. A. A. Osman, J. Usman & N. A. Kadri (Eds.). Springer: Berlin Heidelberg.

Liu, K., & Price, G. W. (2011). Evaluation of three composting systems for the management of spent coffee grounds. *Bioresource Technology, 102*, 7966–7974.

Lu, S. G., Imai, T., Li, H. F., Ukita, M., Sekine, M., & Higuchi, T. (2001). Effect of enforced aeration on in-vessel food waste composting. *Environmental Technology, 22*, 1177–1182.

Mohammad, N., Alam, M. Z., Kabbashi, N. A., & Ahsan, A. (2012). Effective composting of oil palm industrial waste by filamentous fungi: A review. *Resources, Conservation and Recycling, 58*, 69–78.

Richard, T. L., Hamelers, H. V. M. B., Veeken, A., & Silva, T. (2002). Moisture relationships in composting processes. *Compost Science and Utilization, 10*, 286–302.

Schuchardt, F., Darnoko, D., & Guritno, P. (2002). *Composting of empty oil palm fruit bunch (EFB) with simultaneous evaporation of oil mill waste water (POME)*. Nusa Dua, Bali, Indonesia: International Oil Palm Conference.

Singh, R. P., Embrandiri, A., Ibrahim, M. H., & Esa, N. (2011). Management of biomass residues generated from palm oil mill: Vermicomposting a sustainable option. *Resources, Conservation and Recycling, 55*, 423–434.

Suhaimi, M. & Ong, H. K. (2001). *Composting empty fruit bunches of oil palm.* http://www.agnet.org/library/eb/505a/. Accessed Jan 19 2010.

Sundberg, C., Smårs, S., & Jönsson, H. (2004). Low pH as an inhibiting factor in the transition from mesophilic to thermophilic phase in composting. *Bioresource Technology, 95*, 145–150.

Swan, J. R. M., Crook, B. & Gilbert, E. J. (2002). Microbial emissions from composting sites. In: R E. Hester & R. M. Harrison (Eds.), *Environmental and health impact of solid waste management activities* Royal Society of Chemistry.

Thambirajah, J. J., Zulkali, M. D., & Hashim, M. A. (1995). Microbiological and biochemical changes during the composting of oil palm empty-fruit-bunches. Effect of nitrogen supplementation on the substrate. *Bioresource Technology, 52*, 133–144.

Tuomela, M., Vikman, M., Hatakka, A., & Itävaara, M. (2000). Biodegradation of lignin in a compost environment: A review. *Bioresource Technology, 72*, 169–183.

Waldron, K. W. & Nichols, E. (2009). Composting of food-chain waste for agricultural and horticultural use. In: K. Waldron (Ed.), *Handbook of waste management and co-product recovery in food processing.* Cambridge: Woodhead Publishing Limited.

Weidong, W., Xiaofen, W., Jianbin, L., Masaharu, I., et al. (2007). Effect of oxygen concentration on the composting process and maturity. *Compost Science and Utilization, 15*, 184–190.

Yahya, A., Sye, C. P., Ishola, T. A., & Suryanto, H. (2010). Effect of adding palm oil mill decanter cake slurry with regular turning operation on the composting process and quality of compost from oil palm empty fruit bunches. *Bioresource Technology, 101*, 8736–8741.

Zahrim, A. Y., & Asis, T. (2010). Production of non shredded empty fruit bunch semi-compost. *The Institution of Engineers, Malaysia, 71*, 11–17.

Acetone–Butanol–Ethanol Fermentation From Palm Oil Mill Effluent Using *Clostridium acetobutylicum*

A. S. Japar, M. S. Takriff, J. M. Jahim and A. A. H. Kadhum

Abstract Acetone–butanol–ethanol (ABE) fermentation from sterilized supernatant and sterilized sludge of palm oil mill effluent (POME) by *Clostridium acetobutylicum* NCIMB 619 was investigated. The fermentation was carried out anaerobically in batch mode at a temperature of 37 °C for 72 h. ABE recovery from the fermentation media was attempted by liquid–liquid extraction using oleyl alcohol as a solvent. Total solvent concentrations obtained were 0.3985 and 1.1253 g/l in sterilized POME supernatant and sterilized POME sludge, respectively. However, product recovery by liquid–liquid extraction using oleyl alcohol was very not effective where only 13.7 and 2.85 % total solvent were recovered from sterilized POME supernatant and sterilized POME sludge media, respectively.

Introduction

The petroleum base production of acetone, butanol, and ethanol (ABE) requires alternatives due to high price of its raw materials. One possible alternative is to produce ABE via fermentation process using clostridia strains. There are four known strains that can produce these solvents, namely the *Clostridium acetobutylicum, C. beijerinckii, C. saccharobutylicum,* and *C. saccharoperbutylacetonicum* (Shaheen et al. 2000). The ABE fermentation occurs in two distinct phases, namely acidogenic phase and solventogenic phase. During the acidogenic phase, the pH of the medium decreases due to the production of organic acids. As the fermentation enters the stationary growth phase, the cell metabolism undergoes a shift to solventogenic phase where ABE are produced (Jones and Woods 1986). To

A. S. Japar (✉) · M. S. Takriff · J. M. Jahim · A. A. H. Kadhum
Department of Chemical and Process Engineering, Faculty of Engineering & Built Environment, Universiti Kebangsaan Malaysia, UKM Bangi, 43600 Selangor, Malaysia
e-mail: azima_syafaini88@ymail.com

R. Pogaku et al. (eds.), *Developments in Sustainable Chemical and Bioprocess Technology*, DOI: 10.1007/978-1-4614-6208-8_5,

ensure ABE fermentation is economically viable, readily available and cheap substrates are required (Ezeji et al. 2007). The *C. acetobutylicum* has the flexibility of using various starchy substances as fermentation media (Jones and Woods 1986).

Product recovery is an important step in fermentation processes. Liquid–liquid extraction and pervaporation are common purification methods for fermentation processes due to their potential in solvent recovery (Groat et al. 1992). Qureshi and Maddox (1995) reported that among suitable solvent for liquid–liquid extraction without harming the fermentation process include fatty acids esters, dibutyl phthalate, oleyl alcohol, decanol, ethyl hexanol, polyoxyalkylene ethers, and polypropylene glycol.

Malaysia is one of the world's largest producers of palm oil. However, its production generates various wastes chief among which is palm oil mill effluent (POME). On the average, 2.5 ton of POME is generated for every ton of crude palm oil produced. Raw POME is thick, brownish liquid that consists of various suspended components, including short fibers, a spectrum of carbohydrates ranging from hemicellulose to simple sugar, a range of nitrogenous compound from proteins to amino acids and free organic acids (Ugoji 1997). In palm oil mills, POME is generated from three major sources, sterilizer condensate, separator sludge, and hydrocyclone operation where the broken shells are separated from the kernels (Vijayaragharan et al. 2007). Table 1 shows typical characteristic of POME (Ahmad et al. 2006). This entire feature has made POME a potential substrate for ABE fermentation (Somrutai et al. 1996; Kalil et al. 2003). Based on previous studies, ABE fermentation using whole POME as media can produce total solvent between 1.0 and 2.66 g/l (Siti Jamilah Hanim et al. 2010; Masngut et al. 2007 and Kalil et al. 2003). Therefore, the objective of this study was to investigate the production of ABE via fermentation in sterilized POME supernatant and sterilize POME sludge *using C. acetobutylicum NCIMB* 619. In addition, this work investigated the potential of recovering the ABE via liquid–liquid extraction using oleyl alcohol as a solvent.

Table 1 Characteristics of palm oil mill effluent

Parameter	Concentration (mg/L)	Element	Concentration (mg/L)
Oil and grease	4,000–6,000	Phosphorus	180
BOD	20,000–25,000	Potassium	2,270
COD	40,000–50,000	Calcium	439
Total solids	40,500	Boron	7.6
Suspended solid	18,000	Iron	46.5
Total volatile solids	34,000	Manganese	2.0
Ammoniacal nitrogen	35	Copper	0.89
Total nitrogen	750	Magnesium	615
		Zinc	2.3

Materials and Methods

Microbes and Media

C. acetobutylicum NCIMB 619 was used in this study. Laboratory stocks of *C. acetobutylicum* were routinely maintained as spore suspensions of sterile deoxygenated reinforced clostridial medium (RCM) at 4 °C under anaerobic condition. The inoculum was prepared in a 250-ml Schott bottle with deoxygenated RCM medium and 10 % by volume of spore suspension and was incubated at 37 °C for 18 h anaerobically in a stationary mode. All procedures were carried out anaerobically, and inoculum with optical density of 1.3–1.8 at 680 nm wavelength was used in the fermentation. Reinforced clostridial medium (RCM) liquid medium was used in inoculum production, while raw palm oil mill effluent (POME) was used as the fermentation medium as well as in acetone–butanol–ethanol (ABE) production. Raw POME obtained from Sime Darby's East Palm Oil Mill in Carey Island was sedimented overnight in a cool room at temperature of 4 °C. The supernatant and the sludge were then separated and autoclaved at 121 °C for 15 min.

Fermentation Process and Analysis

The fermentation process was carried out anaerobically in 500-mL Duran bottles with working volume of 400 mL at a temperature of 37 °C for 72 h with no agitation. The pH of the sterilized POME supernatant and the sterilized POME sludge was adjusted to 5.8 using 5 M NaOH (Kalil et al. 2003) and deoxygenated by sparging with purified nitrogen gas. The ratio of inoculum to medium was 1:10. The pH value was determined by pH meter. The cell concentration and reducing sugar in the sample were determined using dinitrosalicylic acid (DNS) and volatile suspended solid (VSS) methods, respectively. The concentrations of acetone, butanol, and ethanol were measured by a gas chromatograph that is equipped with flame ionization detector (FID). All samples were analyzed at oven temperature of 40 °C for 8 min then increased to 130 °C at the rate of 4 °C min^{-1} for 2 min. The injector and detector temperature was programmed at 250 °C, and helium was used as a carrier gas.

Liquid–Liquid Extraction

Liquid–liquid extraction was carried to recover ABE from the fermentation media using oleyl alcohol which has been reported as a potential separation solvent for ABE (Boudreau and Gordon 2006). An amount of oleyl alcohol was added into the sample in centrifuge tubes at the ratio of 1:1 (oleyl alcohol:sample). In order to increase the interaction between oleyl alcohol and the samples, the solution in the tube was mixed using vortex mixer at 1,500 rpm for 1 min, and then, the samples were centrifuge at 2,000 rpm for 5 min to complete the phase separation; 1.5 mL were taken from each phase using syringe with fine needles and transferred to sealed GC vials tubes for further analysis in the gas chromatograph.

Results and Discussion

ABE Fermentation

Figure 1a and b shows the growth profiles of the bacteria in both media. There were no lag phases showed in both profiles that indicated the bacteria that could easily adapted to the both POME media. For the sterilized POME supernatant, the sugar concentration was reduced with time, while the sugar concentration in sterilized POME sludge was fluctuated with time that may be due to the high pressure in the cells that leads to the increase in sugar. The sugar concentration decreased after 36 h of fermentation that shows the cells starting to utilize sugar. During the acidogenic phase where the organic acids were produced, the pH of the media decreases from 5.8 to 4.61 for sterilized POME supernatant and 5.8–4.82 for sterilized POME sludge. As the culture enters the solventogenic phase, organic acids together with residual carbohydrate were converted to ABE. During this phase, the pH of the media remains fairly constant implying that only small amount of the organic acids were converted to ABE. This may be due to the low amount of sugar concentration in POME. Previous studies show that low sugar concentration can lead to low production of solvent (Shaheen et al. 2000).

The results of the ABE fermentation are presented in Fig. 2, while the comparison of ABE fermentation in sterilized POME supernatant and sterilized POME

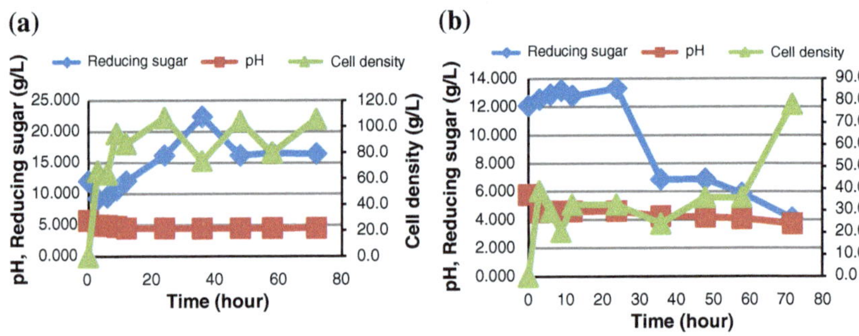

Fig. 1 The growth, pH, and reducing sugar profiles of *C. acetobutylicum* NCIMB 619 in **a** sterilized POME sludge and **b** sterilized POME supernatant

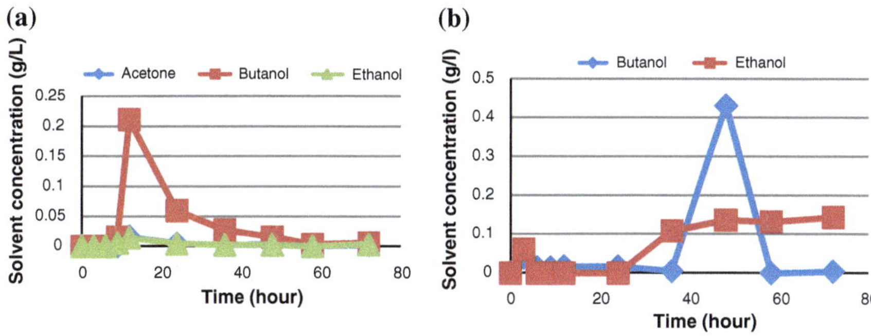

Fig. 2 Solvent productions for **a** sterilized sludge and **b** sterilized supernatant

sludge are summarized in Table 2. Figure 2a shows that the maximum production of solvents occurred after 12 h of fermentation in sterilized POME sludge where concentration of butanol was the highest at 0.21 g/l. However, minimal concentrations of acetone and ethanol were produced in the ABE fermentation using sterilized POME sludge as a medium. Figure 2b shows that in the sterilized POME supernatant, the highest production of solvent occurred after 36 h where the maximum level of butanol reached 0.43 g/l. Maximum concentration of ethanol was 0.15 g/l, and no acetone was produced in the ABE fermentation using sterilized POME supernatant as fermentation media. Previous study (Masngut et al. 2007) using POME as fermentation media in the production of ABE also found that the main product was butanol. Table 2 shows that the total solvents that produced in sterilized POME supernatant were about threefold higher than that in sterilized POME sludge. A typical ABE fermentation produces these solvents in the ratio of 3:6:1 (acetone:butanol:ethanol) (Jones and Woods 1986). The solvent ratio obtained in this study was 0.9:10.6:1 for sterilized POME sludge and 0:0.9:1 for sterilized supernatant. The final ABE ratio varies depending on the type of strain, substrate, and cultural condition (Awang et al. 1988). The total solvent produced in the sterilized POME supernatant was 1.12 g/l which is within the range of the 1.0–2.66 g/l as reported in the previous study (Siti Jamilah Hanim et al. 2010; Masngut et al. 2007 and Kalil et al. 2003). However, the total solvent produced in the sterilized POME sludge was only 0.39 g/l.

Table 2 Comparison of fermentation performance

Parameter	Sterilized POME sludge	Sterilized POME supernatant
Maximum cell concentration (g/l)	107.0	78.3
Maximum acetone concentration (g/l)	0.0163	0
Maximum butanol concentration (g/l)	0.2116	0.4307
Maximum ethanol concentration (g/l)	0.0142	0.1371
Total solvent concentration (g/l)	0.3985	1.1253
A:B:E ratio	0.9:10.6:1	0:0.9:1
$Y_{P/S}$ (g ABE/g substrate)	0.03	0.09
Productivity (g/l/h)	0.01	0.02

ABE Recovery

The recovery of ABE from the fermentation media was done by liquid–liquid extraction using oleyl alcohol as the separation solvent. The total percentages of recovery for sterilized POME supernatant and sterilized POME sludge were 13.17 % and 2.85 %, respectively. There are several possible explanations for the low recovery of the ABE. The recovery percentages are very low for both media due the formation of emulsion that may occur during the extraction process. In addition, low concentration of the ABE in fermentation media could also affect the performance of the liquid–liquid extraction process. Furthermore, the accumulation of mineral salts that may be occurred during the fermentation process could also affect the ABE recovery process (Qureshi and Maddox 1995).

Conclusion

Fermentation process that utilized POME as the main medium to produce solvents proved that palm oil industrial waste can be converted to useful products. In this study, total solvent produced was 1.52 g/l, with butanol as the major product. By means of extraction process using oleyl alcohol as the separation solvents, 2.85 and 13.17 % of ABE were successfully recovered for sterilized sludge and sterilized supernatant, respectively. Sterilized sludge showed the best performance as medium for ABE production, with solvent ratio 0.9:10.6:1. This may be due to the large amount of active bacteria presence in the sludge that helped solvent production.

Acknowledgments The authors wish to thank Ministry of Science, Technology, and Innovation, Malaysia, for funding this project under Grant number 03-01-02-SF0710.

References

Ahmad, A. L., Sumathi, S., & Hameed, B. H. (2006). Coagulation of residue oil and suspended solid in palm oil mill effluent by Chitosan, Alum and PAC. *Chemical Engineering Journal, 118*, 99–105.

Awang, G. M., Jones, G. A., & Ingledew, W. M. (1988). The acetone-butanol-ethanol fermentation. *Critical Reviews in Microbiology, 15*(1), 33–67.

Boudreau, T. M., & Gordon, A. H. (2006). Improved ethanol-water separation using fatty acids. *Process Biochemistry, 41*, 980–983.

Ezeji, T. C., Qureshi, N., & Blaschek, H. P. (2007). Bioproduction of butanol from biomass: From genes to bioreactors. *Current Opinion in Biotechnology, 18*, 1–8.

Groat, W. J., van der Lans, R. G. J. M., & Luyben, K. C. B. A. M. (1992). Technologies for butanol recovery integrated with fermentations. *Process Biochemistry, 27*, 61–75.

Jones, D. T., & Woods, D. R. (1986). Acetone-butanol fermentation revisited. *Microbiological Reviews, 50*(4), 484–524.

Kalil, M. S., Kit, P. W., Yusoff, W. M. W., Sadazo, Y., & Rahman, R. A. (2003). Direct fermentation of palm oil mill effluent to acetone-butanol-ethanol solvent producing clostridia. *Pakistan Journal of Biological Science, 6*(14), 1273–1275.

Masngut, N., Takriff, M. S., Mohammad, A. W., Kalil, M. S., & Kadhum, A. A. H. (2007). Performance of oscillatory flow reactor and stirred tank reactor in solvent fermentation from palm oil mill effluent. *Jurnal Teknologi, 47*, 45–54.

Qureshi, N., & Maddox, I. S. (1995). Continuous production of acetone-butanol-ethanol using immobilized cells of *Clostridium acetobutylicum* and integration with product removal by liquid–liquid extraction. *Journal of Fermentation and Bioengineering, 80*(2), 185–189.

Shaheen, R., Shirley, M., & Jones, D. T. (2000). Comparative fermentation studies of industrial strains belonging to four species of solvent-producing Clostridia. *Journal of Molecular Microbiology and Biotechnology, 2*(1), 115–124.

Siti Jamilah Hanim, M.Y., Mohd Sobri, T., Abdul Amir, H. K., Jamaliah, M. J. & Abdul Wahab, M. (2010). *Fermentasi asiton-butanol-etanol (ABE) dan perolehan pelarut daripada kaldu fermentasi*. Penerbit UKM.

Somrutai, W., Takagi, M., & Yoshida, T. (1996). Acetone-butanol fermentation by Clostridium acetobutyricum ATCC 17777 from a model medium for palm oil mill effluent. *Journal of Fermentation and Bioengineering, 81*(6), 543–547.

Ugoji, E. O. (1997). Anaerobic digestion of palm oil mill effluent and its utilization as fertilizer for environmental protection. *Renewable Energy, 10*(2), 20.

Vijayaragharan, K., Desa, A., & Abdul Aziz, M. E. (2007). Aerobic treatment of palm oil mill effluent. *Journal of Environmental Management, 82*, 24–31.

Initial Study of Thermophilic Hydrogen Production from Raw Palm Oil Mill Effluent (POME) Using Mixed Microflora

N. Ibrahim, J. Md. Jahim and Lim S. S.

Abstract The screening of cultivation conditions for the highest hydrogen production using mixed microflora was performed in serum bottles. The fermentation conditions were studied for temperature, inoculum size, and initial pH, and it shows the significant effect of these factors in enhancing the biohydrogen production. The optimum condition was found at 60 °C with initial pH of 7, and the percentage of inoculum added to POME medium was at 30:70 % (v/v). In the following experiment, batch thermophilic hydrogen fermentation was conducted in 0.5 L medium using the optimum conditions, which were run for 30 h. The highest yield of 4.85 mol H_2/mol glucose was achieved when the fermentation was started with 4 g/L initial glucose concentration.

Keywords Biohydrogen · Thermophilic condition · Palm oil mill effluent · Mixed microflora

Introduction

Biological waste and wastewater treatment by anaerobic digestion is an economically and environmentally sustainable technology (Noike and Mizuno 2000) that has grown substantially over the last two decades. The hydrogen production through anaerobic conversion not only produced renewable energy but also provide an efficient system for wastewater treatment.

N. Ibrahim · L. S. S.
Fuel Cell Institute, Universiti Kebangsaan Malaysia, Bandar Baru Bangi, 43600 Selangor, Malaysia

J. Md. Jahim (✉)
Department of Chemical and Process Engineering, and Built Environment, Universiti Kebangsaan Malaysia, Bandar Baru Bangi 43600 Selangor, Malaysia
e-mail: jamal@eng.ukm.my

R. Pogaku et al. (eds.), *Developments in Sustainable Chemical and Bioprocess Technology*, DOI: 10.1007/978-1-4614-6208-8_6,

Palm oil is one of the world's most rapidly expanding equatorial crops. There is about 1.5 m^3 water used to process one tonne of fresh fruit bunches (FFB), and half of this quantity would end up as waste, called palm oil mill effluent (POME). POME poses a great threat to the environment because of its high biological and chemical oxygen demands (Zhang et al. 2008).

To date, there have been several studies conducted using POME to produce hydrogen and among them are Atif et al. (2005), O-Thong et al. (2008) and Badiei et al. (2012). The first two groups conducted hydrogen production at 60 °C, while the later used mesophilic condition at 37 °C. Currently, the thermophilic bacteria are considered to be more promising microorganisms compared to mesophilic ones as they give higher hydrogen production rate and less variety of fermentation end products (O-Thong et al. 2008).

Thus, this study was conducted to investigate the best cultural conditions such as temperature, pH, and inoculum size for biohydrogen production from POME under thermophilic condition. Hydrogen fermentation using 0.5-L bioreactor was conducted to study the profile of biogas and hydrogen production.

Methodology

Preparation of Seed Sludge

Mixed microflora used in this study was collected from the sludge pit at POME Treatment Plant at West Palm Oil Mill, Pulau Carey, Malaysia. pH and volatile suspended solid (VSS) of the sludge were measured around pH 4–4.5 and 28 g/L, respectively. Acclimatization of the sludge was done in sequencing batch mode by taking out 50 % of the medium and adding 50 % fresh POME into the reactor every 48 h for a month. The sludge later was used as an inoculum for the hydrogen fermentation.

Preparation of Fermentation Substrate

Freshly discharged brownish color raw POME at temperature of 80–90 °C was collected from East Palm Oil Mill, Pulau Carey, Malaysia. It was kept in cold room at 4 °C prior to use. Table 1 indicates the characteristics of POME used throughout this study.

Table 1 Characteristics of POME used as feedstock in this study

Parameter	Value	Parameter (g/L)	Value
Moisture (%)	94.33 ± 0.6	TSS	33.62 ± 1.1
pH	3.96 ± 0.07	VSS	26.88 ± 2.8
COD (g/L)	42.2 ± 2.9	Total sugar	9.03 ± 0.6

Serum Bottle Experiments

A series of serum bottle tests were conducted to determine the optimal cultural condition for hydrogen production. The fermentation was carried out at three different temperatures (50, 60, and 70 °C), four initial pHs (5, 5.5, 6, and 7), and volume ratio of the inoculum to POME medium at 10, 30, 60, 90, and 100 % (v/v). The pH of the medium was adjusted, and it was purged with nitrogen for 3 min to provide anaerobic condition. Once the inoculum was added, they were sealed with rubber stopper and incubated for 24 h. Total volume of biogas produced from each medium was measured by releasing the evolved gas from the bottles with syringes.

Batch Fermentation in 0.5 L Working Volume

Batch fermentation was carried out in a 500-ml Scott Duran® bottle incubated at 60 °C. 30 % (v/v) of the inoculum was transferred to raw POME medium, and the pH of the mixture was adjusted to pH 7.0. Before starting the fermentation, nitrogen was purged for 15 min to provide anaerobic condition. Gaseous and liquid products were taken 3 h and sent for the analysis.

Analytical Method

Reducing sugar content in the media was analyzed using dinitrosalicylic (DNS) assay (Miller 1959). The amount of metabolites such as acetic acid, butyric acid, and ethanol were analyzed using GC SRI 8600C. Total suspended solids (TSS) and volatile suspended solids (VSS) were measured according to the standard methods (APHA 1998). Biogas produced from fermentation was collected using water displacement method where an inverted measuring cylinder was filled with acidic water. Then, the composition of biogas was analyzed using gas chromatography (GC) model SRI 8600C with HID and TCD detectors. The gas carrier for GC is helium (MOX 99.99 %) at a flow rate of 20 mL/min. The pressure for the carrier was set at 29 psi, while the injection port, detector, and oven temperatures were set at 40 °C, 150 °C, and 40 °C, respectively.

Hydrogen Yield and Hydrogen Production Rate

The profile of hydrogen production was subjected to the Gompertz equation for hydrogen yield and hydrogen production estimation, as shown in Eq. (1) (Lay et al. 1999),

$$H(t) = P \cdot \exp\left\{-\exp\left[\frac{R_m e}{P}(\lambda - t) + 1\right]\right. \tag{1}$$

where H is cumulative hydrogen production (mL), λ is the lag phase (hr), P is maximum hydrogen production (mL), R_m is maximum hydrogen production rate (mL/hr), and e is a constant value of 2.72. The values of P, R_m, and λ for each batch were estimated by simulation of experiment data with Eq. (1) using statistic nonlinear regression wizard function of Sigma Plot® (version 10). The molar hydrogen production rate (mmol H_2 L^{-1} h^{-1}) was calculated using the ideal gas law where molar H_2 production rate (mmol H_2 L^{-1} h^{-1}) = volumetric H_2 production rate (ml H_2 L^{-1} h^{-1})/(RT) where $R = 0.0821$ atm $K^{-1}mol^{-1}$ and $T = 298$ K. Hydrogen production yield was calculated as total molaric amount of hydrogen divided into molaric amount of consumed glucose.

Results and Discussions

Cultural Conditions of Thermophilic Anaerobic Mixed Microflora

In the initial study in serum bottle experiment, the best conditions of incubation temperature, initial pH, and right inoculum size were found. The results shown in Fig. 1a were obtained from 3 different temperatures. Temperature at 60 °C gave the highest hydrogen production which agrees with Atif et al. (2005) and O-Thong et al. (2008) as they recorded a similar optimum temperature for hydrogen production.

With the best temperature 60 °C, experiments were continued with different initial pHs and inoculum sizes. The effect of initial pH is known to be crucial in influencing biohydrogen production as in Fig. 1b; it shows that the biogas and hydrogen production increased with different initial pHs ranging from 5.0 to 7.0. The inoculum composition had a significant effect on the biohydrogen production as Fig. 1c shows that 30 % (v/v) inocula gave the highest biogas and hydrogen

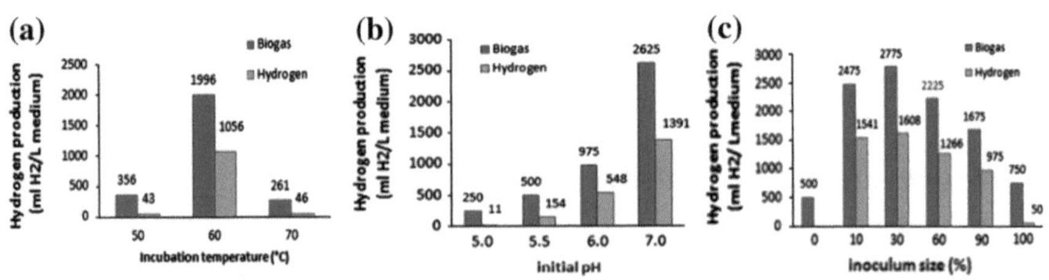

Fig. 1 Cultural condition of thermophilic anaerobic sludge in 20 ml working volume; effect of incubation temperature (**a**), effect of initial pH (**b**), effect of inoculum size (**c**)

production. Therefore, the results obtained showed that the incubation temperature at 60 °C with initial pH7 and 30 % (v/v) inoculum size was optimal for hydrogen production using the mixed microflora.

Batch Fermentation at Optimum Cultural Condition in 0.5 L Working Volume

The fermentations were continued in 0.5 L laboratory-scale fermenter. Biogas produced after 3 h fermentation and stopped after 21 h. The total cumulative biogas produced as shown in Fig. 2a was 2 L and H_2 content was around 50-65 %, with no CH_4 detected. A maximum hydrogen productivity of 5.7 mmol $H_2\ L^{-1}\ h^{-1}$ equivalent with 3.1 L H_2/L POME were achieved between 9th and 15th hour in late exponential phase. Acetic and butyric acids were the main organic acids whose dominant production is closely related to the high hydrogen productivity. Figure 2b shows the profile of metabolites production during fermentation. The amount of butyric acid, acetic acid, and ethanol produced were 2.07 g/L, 1.43 g/L, and 1.07 g/L, respectively. No propionic acid and lactic acid were detected.

Figure 2c shows the profile of pH and substrate consumption throughout fermentation. pH of the fermentation reduced from pH of 7 to 5.3. The decrease in pH was attributed to the accumulation of organic acids. After 3 h of fermentation, the glucose was utilized and the consumption continues until hydrogen production stopped. Total glucose consumption was 75 %, and the yield of mol hydrogen production per mol glucose consumed in this study was 4.85. As fermentation progressed, TSS and VSS decreased and the total reduction was 20 % and 23 %, respectively.

Generally, hydrogen can be produced through two basic biochemical reactions in acidogenic phase. Equations (2) and (3) show 4 mol of hydrogen is produced with 2 mol of acetate, while only 2 mol hydrogen is produced from glucose concomitantly with 1 mol of butyrate (Hawkes et al. 1998). The highest theoretical yield of hydrogen is thus when acetic acid is the only end product, but this is almost never the case because there is always a mixture of acetic acid, butyric acid, or other end products (Levin et al. 2004)

$$C_6H_{12}O_6 + 2H_2O \rightarrow 2C_2H_4O_2 + 2CO_2 + 4H_2 \quad (2)$$

$$C_6H_{12}O_6 \rightarrow C_4H_8O_2 + 2CO_2 + 2H_2. \quad (3)$$

O-Thong et al. (2008) deduced that metabolic pathway for carbohydrate fermentation by thermophilic hydrogen producer for H_2 production associated with ratio mol of butyric (Hbu) to acetic acid (Hac) with respect of Hbu/Hac of 1.0 was shown in Eq. (4),

$$3C_6H_{12}O_6 + 2H_2O \rightarrow 8H_2 + 6CO_2 + 2C_2H_4O_2 + 2C_4H_8O_2 \quad (4)$$

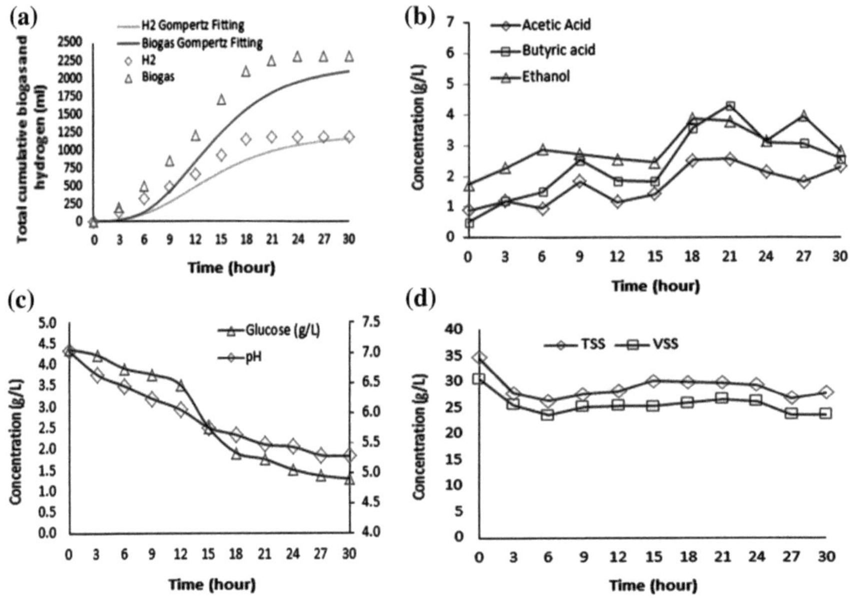

Fig. 2 Batch fermentation of thermophilic anaerobic sludge with POME in 0.5 L fermentation size; cumulative biogas and hydrogen production with Gompertz fitting (**a**), profile of volatile fatty acids (VFAs) and alcohol production (**b**), profile of glucose consumption and pH changes (**c**), profile of TSS and VSS reduction (**d**)

This equation hence resulted in a theoretical H_2 yield of 2.67 mol H_2/mol hexose. Through this study, the Hbu/Hac obtained was around 1.3–2.2. Hence, with respect to average Hbu/Hac of 2.0, the theoretical H_2 yield for this study would be 5.34 mol H_2/mol hexose. The overall experimental hydrogen yield in this batch fermentation with initial glucose concentration of 4 g/L was 4.85 mol H_2/(mol glucose consumed). This value is considered very high but still lower than the theoretical value. The hydrogen yield based on L H_2/L POME with previous study using mixed culture was illustrated in Table 2.

Table 2 Comparison on H_2 production using raw POME with other researchers

Microorganism	Mode/pH/T (°C)	Hydrogen production (L H_2/L POME)	References
Mixed culture	Batch/5.5/60	4.7	Atif et al. (2005)
Mixed culture	Continuous/5.5/55	2.6	Ismail et al. (2010)
Mixed culture	Batch/6/37	0.7	Khaleb et al. (2012)
Mixed culture	Batch/7/60	3.1	This study

Conclusion

The study of the cultural conditions including temperature, inoculum size, and initial pH had shown the significant effect in enhancing biohydrogen production. The thermophilic biohydrogen production conducted in 0.5 L batch fermentation size gave the maximum H_2 production of 3.1 L H_2/(L POME added) equivalent with 4.85 mol H_2/mol glucose and productivity of 5.7 mmol H_2 L^{-1} h^{-1}.

Acknowledgments The authors would like to acknowledge financial support from UKM-Sime Darby Foundation program on Sustainable Development–Zero Waste Technology for the Palm Oil Industry.

References

American Public Health Association (Apha). (1998). *Standard methods for the examination of water and wastewater*. Washington DC, USA.

Atif, A. A. Y., Fakhru'razi, A., Ngan, M. A., Morimoto, M., Iyuke, S. E., & Veziroglu, N. T. (2005). Fed batch production of hydrogen from palm oil mill effluent using anaerobic microflora. *International Journal of Hydrogen Energy, 30*, 1393–1397.

Badiei, M., Jahim, J. M., Anuar, N., Abdullah, S. R. S., Su, L. S., & Kamaruzzaman, M. A. (2012). Microbial community analysis of mixed anaerobic microflora in suspended sludge of ASBR producing hydrogen from palm oil mill effluent. *International Journal of Hydrogen Energy, 37*, 3169–3176.

Hawkes, F. R., Dinsdale, R., Hawkes, D. L., & Hussy, L. (1998). Sustainable fermentative hydrogen production : Challenges for process optimisation. *International Journal of Hydrogen Energy, 27*, 1339–1347.

Ismail, I., Hassan, M. A., Abdul Rahman, N., & Chen, S. S. (2010). Thermophilic biohydrogen production from palm oil mill effluent (POME) using suspended mixed culture. *Biomass and Bioenergy, 34*, 42–47.

Khaleb, N. A., Jahim, J. M., & Ahmad Kamal, S. (2012). Biohydrogen production using hydrolysate of palm oil mill effluent. *Journal of Asian Scientific Research, 2*(11), 705–710.

Lay, J. J., Lee, Y. J., & Noike, T. (1999). Feasibility of biological hydrogen production from organic fraction of municipal solid waste. *Water Research, 33*, 2568–2579.

Levin, D. B., Pitt, L., & Love, M. (2004). Biohydrogen production technology: Prospect and limitations for practical application. *International Journal of Hydrogen Energy, 29*, 173–185.

Miller, G. L. (1959). Use of dinitrosalicylic acid reagent for determination of reducing sugar. *Analytical Chemistry, 31*, 426–428.

Noike, T., & Mizuno, O. (2000). Hydrogen fermentation of organic municipal wastes. *Water Science and Technology, 42*(12), 155–162.

O-Thong, S., Prasertsan, P., Karakashev, D., Dimitar, S., & Irini, V. (2008). Thermophilic fermentative hydrogen production by the newly isolated *Thermoanaerobacterium thermosaccharolyticum* PSU-2. *International Journal of Hydrogen Energy, 33*(4), 1204–1214.

Zhang, Y., Yan, L., Qiao, X., Chi, L., Niu, X., & Mei, X. (2008). Integration of biological method and membrane technology in treating palm oil mill effluent. *Journal of Environmental Sciences China, 20*(5), 558–564.

Production of Cellulases by *Aspergillus* sp. GDX02 in a Solid-State Fermentation Using Oil Palm Empty Fruit Bunch

H. W. Kim, G. J. Lee, D. M. Kim, Y. W. Lee, Y. S. Kim and H-J. Chung

Abstract A reduction in enzyme cost is one of the important factors for making bioethanol production more economically feasible. To this aim, we isolated a fungus, GDX02, to produce cellulases using agricultural wastes, including empty fruit bunch (EFB), under solid-state fermentation (SSF). This study was conducted to determine the best substrate and optimal fermentation conditions for maximum cellulase production using GDX02. Of the different substrates tested, rice straw resulted in the highest enzyme production. As a nitrogen source, a supplement of yeast extract was necessary to achieve high enzyme activity. In particular, FPase activity reached its maximum at a supplement of 7 % yeast extract, whereas β-glucosidase activity experienced little change under all concentrations of yeast extract tested. The GDX02 produced high FPase and β-glucosidase after three-day culture of EFB at 40 % moisture level. The saccharification yield of pretreated EFB by GDX02 cellulase was comparable to that from the commercial enzymes Celluclast and C-Tec2, indicating the great potential use of GDX02 cellulase for cellulosic ethanol production from EFB.

Keywords Cellulase · Oil palm empty fruit bunch · Solid-state fermentation · *Aspergillus*

H. W. Kim · G. J. Lee · D. M. Kim · Y. W. Lee · Y. S. Kim (✉) · H-J.Chung (✉)
GenDocs Inc, Daejeon, Republic of Korea
e-mail: yskim@gendocs.co.kr

H-J.Chung
e-mail: hwajee@gendocs.co.kr

R. Pogaku et al. (eds.), *Developments in Sustainable Chemical and Bioprocess Technology*, DOI: 10.1007/978-1-4614-6208-8_7,

Introduction

Lignocellulose, an important source for the production of both renewable energy and chemicals, is composed of cellulose and hemicellulose that can be hydrolyzed into glucose or xylose by cellulases and hemicellulases, respectively. Cellulases are a mixture of endo-D-glucanase, exo-D-glucanase, and β-glucosidase that synergistically convert cellulose into glucose. The cellulases are mostly produced by fermentation of the fungi *Trichoderma* and *Aspergillus* under a condition of either submerged or solid media. Solid-state fermentation (SSF) is a process wherein the solid substrate provides both the nutrients and physical support during the culture. SSF has been widely employed in the production of industrial enzymes due to its several advantages over submerged fermentation, including higher concentration of enzymes, high volumetric productivity, lower operating expenses, and lower sterility demand. Besides, SSF generates less wastewater because of lower water requirements.

Agro-industrial by-products and crop residues can be used as carbon sources for cellulase production in SSF at a lower cost. Oil palm empty fruit bunch (EFB) is an agricultural by-product rich in cellulose and hemicellulose, typically containing 35 to 40 % cellulose and 20 to 25 % hemicellulose. It is abundantly produced in Indonesia and Malaysia. Palm oil mills in Malaysia generate 2.4 million tons of EFB annually. EFB was often used as fuel to generate steam or electricity, but most of EFB is discarded as waste.

There have been a few reports describing SSF production of cellulases using EFB and various microbes such as *Trichoderma hazianum* (Alam et al. 2009) and *Botryosphaeria* sp. (Bahrin et al. 2011). In addition, a pretreated EFB was also used as a carbon source in submerged fermentation (Umikalson et al. 1997; Ariffin et al. 2008; Shahriarinour et al. 2011).

The objective of this study was to identify the promising fungal species exhibiting a high level of production of cellulases and to optimize and evaluate cellulase production during SSF of EFB.

Methodology

Solid-State Fermentation

SSF was carried out using 40 g of substrate (EFB or rice straw) and yeast extract as a nitrogen source in a 400 ml glass bottle (81 × 132 mm). Initial moisture content was adjusted to between 40 and 80 % (w/w) with deionized water before being sterilized at 121 °C for 15 min. Four agar blocks (15 × 15 mm) of fungi grown for one to two weeks ($1 \times 10^6 \sim 10^8$ conidia) were inoculated into each bottle and then incubated statically at 30 °C for two to ten days.

Enzyme Extraction

The enzymes were extracted from the fermented substrate with 6 ml (per gram of biomass) of 0.05 M sodium citrate buffer (pH 4.8) by shaking (250 rpm) at 30 °C for 1 h. Eluted enzyme mixture was then separated from the fungal biomass by centrifugation (10,000 × g for 10 min). The clarified supernatant was used as the source for crude enzymes.

Enzyme Assay and Protein Determination

The activities of total cellulase, endo-glucanase, and *β*-glucosidase were determined. Total cellulase activity (filter paper activity, FPase) was assayed by incubating 500 μl of crude enzyme with 1 ml of 0.05 M sodium citrate buffer containing Whatman No. 1 filter paper (50 mg, 1 × 6 cm) at 50 °C for 1 h. The amount of released reducing sugars was determined by the 3, 5-dinitrosalicylic acid (DNS) method. Endo-glucanase activity (CMCase) was assayed in the total reaction mixture of 200 μl containing 50 μl of crude enzymes, 100 μl of 0.1 M potassium phosphate buffer (pH 6.0), and 50 μl of 1 % (w/v) CMC solution. The mixture was incubated at 50 °C for 30 min, and the amount of released reducing sugars was determined by the DNS method. The activity of *β*-glucosidase (CBase) was determined by measuring hydrolyzed glucose from cellobiose. A reaction mixture of 1 ml containing 500 μl of crude enzyme and 15 mM cellobiose solution in sodium citrate buffer was incubated at 50 °C for 30 min. After incubation, 10 μl of reaction mixture was mixed with 1 ml of glucose kit reagent (Young-Dong Diagnostics, Yong-in, Korea) and incubated at 37 °C for 10 min. Then, the amount of glucose in the final reaction solution was measured with a UV–Vis spectrophotometer at 600 nm with glucose as the standard. Total protein in the crude enzyme extract was determined by the Bradford method with bovine serum albumin (BSA) as the standard. One unit (U) of enzyme activity was defined as the amount of enzyme required to release 1 μ mol of products (reducing sugar or glucose) from the appropriate substrates per minute under the assay conditions. In this study, the enzyme activities are expressed as units per gram of EFB (U/g).

Saccharification of EFB

One gram of pretreated EFB was digested with 10 FPU of concentrated enzymes from GDX02, Celluclast, or C-Tec2 (Novozymes, the Netherlands) in 100 ml of 0.05 M sodium citrate buffer. This mixture was incubated at 50 °C for three days with gentle shaking. Finally, the glucose yield (%) was analyzed using high-performance liquid chromatography.

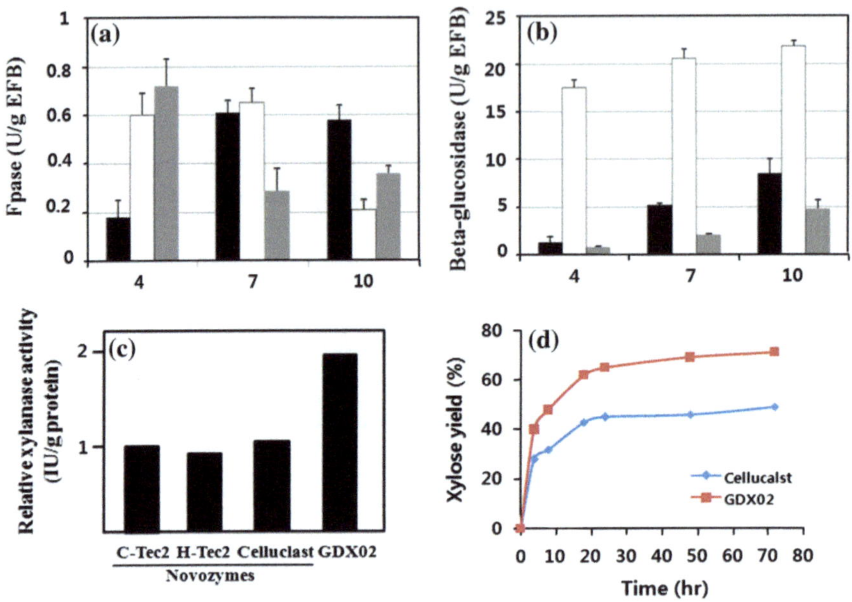

Fig. 1 Cellulase activities from GDX01, GDX02, and *T. longibrachiatum*. **a** FPase. **b** β-glucosidase. *Filled bar* GDX01; *Open bar* GDX02; *Gray bar T. longibrachiatum*. GDX02 produces a high level of xylanase. **c** Comparison of xylanase activity between commercial and GDX02 cellulase. **d** Xylose yield after saccharification

Results and Discussion

Cellulase Production

Several fungi were tested on the basis of cellulolytic activity in SSF to determine the high cellulase producer(s). One of the fungi belonging to *Aspergillus* sp. demonstrated high cellulase activities, especially with regard to β-glucosidase, and it was named GDX02 (Fig. 1a and b). GDX02 produced cellulases at quite an early stage of SSF compared with GDX01 (data not shown). Obviously, GDX02 is a good cellulase producer in terms of the high production of both β-glucosidase (Fig. 1b) and xylanase (Fig. 1c).

The xylanase activity of GDX02 cellulase was two times higher than that of commercial enzymes (Fig. 1c). GDX02 cellulase produced xylose to 70 % during the saccharification of NaOH-pretreated EFB, and its yield was 25 % higher than that of Celluclast (Fig. 1d).

Optimization of SSF

Many factors affect enzyme production in SSF. Among them, we investigated the effects of biomass, the concentration of nitrogen source, and initial moisture content. First, we compared rice and EFB on their cellulase-producing ability. The best results were obtained when rice straw was used as the solid substrate. On the third day of SSF, GDX02 cultured with the rice straw exhibited the highest FPase and β-glucosidase activities. In particular, there was a big difference in FPase activity between rice straw and EFB. This result differs from that using the GDX01 fungus, which showed a similar amount of FPase production on both rice straw and EFB (data not shown).The initial moisture content of growth medium is also one of the important factors affecting enzyme production during SSF. A report indicated that the optimal initial moisture content depends on the microorganism, biomass, and type of end product (Kalogeris et al. 2003). The highest FPase and β-glucosidase activities were obtained after three-day culture of EFB at 40 % moisture level. High moisture levels (above 70 %) rather reduced GDX02 cellulase production.

Addition of different nitrogen sources, whether in organic or inorganic form, to the solid substrate has been shown to have variable effects on cellulose production. NH_4NO_3 enhanced the cellulose production of *T. reesei* under the condition of SSF (Singhania et al. 2006). Several studies reported that organic nitrogen sources resulted in higher production of cellulases than did inorganic nitrogen sources (Jeya et al. 2010; Deswal et al. 2011).

The effect of yeast concentrations on cellulose production during SSF was investigated. For FPase production, the optimum concentration of yeast extract was 7 %. Without yeast extract, GDX02 could not produce cellulase, indicating that the nitrogen source was essential for cellulase production by GDX02. A concentration (above 7 %) caused a decrease in FPase production due to the growth inhibition of GDX02. Interestingly, β-glucosidase production exhibited a different pattern. There was no difference in β-glucosidase production in all concentrations of yeast extract examined. This result is different from that using GDX01, in which β-glucosidase activity was maximum at 5 % yeast extract (data not shown).

Glucose, the end product of saccharification of biomass, has been known to inhibit β-glucosidase activity. To examine the effect of glucose on GDX02-produced β-glucosidase, the enzymatic saccharification of pretreated EFB was performed both with and without glucose. C-Tec2 was the most resistant to glucose (almost 50 % inhibition), whereas GDX02 cellulase was the least resistant to glucose (Fig. 2a).

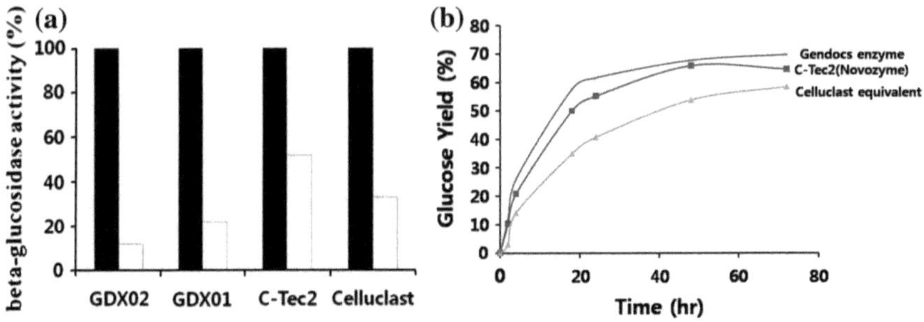

Fig. 2 **a** The inhibitory effect of glucose on beta-glucosidases. *Filled bar* 0 g/L; *open bar* 10 g/L glucose. **b** Saccharification of NaOH-pretreated EFB

Saccharification of EFB Using GDX02 Cellulase

GDX02 cellulase was applied for saccharification of NaOH-pretreated EFB to evaluate its hydrolysis efficiency (Fig. 2b). Compared with commercial enzymes C-Tec2 and Celluclast, the saccharification rate of GDX02 cellulase was much faster than that of Celluclast and was even equivalent to that of C-Tec2.

Conclusions

Results from this study have indicated that GDX02 produced high activities of β-glucosidase and xylanase in SSF using EFB and rice straw as substrate. Xylose yield after saccharification by GDX02 was 25 % higher than that produced by Celluclast. Cellulase production by GDX02 was improved by the addition of a 7 % yeast extract under the condition of 40 % initial moisture content of growth medium. Glucose inhibition of β-glucosidase activity of GDX02 could be overcome by the introduction of a simultaneous saccharification and fermentation (SSF) processes. In conclusion, the fungus GDX02 can be used for large-scale cellulase production using the agricultural by-products in SSF, and GDX02-produced cellulase is capable of converting the lignocellulosic biomass to bioethanol.

Acknowledgments This work was supported by a grant to Y-W.L (20103010090020) from the New & Renewable Energy Technology Development Program of the Korea Institute of Energy Technology Evaluation and Planning funded by the Korean Ministry of Knowledge Economy and by a grant to H-J.C (30801005) from the R&D Convergence Center Support Program, Ministry for Food, Agriculture, Forestry and Fisheries.

References

Alam, M. Z., Mamun, A. A., Qudsieh, I. Y., Muyibi, S. A., Salleh, H. M., & Omar, N. M. (2009). Solid state biodegradation of oil palm empty fruit bunches for cellulase enzyme production using a rotary drum bioreactor. Biochemical Engineering Journal, 46, 61–64.

Ariffin, H., Hassan, M. A., Shah, U. K. M., Abdullah, N., Ghazali, F. M., & Shirai, Y. (2008). Production of bacterial Endoglucanase from pretreated oil palm empty fruit bunch by *Bacillus pumilus* EB3. *Journal of Bioscience and Bioengineering, 106*, 231–236.

Bahrin, E. K., Seng, P. Y., & Abd-Aziz, S. (2011). Effect of oil palm empty fruit bunch particle size on cellulase production by *Botryosphaeria sp.* under solid state fermentation. *Australian Journal of Basic and Applied Science, 5*, 276–280.

Deswal, D., Khasa, Y. P., & Kuhad, R. C. (2011). Optimization of cellulase production by a brown rot fungus *Fomitopsis sp.* RCK2010 under solid state fermentation. *Bioresource Technology, 102*, 6065–6072.

Jeya, M., Joo, A., Lee, K. M., Sim, W. I., Oh, D. K., Kim, Y. S., et al. (2010). Characterization of endo—*β*- 1,4-glucanase from 1 novel strain of *Penicillium pinophilum* KMJ601. *Applied Microbiology and Biotechnolgy, 85*, 1005–1014.

Kalogeris, E., Iniotaki, F., Topakas, E., Christakopoulos, P., Kekos, D., & Macris, B. J. (2003). Performance of an intermittent agitation rotating drum type bioreactor for solid-state fermentation of wheat straw. *Bioresource Technology, 86*, 207–312.

Shahriarinour, M., Ramanan, R. N., Wahab, M. N. A., Mohamad, R., Mustafa, S., & Ariff, A. B. (2011). Improved cellulase production by *Aspergillus terreus* using oil palm empty fruit bunch fiber as substrate in a stirred tank bioreactor through optimization of the fermentation conditions. *Bio Resources, 6*, 2663–2675.

Singhania, R. R., Sukumaran, R. K., Pillai, A., Szakacs, G., & Pandey, A. (2006). Solid-state fermentation of lignocellulosic substrates for cellulase production by Trichoderma reesei NRRL 11460. *Indian Journal of Biotechnology, 5*, 332–336.

Umikalson, M. S., Ariff, A. B., Zulkifli, H. S., Tong, C. C., Hassan, M. A., & Karim, M. I. A. (1997). The treatment of oil palm empty fruit bunch fibre for subsequent use as substrate for cellulase production by *Chaetomium globosum* kunze. *Bioresource Technology, 62*, 1–9.

Acclimatization Process of Microorganisms from Activated Sludge in Kenaf-Retting Wastewater

Z. Zawani, L. Chuah-Abdullah, F.-R. Ahmadun and K. Abdan

Abstract Acclimatization process of microorganisms in Kenaf-retting wastewater was studied using return activated sludge (RAS). RAS was collected from near sewage treatment plant. Initial concentration of chemical oxygen demand (COD) in the wastewater and mixed liquor suspended solids (MLSS) of activated sludge were 2,300 and 1,000 mg/l, respectively. The MLSS and COD concentrations were stabilized at day-25th. The highest substrate removal was recorded at 85 %, and it is found that activated sludge bacteria are well adapted in the new environment. At the end of the experiment, microorganisms, namely *Achromobacter sp, Bacillus sp* and *Acinetobacter sp*, were found dominated in the wastewater.

Introduction

Kenaf is known as a renewable source and often used in developing new environmental-friendly products. It consists of multiple useful components (e.g., stalks, leaves and seeds), and within each of these plant components, there are various usable portions (e.g., fibres and fibre strands, proteins, oils and allelopathic chemicals). The most important part in kenaf is bast, which is a long fibre strands

Z. Zawani · L. Chuah-Abdullah · F.-R. Ahmadun
Chemical and Environmental Engineering, Department of Engineering,
Universiti Putra Malaysia, 43300 UPM Serdang, Selangor, Malaysia

L. Chuah-Abdullah (✉)
Institute of Tropical Forestry and Forest Products (INTROP),
Universiti Putra Malaysia, 43300 UPM Serdang, Selangor, Malaysia
e-mail: chuah@eng.upm.edu.my

K. Abdan
Biological and Agricultural Engineering, Department of Engineering,
Universiti Putra Malaysia, 43300 UPM Serdang, Selangor, Malaysia
e-mail: khalina@eng.upm.edu.my

R. Pogaku et al. (eds.), *Developments in Sustainable Chemical and Bioprocess Technology*, DOI: 10.1007/978-1-4614-6208-8_8,

that are composed of many individual smaller fibres, and it is contained in the bark of kenaf stalk. The bast can be obtained from retting process. Retting is a fibre separation process, and it can be done in several ways like dew, chemical or water-retting. The process is a slow degradation process that can take up a few days to complete. Water-retting generally is applied in industrial sector due to its effectiveness and simplicity of the process itself. However, this procedure utilizes huge amount of water. Wastewater that generated from the retting process produced greenish colour, high in suspended solids and odour. The condition of wastewater is unacceptable from an aesthetic value. Generally, the wastewater will be released to watercourse without any proper treatment. Stringent regulations of water quality, public awareness on environmental issues and a call for water reclamation encourage an immediate action to resolve this issue.

Aerobic technology has been applied in wastewater treatment plant for a very long time. Being one of the biological treatment processes, aerobic process involved decompositions of organic and inorganic materials with the presence of oxygen molecule and aerobic microorganism. Activated sludge (AS) from aerobic process in wastewater treatment plant consists of various types of microorganisms (MOs). Therefore, AS is often selected as a candidate for adaptation process in new wastewater environment. The studies on acclimatization process are important in introducing the MOs to unknown or foreign pollutants, which at some point MOs can enhance degradation of pollutants compound effectively. Furthermore, research on the acclimatization process is also necessary to improve WW technologies and the acclimatized MOs will help to increase the efficiency of biological treatment process.

Some researchers have studied the acclimatization process. Most of them examined the capability of MOs from activated sludge process (ASP) to adapt either with new environment or under extreme condition. In 2007, Özbelge et al. were studied about an exposure of heavy metal (HM) on MOs from activated sludge process. They found that high level of HM exposure to the acclimatized MOs can affect the ASP performance and it is difficult to predict the response of microorganisms (MOs) in the AS as different MOs have different sensitivity to the factor imposed to them. In contrast, other researchers studied on the capability of MOs and mechanisms that may occur during acclimatization period. Wiggins et al (1987) stated that several investigators have proposed that acclimatization process is a result of the time needed for enzymes to be induced (Stephenson et al. 1984; Torstensson et al. 1975), while others suggested time needed by MOs for mutation or genetic exchange to occur (Schmidt et al. 1983; Walker and Newman 1956). Mutation or appearance of new genotypes by genetic exchange may take place in acclimatization process (Wiggins et al 1987).

The main purpose of this study is to investigate acclimatization process of activated sludge in the kenaf-retting wastewater (KRW).

Table 1 Characteristics of wastewater used in the experiment

Parameter	Value (mg/l)
pH	4.0
Total suspended solids (TSS)	550
Total solids (TS)	2,400
Colour	2,390
Turbidity	500
Phosphorus	7.7
Nitrogen	13.5
Chemical oxygen demand (COD)	2,300.0
Biological oxygen demand (BOD)	20.0

Methodology

Wastewater

Wastewater used in this experiment was synthetically prepared. Kenaf bast fibres were retted in water for ten (10) days with the ratio of fibre to water 1 g: 40 ml as suggested by Hongqin and Chongwen (2010). On day 10, the fibres were removed from the container and the wastewater was kept in the container prior to the use. The characteristics of sample are tabulated in Table 1.

Inoculums

For this experiment, activated sludge was obtained from University Putra Malaysia (UPM) sewerage treatment plant. The sludge was collected from return activated sludge (RAS) point and preserved in the cold room.

Experimental Set-Up

Sequencing batch reactor (SBR) system is adapted in order to acclimatize the MOs collected from the treatment plant. The working volume of the system is 1.5 L with the total volume of 2 L. Peristaltic pump (Watson-Marlow Bredel 323 Dz) equipped with silicon tubing was used for feeding and withdrawing the sample. Aeration from air pump and mixing were provided in the reactor. The concentration of biomass in this experiment was set to 1,000 mg/L. Temperature and pH of the system were kept constant at 28.0 ± 0.2 °C and pH 7.0 ± 0, respectively. The pH of medium and samples were adjusted using 0.1 N sulphuric acid (H_2SO_4) and 0.1 N sodium hydroxide (NaOH).

The SBR was operated in 24-h cycle period in four stages; feed and decant time is 1 h (which total hours is 2 and there was no air provided during this period—anoxic condition), reaction time is 20 h (with air supply) and settle/idle time is 2 h (without air supply). The experiment was continuous running until the concentration of COD in treated sample did not show further significant reduction.

Sample Analysis

The treated wastewater was analysed for concentrations COD and mixed liquor suspended solids (MLSS). All parameters followed standard methods (D. Eaton, H. Franson, & American Public Health Association 2005).

The acclimatized sludge samples were cultured and isolated for identification of microorganism. Then, the samples were sent to accredited laboratory for identification process.

Results and Discussion

From the experiment, it is found that MOs in the system required 25 days to acclimatize in the wastewater (as shown in Fig. 1). From the graph, it is revealed that on day 25th, the MOs start to stabilize and did not change significantly until day 30th. At this stage, biomass concentration was recorded at 2,900 mg/L with the increase in 65.5 %. Since day 1st, the biomass starts to increase gradually, and this condition may occur due to diversity of MOs present in the sludge. The variations in MOs in the wastewater have made adaptation process easy.

The highest removal of COD was recorded at 85 %, as shown in Fig. 2. As can be seen from both graphs, degradation level of organic compound and the growth of biomass were increased gradually since first day. However, after 25 days, the degradation process became slow which shows no significant change in concentration of COD. The same condition occurred in biomass growth. There is no

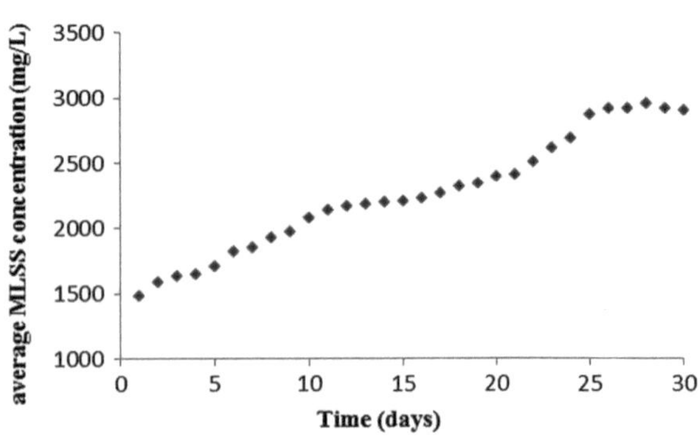

Fig. 1 The average biomass concentration recorded in acclimatization process

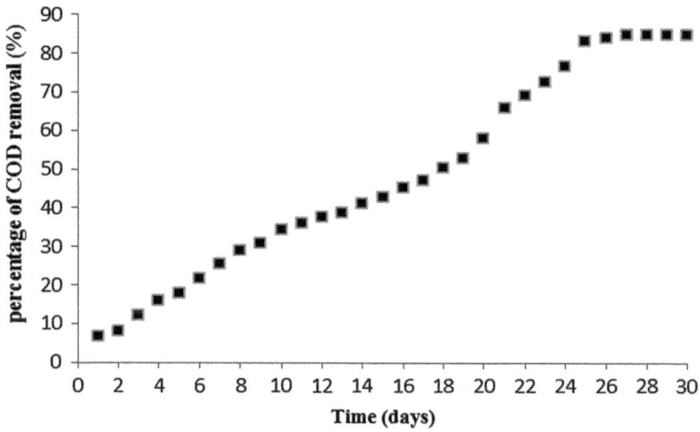

Fig. 2 The COD removal efficiency in the acclimatization process

significant change after 30 days. The main species of bacteria that survived and responsible in degrading pollutant compounds, there are *bacillus sp., achromobacter sp.* and *acinetobactor sp.*

Several events may occur in the acclimatization process as proposed by previous researchers. Frequently, it is suggested that adaptation process indicates the time for induction of enzymes after exposure of the indigenous populations to the compound of interest (Wiggins et al. 1987). The time required for enzyme induction is usually minutes or hours (Richmond 1968). The exposure to unknown and unfamiliar organic pollutant may give obstacle to MOs in degrading the pollutant. However, a long period of exposure to the wastewater and the diversity of microorganism present in the AS may alleviate acclimatization process.

Another factor that may account for the acclimatization process is mutation or the appearance of new genotypes by genetic exchange. In 1983, Schmidt et al. explained that in their studies, new genotype by plasmid transfer appeared during acclimation of microbial communities to chlorophenols. However, the latter suggestion shows that the possibilities of new genotype appearance remain unidentified.

Conclusions

Activated sludge from sewage treatment plant has potential to be used as inoculums for the biological treatment of wastewater due to its assortment. Wastewater used in the experiment was prepared using the same procedure done by farmers and industry. From the experiment, it is found that microorganisms took 25 days to adapt in the kenaf-retting wastewater . The percentage removal of COD and the number of biomass growth recorded in the experiment shows that AS is capable to live and help degrading the organic pollutant that is present in the wastewater.

References

Eaton, A. D., Franson, M. A. H., & American Public Health Association (Eds.) (2005). *Standard Methods for the Examination of Water and Wastewater*. Washington: Amer Public Health Assn

Lewis, D. L., Kollig, H. P., & Hodson, R. E. (1986). Nutrient limitation and adaptation of microbial populations to chemical transformations. *Applied and Environment Microbiology, 51*, 598–603.

Özbelge, T. A., Özblege, H. Ö., & Altinten, P. (2007). Effect of acclimatization of microorganisms to heavy metals on the performance of activated sludge process. *Journal of Hazardous Materials, 142*, 332–339.

Richmond, M. H. (1968). Enzymic adaptation in bacteria: Its biochemical and genetic basis. *Essays in Biochemistry, 4*, 105–154.

Schmidt, E., Hellwig, M., & Knackmuss, H. J. (1983). Degradation of chlorophenols by a defined mixed microbial community. *Applied and Environmental Microbiology, 46*, 1038–1044.

Stephenson, T., Lester, J. N., & Perry, R. (1984). Acclimatisation to nitrilotriacetic acid in activated sludge process. *Chemosphere, 13*, 1033–1040.

Torstensson, N. T. L., & Stark, J. (1975). The effect of repeated applications of 2, 4-D and MCPA on their breakdown in soil. *Weed Research, 15*, 159–164.

Walker, H. H., & Newman, A. S. (1956). Microbial decomposition of 2, 4-dichlorophenoxyacetic acid. *Applied Microbiology, 4*, 201–206.

Wiggins, B. A., Jones, S. H., & Alexander, M. (1987). Explanations for the acclimation period preceding the mineralization of organic chemicals in aquatic environment. *Applied and Environmental Microbiology, 53*(4), 791–796.

Yu, H., & Yu, C. (2010). Influence of various retting methods on properties of kenaf fiber. *Journal of the Textile Institute, 101*(5), 452–456.

Construction of ASMC by Dielectrophoresis Using Wirecloth Electrode for the Treatment of Wastewater

Z. Z. Abidin, Y. C. Wai, N. Haffifudin and F. Ahmadun

Abstract Nowadays, the concern for biological treatment of wastewater has been increased by momentum. The utilization of wirecloth to construct artificially structured microbial consortia (ASMC) for the treatment of synthetic wastewater had been studied in this research. Microorganisms, which grow in pharmaceutical wastewater, were isolated by culture method. The experiment parameters include temperature and pH had been studied. Microorganisms were attracted to the wirecloth by using dielectrophoresis and immobilized with polyethylenimine (PEI). The treatment analysis showed the selected microorganisms consumed the organic materials which significantly reduced the COD in the synthetic wastewater. Other analyses such as BOD, TS, VSS, TSS were also investigated. One of the advantages of using wirecloth electrode is able to significantly shorten the formation time of biofilm which normally take up to few months. SEM observation revealed the patterns of biomass immobilization on wirecloth. Besides, immobilized wirecloth showed a higher efficiency compared to suspended growth with the same synthetic wastewater applied. The attached grow on the wirecloth, suggested as the main role in COD removal.

Introduction

Historically, environmental microbiology can be traced to studies of municipal waste treatment and disposal, but during the last 10 years, biofilms have become an important object of microbiological inquiry as a critical element in the preservation of quality in water systems as well as a key component of biological reactions in wastewater treatment (Flemming et al. 2000). The basic concept of

Z. Z. Abidin (✉) · Y. C. Wai · N. Haffifudin · F. Ahmadun
Department of Chemical and Environmental Engineering, University Putra Malaysia, 43400 Serdang, Selangor, Malaysia
e-mail: zurina@upm.edu.my

R. Pogaku et al. (eds.), *Developments in Sustainable Chemical and Bioprocess Technology*, DOI: 10.1007/978-1-4614-6208-8_9,

biofilms formation is the intrinsic tendency of microorganisms to adhere or growth on surfaces in contact with aqueous systems (Alimahmoodi et al. 2012). Many natural and engineered systems are influenced by biofilms, microorganisms firmly attached to surfaces, and the effects of biofilms may vary from desirable, through undesirable, to a very bad situation, depending on the specific locations where the biofilms are deposited (Lewandowski and Beyenal 2007). One of the significant elements in biofilms is naturally occurred microbial consortia where it is used widely in industry, for example in wastewater treatment, bioremediation, metal leaching, silage production, and various food fermentations (James 1993). Besides the application stated above, there are more strategies that we can apply on the usage of biofilms where biological control will be safer if we compare to chemically modification for our environment sustainability (Woolard 1997). Microbial consortia in nature tend to exist in highly organized macroscopic structures with extensive internal organization where the macroscopic structuring also provides protection against exogenous toxic substrates (Alp et al. 2002).

The use of dielectrophoresis (DEP), the induced movement of particles in non-uniform electric fields, is currently one of the successful methods used in cell separation, biofilms, or artificially structured microbial consortia (ASMC) formation (Abidin et al. 2007a,b; Andrews et al. 2006). DEP occurred when a particle existed in a non-uniform electric field, which induced it to form an electric dipole. A stronger side of electric dipole will result a net force on the particle and caused it to move. The direction of the particle depends on the electric properties of the medium (Pohl et al. 1978; Pohl and Pollock 1978).

In this chapter, microorganisms are attracted by using DEP and immobilized by an immobilizing agent (e.g., polyethylenimine) where it can significantly reduced the formation time of biofilms formed on the wirecloth. The environmental conditions are varying, in order to investigate the removal efficiency of COD during the treatment. The durability and condition of wirecloth are observed after the treatment.

Materials and Methods

Microorganisms

Bacillus sp., *Rhodococcus* sp., and *Micrococcus* sp. (isolated from pharmaceutical wastewater, obtained from Norhafizah binti Hafifudin, Universiti Putra Malaysia) were separately grown overnight in Oxoid nutrient broth ('Lab-Lemco' powder 1 g/L, yeast extract 2 g/L, peptone 5 g/L, and sodium chloride 5 g/L) at 35 °C in an incubator shaker with a speed of 150 rpm. Each type of the cells was centrifuged and washed 4 times to reduce the conductivity. Cell concentrations were measured by obtaining the optical density (OD) from UV spectrophotometer

(Thermo Genesys 10 uv) by using wavelength of 600 nm. Each type of cell concentration (OD = 1.2) was mixed together for immobilizing process.

Wirecloth

Wirecloth was produced by using textile technology in Swiftec Sdn. Bhd., Kajang. The wirecloth was plain weaved using polyester yarns (Hualon) and stainless steel wires of 100 μm diameter (Ormiston Wire Ltd, Middlesex, UK).

Composition of the Medium

The synthetic wastewater had the following composition in (mg/L): glucose (120), peptone (90), yeast extract (12), NH_4Cl (100), KH_2PO_4 (20), $NaHCO_3$ (300), $MgSO_4{\cdot}7H_2O$ (25), $CaCl_2\_2H_2O$ (3), MnSO4$\cdot_5H_2O$ (2), and FeCl3$\cdot_6H_2O$ (0.1) (Jung et al. 2005). Slight modification had been made to the composition to simulate local domestic wastewater.

Immobilization of Microorganism (Formation of Biofilm)

To immobilize microorganisms on the wirecloth by DEP, the electrodes in the wirecloth were activated by AC signal of 20 V_{pk-pk} and a frequency of 1 MHz using a Thurlby-Thandar TG120 function generator. Next, the mixed microorganisms were added on the wirecloth, and the attraction of the microorganisms on the electrode was observed by microscope (Leica Q500IW). The immobilizing agent PEI (0.005 %) was introduced to form attached artificial consortia. After 10 min, the function generator was switched off and gently washed with deionized water to rinse off the unbound microorganisms (Abidin et al. 2007a). The wirecloth with immobilized microorganisms was ready to transfer into the reaction medium.

Biofilms Reactor and Treatment System

The laboratory-scale reactor was fabricated in laboratory by glass material (Perplex). It consists of simple feeding inlet, aeration inlet, sample outlet, and sludge outlet. The total volume of the reactor was 4 l. The synthetic wastewater was introduced from the feeding inlet before the immobilized wirecloth was put into the reactor. The aeration was blown from the bottom of the reactor to ensure a better mixing during the reaction phase. The environment conditions were varied as the parameter of this experiment. The COD removal rate as treatment efficiency was recorded each day during the treatment period.

Analytical Method

The performance of the wirecloth and formation of biofilms were accessed by monitoring the concentration of COD (closed reflux titrimetric method) throughout the reactor cycle operation. The pH of the initial and final treatment medium was recorded. The procedure of COD titration method was followed by standard method (APHA 1998).

Results and Discussion

Immobilization and Biofilms

Figure 1 showed a part of wirecloth while electric field was applied on the electrode. The horizontal opaque rod-shaped electrodes were separated by the multifilament polyester yarns (thinner semitransparent lines shown in Fig. 1), no electrical contact between each of the electrode, microorganisms showed positive DEP effect on the electrode. Immobilizing agent PEI was added just after positive DEP achieved. The electric field was turned off after 10 min of immobilizing process. The microorganisms that were not successfully bond to wirecloth were washed away by a gently flow of deionized water. Besides electrodes, the microorganisms were also attracted on the polyester yarns, and the field distortion was observed and explained previously (Abidin et al. 2007a, Abidin and Markx 2005).

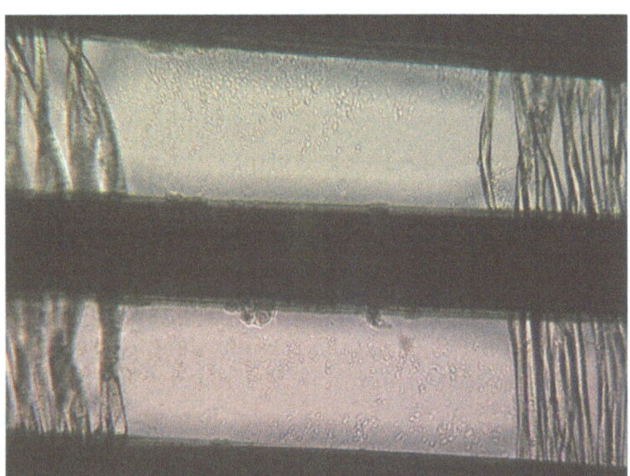

Fig. 1 Positive DEP effect

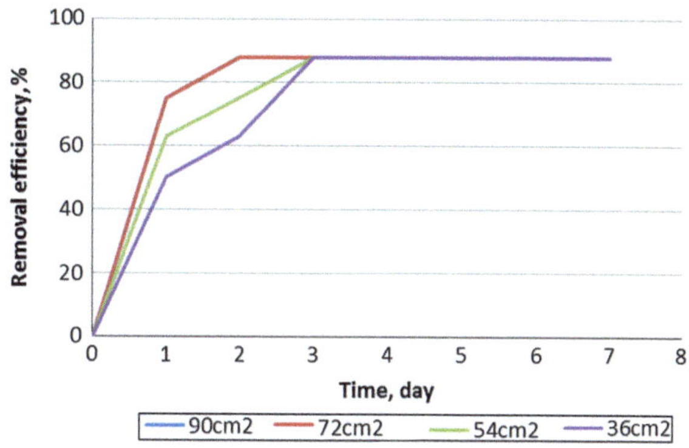

Fig. 2 Removal efficiency, % of COD versus time, day

Effect of Wirecloth's Surface Area

The performance of the treatment was evaluated by the removal efficiency of COD. Four wirecloths with different surface area (36, 54, 72, 90 cm^2) were used. Figure 3 shows the results of the treatment. Samples were taken each day during the treatment period. The conditions for temperature (26 °C), agitation (125 rpm), and aeration (250 cc/min) were set as constant throughout the period. The volume used was 1000 ml for each type of wirecloth, and the removal efficiency of COD for all of the wirecloths achieved about 90 % after 2–3 days of treatment time. This showed the attached microorganisms worked and adapted well in the medium after immobilizing on the wirecloth. Less sludge was produced during the period and both 72 and 90 cm^2 wirecloth overlapped in Fig. 2 which showed the fastest removal efficiency within 2 days.

Effect of pH

Figure 3 showed the removal efficiency of COD on different pH (2, 4, 6, 8, and 10). When different pH were applied to the system, microorganisms took time to acclimatize. The microorganisms were successfully adapted and grown on the wirecloth even in extreme acidic and alkaline condition, but the disadvantage was that they need more time to acclimatize. The removal efficiency was fast for pH 6 and pH 8 which means the environmental conditions are better for the microorganisms. All of the removal efficiencies were reached almost 90 %, but it was lower for pH 10 (about 80 %). The wirecloth showed no corrosive defect by observing through microscope after the experiments.

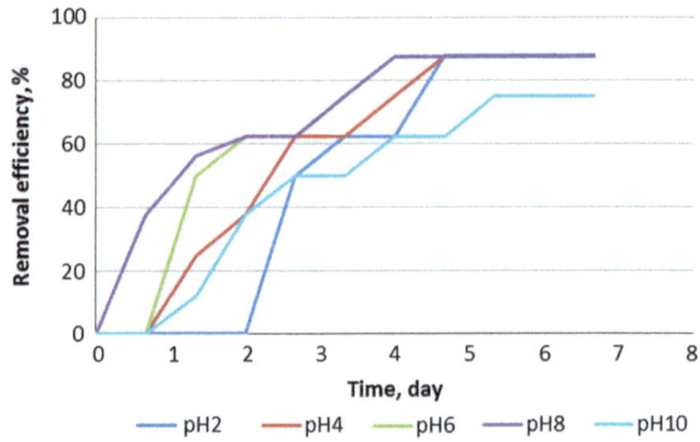

Fig. 3 Removal efficiency, % of COD against time, day

Conclusions

The experimental data showed the capability of wirecloth to use in the treatment of low-biodegradable chemical wastewater. The wirecloth can be used to attract more than one type of microorganisms to achieve a better treatment results. Less control of the initial concentration of microorganisms needed as the unbounded microorganisms will be washed away which the artificial consortia will remain on the wirecloth with a positive DEP effect. The treatment efficiency for both parameters above showed more than 80 % of COD was removed as this proven the effectiveness of the biofilms formed on wirecloth. The flexibility of wirecloth allowed it to fit in most places which could consider as an addition benefit to use as a media on the treatment of wastewater.

Acknowledgments This research was funded by Universiti Putra Malaysia, and support from MOSTI Science Fund (Project number UPM-03-01-04-SF0842) was gratefully acknowledged.

References

Abidin, Z. Z., Downes, L., & Markx, G. H. (2007a). Large scale dielectrophoretic construction of biofilms using textile technology. *Biotechnology and Bioengineering, 96*, 1222–1225.

Abidin, Z. Z., Downes, L., & Markx, G. H. (2007b). Novel electrode structures for large scale dielectrophoretic separations based on textile technology. *Journal of Biotechnology, 130*, 183–187.

Abidin, Z. Z., & Markx, G. H. (2005). High-gradient electric field system for the dielectrophoretic separation of cells. *Journal of Electrostatics, 63*, 823–830.

Alimahmoodi, M., Yerushalmi, L., & Mulligan, C. N. (2012). Development of biofilm on geotextile in a new multi-zone wastewater treatment system for simultaneous removal of COD, nitrogen and phosphorus. *Bioresource Technology, 107*, 78–86.

Alp, B., Stephens, G. M., & Markx, G. H. (2002). Formation of artificial, structured microbial consortia (ASMC) by dielectrophoresis. *Enzyme and Microbial Technology, 31*, 35–43.

Andrews, J. S., Mason, V. P., Thompson, I. P., Stephens, G. M., & Markx, G. H. (2006). Construction of artificially structured microbial consortia (ASMC) using dielectrophoresis:

examining bacterial interactions via metabolic intermediates within environmental biofilms. *Journal of Microbiological Methods, 64*, 96–106.

APHA. (1998). *Standard methods for the examination of water and wastewater.* Washington: DC, American Public Health Association.

Woolard, C.R. (1997). The advantages of periodically operated biofilm reactors for the treatment of highly variable wastewater. Water Science and Technology, *35*, 199–206.

Flemming, H.-C., Szewzyk, U., & Griebe, T. (Eds.). (2000). *Biofilms investigative methods & application.* USA: Technomic Publishing Company, Inc.

James, D. B. (1993). Bacterial biofilms. *Current Opinion in Biotechnology, 4*, 197–204.

Lewandowski, Z., & Beyenal, H. (2007). *Fundamentals of biofilm research.* USA: Taylor & Francis.

Pohl, H., Pollock, K., & Crane, J. (1978). Dielectrophoretic force: a comparison of theory and experiment. *Journal of Biological Physics, 6*, 133–160.

Pohl, H. A., & Pollock, K. (1978). Electrode geometries for various dielectrophoretic force laws. *Journal of Electrostatics, 5*, 337–342.

Jung, Y., Koh, H., Shin, W., & Sung, N. (2005). *Wastewater Treatment Using Combination of MBR Equipped with Non-Woven Fabric Filter and Oyster-Zeolite Column* [Online]. Busan. Available: www.ceric.net/wonmun3/ksee/04708996.pdf [Accessed 10 June 2012].

Catalytic Pyrolysis of Waste Chicken Fats Using Zeolite Catalysts

M. Y. Liew, A. Salmiaton, W. A. K. G. Wan Azlina and R. Yunus

Abstract Chicken fats become a major poultry waste from broiler industries and the production rate is increasing with increasing of human population. The conversion of waste chicken fats containing major triglycerides into potential hydrocarbon chemicals and bio-oils was investigated under pyrolysis process. Catalytic pyrolysis of chicken fats with zeolite catalyst, ZSM-5, was conducted using laboratory-scale distillation set-up under nitrogen flow of 400 °C. In the process, thermal cracking plays an important role in breaking triglyceride into shorter hydrocarbon chain under high temperature. Pyro-oil produced was further analysed by gas chromatography–mass spectroscopy in order to identify its chemical properties. From the chemical analysis, different short hydrocarbon chain products from C_7–C_{24} were indicated. Major chemical composition of pyro-oil ascertained was aliphatics, followed by carboxylic acid, alcohol, ketone, ester, aromatic, anhydride, ether and aldehyde. The effect of ZSM-5 catalysts towards product yield and composition was also explored.

M. Y. Liew · A. Salmiaton (✉) · W. A. K. G. Wan Azlina
Department of Chemical and Environmental Engineering, Faculty of Engineering, Universiti Putra Malaysia, 43400 Serdang, Selangor, Malaysia
e-mail: mie@eng.upm.edu.my

M. Y. Liew
e-mail: meiyi0321@hotmail.com

W. A. K. G. Wan Azlina
e-mail: wanaz@eng.upm.edu.my

R. Yunus
Institute of Advanced Technology, Universiti Putra Malaysia, 43400 Serdang, Selangor, Malaysia
e-mail: robiah@eng.upm.edu.my

R. Pogaku et al. (eds.), *Developments in Sustainable Chemical and Bioprocess Technology*, DOI: 10.1007/978-1-4614-6208-8_10,

Introduction

Fossil fuels represent more than 80 % of total energy in the world and act as the primary resource nowadays. But the scarcity of this conventional energy has become a problem due to the rising of energy demand (Demirbas 2008). The application of renewable energy from biomass conversion via chemical process can be a potential new energy resource. Pyrolysis is a process of thermal decomposition of materials at high temperatures with the absence of oxygen (Jayasinghe and Hawboldt 2012). The production of bio-oil from pyrolysis involves rapid heating to breakdown long hydrocarbon chain species producing smaller hydrocarbon products that have similar fuel properties.

Bio-oil derived from animal fats such as lard, poultry fats and fish oil showed diesel-like fuel properties with major components of alkanes, alkenes, aldehydes, ketones, aromatics and carboxylic acids (Adebanjo et al. 2005; Tian et al. 2008; Wisniewski Jr et al. 2010). Gürü et al. (2010) investigated the performance and emission characteristics of bio-oil on running engines and compared to those of fossil fuel and fatty acid methyl esters. The authors found that bio-oil with low sulphur content was performed well with diesel fuel blends.

Catalytic pyrolysis is a more advanced technique to maximise the production of pyrolytic products and increase the selectivity on specific compositions of compounds (Butler et al. 2011). In one research conducted by Carlson et al. (2009), ZSM-5 catalyst showed good performance in producing higher aromatic compounds. Higher aromatics improved the properties of higher heating value due to relatively reduced oxygenated compounds.

Methodology

Materials

Waste chicken fats as raw materials were collected from a public market located at Seri Kembangan, Selangor for torrefaction. The chicken fats stripped off directly from different chicken species were stored in a fridge of 4 °C after collected. Torrefaction of chicken fats was carried out in a reflux set-up at 200 °C under 300 ml/min of nitrogen flow rate in order to create an oxygen-free condition. About 80 % of torrefied biomass was extracted and was prepared for pyrolysis.

Ultimate analysis of torrefied biomass was performed by LECO CHNS-932 elemental analyser as shown in Table 1. Higher heating value (HHV) was theoretically calculated according to Dulong's formula using the percentage of C, H, O and S elements from the analysis (Rajput 2010).

ZSM-5 catalyst used was a commercialised catalyst purchased from Qingdao Wish Chemical Co., Ltd. It was synthesised from water glass and aluminium

Table 1 Ultimate analysis of torrefied biomass

Ultimate analysis (wt% daf)				
C	H	N	S	O*
77.21	11.96	0.84	0.28	9.71
Heating value (HHV, MJ/kg)			41.64	

*By difference

sulphate and suitable employed in the production oil catalytic cracking reaction (den Hollander et al. 2002).

Pyrolysis of Torrefied Biomass

The torrefied biomass was pyrolysed in a distillation set-up with heat source supplied from heating mantle, nitrogen gas connection and thermocouple. The nitrogen gas was pre-purged for 15 min with 300 ml/min of nitrogen flow rate to create oxygen-free environment prior to pyrolysis process. Non-catalytic and catalytic pyrolysis using ZSM-5 catalyst were conducted in the same pyrolysis condition to investigate the effectiveness of the catalyst. The condition was set at temperature of 400 °C for 30 min under 300 ml/min nitrogen flow rate and 350 rpm stirring rate.

GC–MS and FTIR Analysis

Chemical properties of pyro-oil produced were analysed using gas chromatography-mass spectroscopy (GC–MS, Shimadzu GCMS-QP2010 Plus). A capillary column SGE BPX5 (length 30 m, ID 0.25 mm and film thickness 0.25 μm) was used in the GC analysis. Helium gas was employed as a carrier gas at 1.0 ml/min with a pressure set at 53.6 kPa. Injector and interface temperature were operated at 250 and 320 °C, respectively. Injection volume of pyro-oil was 0.1 μl in split mode ratio of 1:50. Oven temperature was pre-run at 50 °C for 2 min, increased to 300 °C at heating rate of 8 °C/min and maintained for 10 min.

Results and Discussion

Product Yield

The products obtained from pyrolysis of torrefied biomass were divided into three fractions namely char, pyro-oil and gas (calculated by difference). Figure 1 portrayed the product yield of catalytic pyrolysis of waste chicken fats.

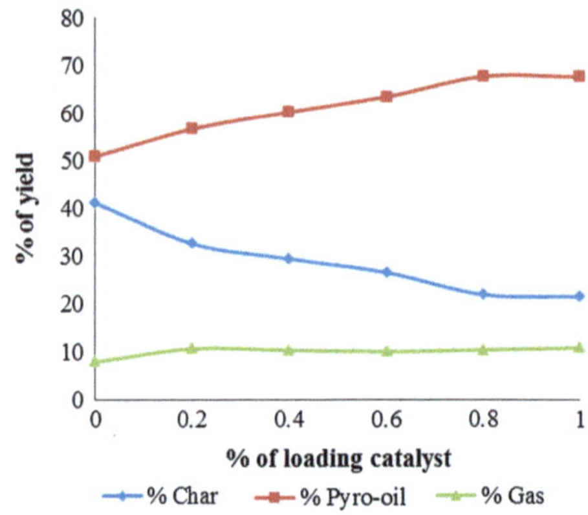

Fig. 1 Yields of catalytic pyrolysis at different percentage of loading catalyst

The effectiveness of ZSM-5 towards yield production was determined by different percentage of catalyst loading to the torrefied biomass mixture pyrolysed at 400 °C. The pyro-oil yield was increased from 50.92 to 67.60 % as the catalyst loading increased from 0.2 to 1.0 %. It was observed that the yield of char decreased from 41.19 to 21.61 % and the gas production was slightly increased from 7.89 to 10.79 % with the increase in catalyst loading. The maximum yield of pyro-oil was occurred when the biomass was pyrolysed with 0.8 % ZSM-5 catalyst, and the yield was slightly reduced with 1.0 % catalyst loading. It can be explained that at the condition above optimum, pyrolysis was more favoured to gasification reaction when higher amount of catalyst was employed (Pütün et al. 2006; Pütün 2010).

Pyro-Oil Characterisation

Various compositions of pyro-oils were observed in the GC–MS analysis and their chromatograms were shown in Fig. 2. Different compositions of non-catalytic pyro-oil and ZSM-5 catalytic pyro-oil were grouped based on carbon numbers and chemical groups in order to determine thermal cracking of hydrocarbon and the selectivity of catalyst (Pütün et al. 2006). Both pyro-oil contained a series of organic compounds with carbon number from C_7 to C_{24} (Fig. 3a). The most abundant compound for both was aliphatic in C_{15}, distributed around 21.59 and 23.11 %, respectively. Catalytic pyro-oil showed higher diversity in C_7–C_{15} (67.12 %) than non-catalytic pyro-oil (61.84 %) due to efficient of catalyst in thermal degradation of higher hydrocarbon into smaller product (Butler et al. 2011). C_{16} and C_{17} products from non-catalytic pyro-oil were significantly reduced from 7.64 to 1.93 % and 15.7 to 5.86 % after the loading of ZSM-5 catalyst, respectively.

Different kinds of compounds were classified based on chemical group (Fig. 3b). The main group for both pyro-oil was aliphatic, followed by carboxylic acid and alcohol with some in ketone and ester and traces in aromatic, anhydride,

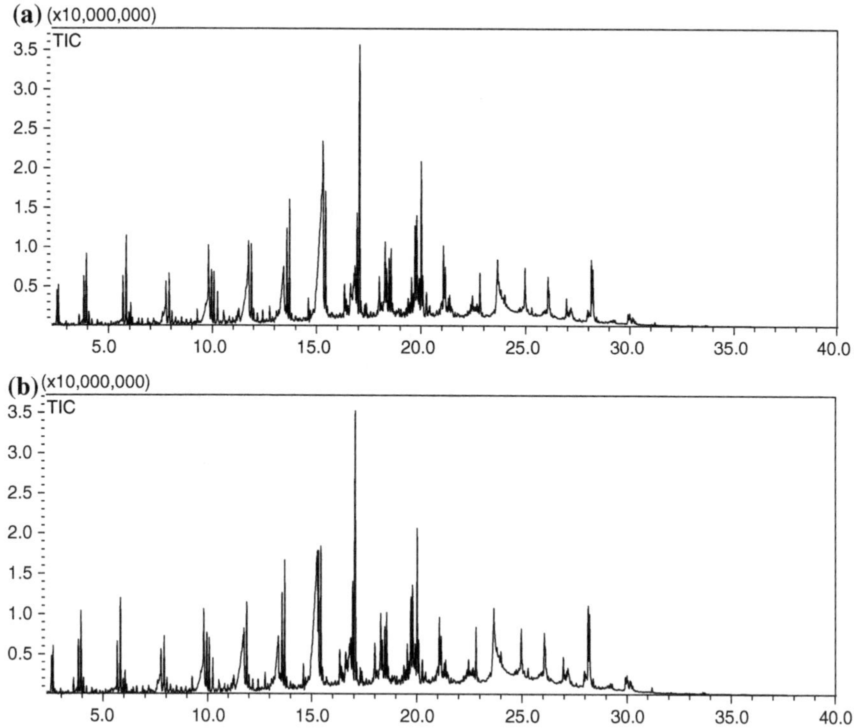

Fig. 2 GC–MS Chromatograms of **a** non-catalytic pyro-oil and **b** ZSM-5 catalytic pyro-oil

ether and aldehyde. Catalytic pyro-oil showed high selectivity in aliphatic compounds as its distribution increased from 55.34 % (non-catalytic pyrolysis) to 61.00 % (catalytic pyrolysis). As reported by Pütün et al. (2006), the advantage of zeolite catalyst was apparently improved the yield of aliphatic fraction. ZSM-5 catalytic pyrolysis favoured the production of alcohols as well, from 9.50 % (non-catalytic pyrolysis) increased to 12.15 % (catalytic pyrolysis). Other than aliphatic and alcohol groups, catalytic pyrolysis was selectively performed on other minor components (anhydrides, aromatics, ester and ketones) with slight increment from 0.05 to 1.11 %. But, ZSM-5 catalyst was not selective in production of carboxylic acids (the second major component) and its yield was decreased from 32.13 to 21.71 %.

Conclusions

Catalytic pyrolysis of torrefied chicken fats using ZSM-5 catalysts showed an increase in pyro-oil and gas production while decrease in char production. The maximum pyro-oil yield was 32.89 % at 0.8 % catalysts loading. From the

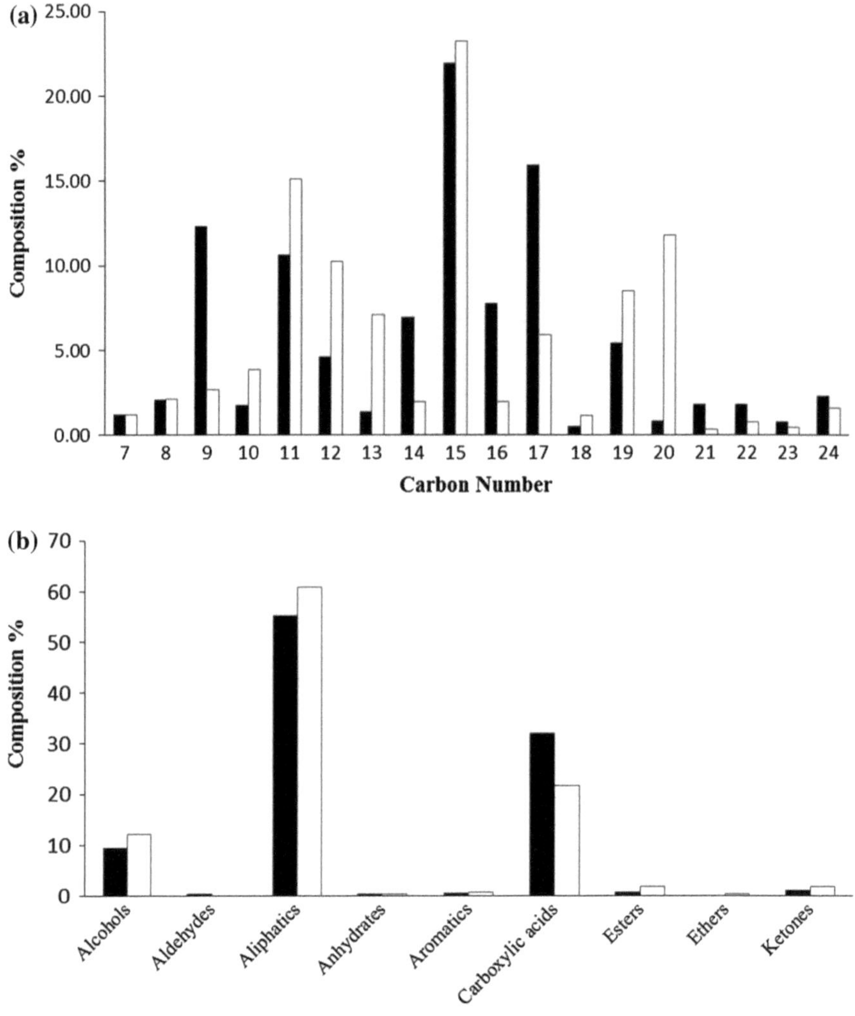

Fig. 3 Composition distribution of **a** carbon number and **b** chemical group for non-catalytic pyro-oil (*black bar*) and ZSM-5 catalytic pyro-oil (*white bar*)

GC–MS analysis, different kinds of organic compounds from carbon number C_7–C_{24} were found. Aliphatic C_{15} for non-catalytic pyro-oil and catalytic pyro-oil was dominant among other compounds. By comparison, catalytic pyro-oil had higher selectivity in chemical group of alcohols, aliphatics, anhydrides, aromatics, ester and ketones.

Acknowledgments The authors would like to thank Universiti Putra Malaysia in particular the Department of Chemical and Environmental Engineering, Faculty of Engineering, for the facility, research funding and all those who gave cooperation and help throughout the whole project.

References

Adebanjo, A. O., Dalai, A. K., & Bakhshi, N. N. (2005). Production of diesel-like fuel and other value-added chemicals from pyrolysis of animal fat. *Energy and Fuels, 19*, 1735–1741.

Butler, E., Devlin, G., Meier, D., & Mcdonnell, K. (2011). A review of recent laboratory research and commercial developments in fast pyrolysis and upgrading. *Renewable and Sustainable Energy Reviews, 15*, 4171–4186.

Carlson, T., Tompsett, G., Conner, W., & Huber, G. (2009). Aromatic production from catalytic fast pyrolysis of biomass-derived feedstocks. *Topics in Catalysis, 52*, 241–252.

Demirbas, A. (2008). *Biodiesel: A realistic fuel alternative for diesel engines*. London: Springer.

Den Hollander, M. A., Wissink, M., Makkee, M., & Moulijn, J. A. (2002). Gasoline conversion: Reactivity towards cracking with equilibrated FCC and ZSM-5 catalysts. *Applied Catalysis, A: General, 223*, 85–102.

Gürü, M., Koca, A., Can, Ö., Çinar, C., & Şahin, F. (2010). Biodiesel production from waste chicken fat based sources and evaluation with Mg based additive in a diesel engine. *Renewable Energy, 35*, 637–643.

Jayasinghe, P., & Hawboldt, K. (2012). A review of bio-oils from waste biomass: Focus on fish processing waste. *Renewable and Sustainable Energy Reviews, 16*, 798–821.

Pütün, E. (2010). Catalytic pyrolysis of biomass: Effects of pyrolysis temperature, sweeping gas flow rate and MgO catalyst. *Energy, 35*, 2761–2766.

Pütün, E., Uzun, B. B., & Pütün, A. E. (2006). Fixed-bed catalytic pyrolysis of cotton-seed cake: Effects of pyrolysis temperature, natural zeolite content and sweeping gas flow rate. *Bioresource Technology, 97*, 701–710.

Rajput, R. K. (2010). *Thermal Engineering*. New Delhi: Laxmi Publications Pvt Limited.

Tian, H., Li, C., Yang, C., & Shan, H. (2008). Alternative processing technology for converting vegetable oils and animal fats to clean fuels and light olefins. *Chinese Journal of Chemical Engineering, 16*, 394–400.

Wisniewski, JR, A., Wiggers, V. R., Simionatto, E. L., Meier, H. F., Barros, A. A. C., & Madureira, L. A. S. (2010). Biofuels from waste fish oil pyrolysis: Chemical composition. Fuel, *89*, 563-568.

Investigation of Pyrolysis Parameters on the Yield and Quality of Bio-Oil from *Jatropha curcas* Wastes

S. A. Jourabchi, S. Gan and H. K. Ng

Abstract *Jatropha curcas* pressed cake leftover after oil extraction for biodiesel production is considered as waste. Following physicochemical characterisation tests, this biomass waste was pyrolysed over a temperature range of 573.15–1073.15 K and a nitrogen linear speed range of 7.8×10^{-5}–6.7×10^{-2} m/s to produce bio-oil. The heating rate and biomass grain size were fixed to constants of 50 K/min and <2 mm, respectively. The bio-oil products after yield measurements were tested for three major specifications, i.e. gross calorific value, water content and acidity, based on ASTM procedures for standard specification of pyrolysis liquid biofuels. In addition, empirical correlations between the bio-oil yield and pyrolysis parameters have been introduced for the experiments using nonlinear multiple regression method.

Introduction

Increasing world population and industrialisation have resulted in both the increasing use of fossil fuels and the impending depletion of fossil fuel reserves worldwide. Additionally, the increasing release of harmful emissions especially carbon dioxide (CO_2) from the increased usage of fossil fuels has led to the need for more sustainable and environment-friendly alternative fuels such as biodiesel, bioethanol and biomass.

S. A. Jourabchi · S. Gan (✉)
Department of Chemical and Environmental Engineering, The University of Nottingham Malaysia Campus, Jalan Broga, 43500 Semenyih, Selangor Darul Ehsan, Malaysia
e-mail: suyin.gan@nottingham.edu.my

H. K. Ng
Department of Mechanical, Materials and Manufacturing Engineering, The University of Nottingham Malaysia Campus, Jalan Broga, 43500 Semenyih, Selangor Darul Ehsan, Malaysia

R. Pogaku et al. (eds.), *Developments in Sustainable Chemical and Bioprocess Technology*, DOI: 10.1007/978-1-4614-6208-8_11,

Presently, bio-oil or tar produced from the pyrolysis of biomass can be used directly or after further physicochemical processes to heat up boilers, or even drive diesel engines or turbines (Balat et al. 2009; Mohan et al. 2006). During the combustion process, its net zero CO_2 as well as lesser nitrogen oxides (NO_x) or sulphur oxides (SO_x) emissions compared to fossil fuels makes it a potential liquid fuel replacement (Putun et al. 2005).

The focus of this research is to determine the pyrolysis parameters required to achieve optimum quantity and quality of bio-oil production from Jatropha curcas (J. curcas) pressed cake after oil extraction of its seeds. Currently, *J. curcas* is a drought free and inedible crop which can be planted economically in tropical and subtropical regions. Its seed is a promising source of oil, especially for the purpose of commercial biodiesel production. A thermo-gravimetric analysis (TGA) shows that the *J. curcas* pressed cake contains less than 4 wt% moisture and over 70 wt% volatiles, both of which are indicative of the potential bio-oil content. This analysis also shows that 70 % of the total volatiles separation occurs at temperatures below 650 K, which is relatively low and hence aids in reducing the energy consumption during the pyrolysis process. To achieve maximum bio-oil production, a laboratory-scale pyrolysis reactor and a quencher are designed and fabricated. A corresponding design of experiment is performed in this work to determine the optimum pyrolysis parameters.

Methodology

Material

J. curcas pressed cake waste was obtained from a biodiesel production company in Malaysia (Bionas Group) to be used as the feedstock for the pyrolysis project. Table 1 shows the physicochemical properties that were measured for the *J. curcas* waste.

The amount of moisture content both from the TGA and from the oven drying method has matching results of below 4 wt% (Sluiter et al. 2008). This is relatively low, considering that under Malaysian climate condition, the average humidity is approximately 80 % (Malaysian Meteorology Department 2012). By having a moisture content of below 10 %, the drying process can be skipped completely which is an advantage since energy is saved for the total bio-oil production process (Mohan et al. 2006). Comparably high gross calorific value, high carbon and volatiles content indicate the potential of *J. curcas* pressed cake as a source of biofuel.

The sample was prepared for pyrolysis by grinding into a powder form. The pressed cake was firstly converted into granules using a Retsch SM100 Comfort grinder machine and sieved using a grinding sieve size of 0.2 mm in a Retsch ZM200 grinder.

Table 1 Physicochemical properties of *Jatropha curcas* pressed cake

Item	Unit	Amount	Method
Moisture content	wt%	3.40	Oven drying
Gross calorific value	MJ/Kg	19.11	Bomb calorimeter
Moisture content	wt%	3.31	Thermo-gravimetric analysis (TGA)
Volatiles content		70.98	
Fixed carbon		19.72	
Ash content		5.99	
Carbon (C)	wt%	57.85	Energy-dispersive X-ray spectroscopy
Oxygen (O)		37.94	
Potassium (K)		1.28	
Aluminium (Al)		1.15	
Phosphorus (P)		0.75	
Magnesium (Mg)		0.43	
Calcium (Ca)		0.29	
Sulphurous (S)		0.22	
Silicon (Si)		0.09	

Experimental Tests

The fixed-bed pyrolysis rig was designed with the aid of Pro/Engineering software and was mainly fabricated at the University of Nottingham Malaysia Campus workshops. The fixed-bed stainless steel reactor has an inner diameter of 52.5018 mm connected to a stainless steel condenser to quench the volatiles released. The condenser utilises cooling water from a refrigerated bath circulator type 632D. By matching the reactor geometry, a Carbolite tubular furnace type CTF 12/65/550 with a maximum heating rate of 50 K/min is used as heater. The biomass was pyrolysed over a temperature range of 573.15–1073.15 K and a nitrogen linear speed range of 7.8×10^{-5}–6.7×10^{-2} m/s to produce bio-oil. The heating rate and biomass grain size were fixed to constants of 50 K/min and <2 mm, respectively. For consistency, 50 ± 1 g samples were used in all test runs. The tests were repeated three times. The residence time for each test was a maximum of 90 min. A run was terminated if no visible outlet gas and bio-oil drops were observed in the collector. Each experimental test point was also tested using a Mettler-Toledo TGA/DSC 1 unit to simulate the heat and mass transfer of the actual experiment.

Results and Discussion

Bio-Oil Yield

Figure 1 shows that the bio-oil yield obtained from the set-up increases when the velocity of nitrogen flowing through the bed reduces.

For a fixed nitrogen flow rate and hence velocity, when the reaction temperature increases, the bio-oil yield increases to a maximum amount at approximately 823.15 K. Above this temperature, a slight decrease is experienced followed by a final increase.

At temperatures below 573.15 K, no noticeable bio-oil product was observed because the heat was enough to crack the hemicellulose only and produce mainly carbon monoxide (CO) and CO_2 gases (Putun et al. 2005). As temperature increases, the cellulose breaks down first. Then, at higher temperatures, the lignin starts to crack down into char, water and heavy oil. This justifies the decrease in char production in favour of higher bio-oil product and almost constant amount of gas release. Beyond 823.15 K, further gasification occurs, and fine and light gases such as hydrogen are generated resulting in lower bio-oil production compared to lower temperatures.

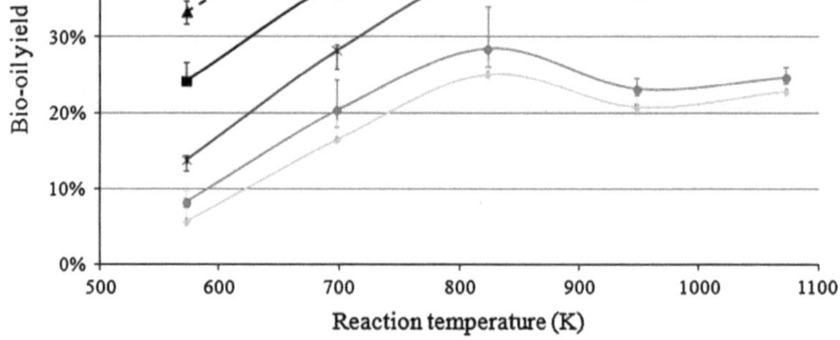

Fig. 1 Bio-oil yield against reaction temperature for different nitrogen velocities

Product Characterisation

The gross calorific value, water content and acidity of the bio-oil products are shown in Fig. 2. The gross calorific values of the bio-oil products were tested according to ASTM D240 method using a Parr 6100 Calorimetric Bomb unit. The gross calorific values neglecting the water content within the products were also calculated. The specification for gross calorific value of pyrolysis liquid biofuels to be used in burners is a minimum of 15 MJ/kg (ASTM D7544-10 2009). Not only do the unprocessed bio-oils meet this specification between the temperatures of 670 and 870 K, but the dry bio-oils generally also have gross calorific value above 22 MJ/kg which is comparable to biodiesel. Both the gross calorific value and water content curves are almost symmetrical which shows that the main factor affecting the heat of combustion is the water content of bio-oil.

The ASTM E203 method was used in the measurement of water content using a Mettler-Toledo V20 Karl Fischer titrator unit. Based on the initial testing results, it was confirmed that the bio-oil contains ketones and aldehydes; hence, methanol could not be used as a solvent in the Karl Fischer titrator because of its reaction with ketones producing water. The Combisolvent Keto was used instead of methanol to prevent this reaction. The maximum allowable water content in liquid biofuels is limited to 30 wt% (ASTM D7544-10 2009). The samples that have been tested meet the same water content specifications for reaction temperatures between 780 and 850 K (Fig. 2).

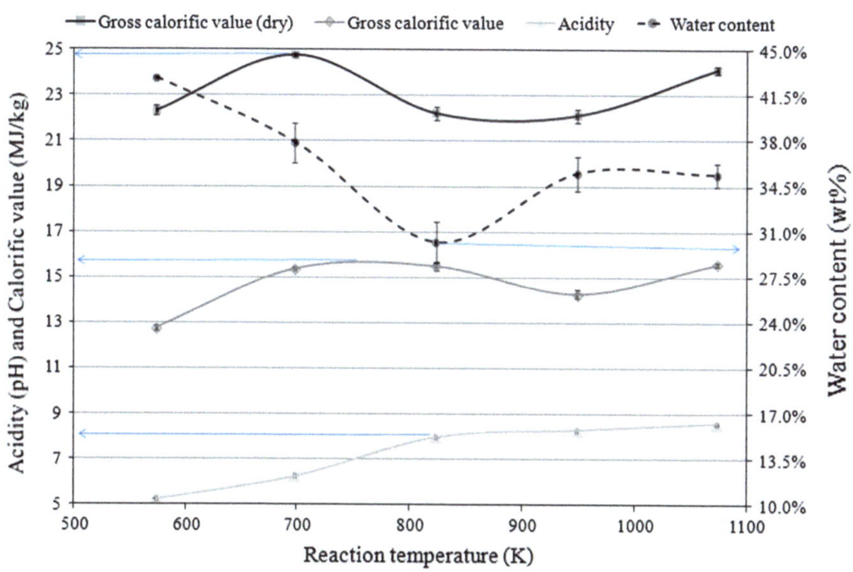

Fig. 2 Characteristics of bio-oil versus reaction temperature for nitrogen velocity of 0.0078 cm/s

Although one of the drawbacks of bio-oil is its typical acidity, the results of ASTM E70 tests on the *J. curcas* bio-oil samples show that the pH values range from 5 to 8.7 at all reaction temperatures (Fig. 2). The pH measured at the reaction temperature of 823.15 is 7.96 which can be considered as neutral.

Empirical Correlations

By averaging the test results for bio-oil yield, a total number of 25 testing points have been used in developing empirical correlations. For each test, there are two variables, i.e. reaction temperature and nitrogen linear speed in the bed. By looking at Fig. 1, it can be deduced that the partial correlation between reaction temperature and bio-oil yield can be modelled as a cubic polynomial. On the other hand, the partial correlation between nitrogen linear speed in the bed and bio-oil yield can be categorised as a quadratic polynomial. Generalising these partial correlations into a multiple correlation by summation of these two polynomials, the following equation can be written as follows:

$$B = c + n_1N + n_2N^2 + t_1T + t_2T^2 + t_3T^3 \tag{1}$$

In Eq. (1), B, N and T are the bio-oil yield, nitrogen linear speed and reaction temperature, respectively. By using nonlinear multiple regression method, the above correlation constants can be found by solving the 6 by 6 simultaneous equations introduced in this method to result in Eq. (2) below (Spiegel and Stephens 1999).

$$B = -58.31 - 9.95\ N + 0.95\ N^2 + 176.98\ T + 0.01\ T^2 - 68.8599\ T^3 \tag{2}$$

B is in wt%, N is in cm/s and T is in K/1,000. The modelling results from the proposed correlation have been compared to experimental results and over the experimental range tested, the maximum error is below 5 % for bio-oil yield.

Conclusions

A fixed-bed pyrolysis rig has been used to produce bio-oil from *J. curcas* pressed cake. The experimental results have shown that both reaction temperature and nitrogen velocity strongly influenced the yield of bio-oil. Above 50 wt% of the waste could be cracked down into bio-oil at a reaction temperature of 823.15 K, heating rate of 50 K/min and nitrogen linear velocity of 7.8×10^{-5} m/s in the bed. Under the same testing condition, the bio-oil produced has less than 30 wt% water content, a gross calorific value of 15.51 MJ/kg and a pH of 7.96. These properties demonstrate that this bio-oil can be used in burners without any modifications provided that other specifications of this fuel match with the ASTM

D7554-10 biofuel standard. An empirical correlation has been introduced to estimate the bio-oil yield over the range of experimental parameters with less than 5 % errors in its predictions.

Acknowledgments This work was supported by the Ministry of Agriculture, Malaysia, under the Science Fund 05-02-12-SF1004. The Faculty of Engineering at the University of Nottingham, Malaysia Campus, is also acknowledged for its support towards this project. The authors would also like to thank the Bionas Group (Malaysia) for providing the *Jatropha curcas* waste.

References

American Society for Testing and Materials. (2009). Standard specification for pyrolysis liquid biofuel. ASTM D7544–10.

Balat, M., & Demirbas, M. F. (2009). Bio-oil from pyrolysis of black alder wood. *Energy Sources, 31* Part A, 1719–1727.

Malaysian Meteorology Department. (2012). *Meteorology Metadata station page*. [http://ww.met.gov.my]. [Accessed 04 July 2012].

Mohan, D., Pittman, C. U, Jr, & Steele, P. H. (2006). Pyrolysis of wood/biomass for bio-oil: a critical review. *Energy & Fuels, 20*, 848–889.

Putun, A. E., Uzun, B. B., Apaydin, E., & Putun, E. (2005). Bio-oil from olive oil industry wastes: Pyrolysis of olive residue under different conditions. *Fuel Processing Technology, 87*, 25–32.

Sluiter, A., Hamas, B., Hyman, D., Payne, C., Ruiz, R., Scarlata, C., Sluiter, J., Templeton, D., Wolfe, J. (2008). Determination of total solids in biomass and total dissolved solids in liquid process samples. *National Renewable Energy Laboratory,* NREL/TP-510-42621.

Spiegel, M. R., & Stephens, L. J. (1999). Multiple and partial correlation. *Statistics* (3rd edn). New York: Schum's outlines, Mc Graw-Hill publishing.

Electrochemical Oxidation of PAHs in Aqueous Solution

A. Yaqub, M. H. Isa and S. R. M. Kutty

Abstract In this work, oxidation of PAHs in aqueous solution was studied using Ti/SnO_2 electrode. Electrochemical oxidation was performed at a current density of 0.33 mA/cm^2 at initial pH 3, 6, and 9. Brush coating thermal decomposition method was used to coat SnO_2 on titanium substrate. Surface morphology and characterization of coating material were studied using FESEM. GC–MS was used to quantify PAHs. PAHs containing 2–3 benzene rings were completely degraded in the presence of electrolyte (NaCl) at all pH condition. Total PAH removal was 87, 93, and 87 % at pH 3, 6, and 9, respectively. Higher current efficiency was shown using NaCl as electrolyte.

Introduction

Polycyclic aromatic hydrocarbons (PAHs) can enter water bodies through industrial effluent discharges and accidental oil spills. Acute or short-term health effects related to occupational exposures to high levels of pollutant mixtures containing PAHs have resulted in symptoms such as eye irritation, nausea, vomiting, diarrhea, and confusion. In laboratory studies, animals exposed to some PAHs over long periods have developed lung cancer from inhalation, stomach cancer from oral ingestion of PAHs in food, and skin cancer from skin contact (PHFS 2009).

A. Yaqub · M. H. Isa (✉) · S. R. M. Kutty
Civil Engineering Department, Universiti Teknologi PETRONAS,
31750 Tronoh, Perak, Malaysia
e-mail: hasnain_isa@petronas.com.my; hasnain_isa@yahoo.co.uk

A. Yaqub
e-mail: environment_green@yahoo.com

S. R. M. Kutty
e-mail: shamsulrahman@petronas.com.my

R. Pogaku et al. (eds.), *Developments in Sustainable Chemical and Bioprocess Technology*, DOI: 10.1007/978-1-4614-6208-8_12,

Table 1 Characteristics of priority PAHs (Pakpahan et al. 2012a, b)

PAHs	Abbreviation	Chemical formula	Mol. weight	Benzene rings	Solubility (mg/l)	MCL* (mg/l)
Naphthalene	Nap	$C_{10}H_8$	128.17	2	32.0	–
Acenaphthylene	Acy	$C_{12}H_8$	152.20	3	3.93	–
Acenaphthene	Ace	$C_{12}H_{10}$	154.20	3	3.40	–
Fluorine	Flu	$C_{13}H_{10}$	166.22	3	1.9	–
Phenanthrene	Phe	$C_{14}H_{10}$	178.23	3	1.3	–
Anthracene	Ant	$C_{14}H_{10}$	178.23	3	0.07	–
Fluoranthene	Fla	$C_{16}H_{10}$	202.26	4	0.265	–
Pyrene	Pyr	$C_{16}H_{10}$	202.26	4	0.132	–
Benz(a)anthracene	BaA	$C_{18}H_{12}$	228.28	4	0.014	0.0001
Chrysene	Chr	$C_{18}H_{12}$	228.28	4	0.002	0.0002
Benzo(b)fluoranthene	B*b*F	$C_{20}H_{12}$	252.31	5	0.0012	0.0002
Benzo(k) fluoranthene	B*k*F	$C_{20}H_{12}$	252.31	5	0.0005	0.0002
Benzo(a)pyrene	B*a*P	$C_{20}H_{12}$	252.31	5	0.11	0.0002
Indeno(1,2,3-cd) pyrene	Ind	$C_{22}H_{12}$	276.33	6	–	0.0004
Dibenz(a,h) anthracene	DbA	$C_{22}H_{14}$	278.34	6	0.005	0.0003
Benzo(g,h,i) perylene	BPer	$C_{22}H_{12}$	276.33	6	0.0026	–

*Maximum contaminant level: USEPA. Standards and regulations for polycyclic aromatic hydrocarbons

The USEPA has classified 16 PAHs as priority pollutants due to their toxic and carcinogenic–mutagenic properties (Pakpahan et al. 2012a, b). Table 1 shows characteristics of the priority PAHs.

PAHs are recalcitrant and are not efficiently removed by conventional biological treatment (Ferrarese et al. 2008). Electrochemical process is a useful tool for the oxidation of a variety of organics present in the liquid waste (Bashir et al. 2009; Mohajeri et al. 2010). Mineralization of organics depends upon high electrocatalytic activity of the anode material and reaction conditions (Montilla et al. 2005; Yaqub et al. 2012). Dimensionally stable anodes (DSAs) have been extensively used for the degradation/oxidation of organics in wastewater. DSAs consist of a titanium substrate covered with conductive metallic oxides coating such as IrO_2, RuO_2, PbO_2, and SnO_2, etc. used as electrocatalyst. The advantage of these metal oxide anodes lies in their good electro-catalytic activities against organic pollutants, high durability, and low production cost (Santos et al. 2006). In this study, SnO_2-coated Ti anode was used to degrade PAHs in aqueous solution. The effect of electrolyte at various pH values was also studied.

Methodology

Electrochemical Cell

All the experiments were conducted under galvanostatic condition in an electrochemical cell that consisted of Ti/SnO_2 anode and a Ti plate cathode connected to a DC power supply. Synthetic PAH solutions were used. Experimental conditions were pH 3, 6, and 9, electrolyte (NaCl) concentration 1 g/l, surface area of electrode 6 cm^2, electrode separation distance 2 cm, current density 0.33 mA/cm^2, and reaction time 4 h.

Electrode Preparation

Anode was prepared by brush coating thermal decomposition method described by Sun et al. (2009). $SnCl_4.5H_2O$ precursor salt solution in ethylene glycol was used. It was applied (about 20 times) to a degreased and etched Ti plate. After each application, the solvent was removed by drying at 100 °C hot air, the plate was dried at 550 °C for 10 min, and finally calcinated at 550 °C for 2 h.

Analysis

Field Emission Scanning Electron Microscopy was used for surface morphology recognition and energy dispersive analysis of X-rays (EDAX) of coating material. GC–MS with column length 30 m and film thickness 0.25 μm was used to determine PAH. The oven temperature was increased from 60 to 175 °C at 6 °C/min, then increased at 3 °C/min until 240 °C, and finally held at 300 °C for 7 min. Injector and transfer line temperatures were 280 and 300 °C, respectively (Pérez et al. 2001). Selected ion monitoring (SIM) mode was employed for data acquisition. PAHs were extracted from aqueous solution using dichloromethane. Each PAH was separately quantified using a five-point calibration of mixed standard solutions in the range from 5 to 25 μg/L.

Calculations

Instantaneous current efficiency (ICE) and energy consumption (EC) were determined as:

$$\text{ICE} = \frac{\left[(\text{COD})_i - (\text{COD})_f\right]}{8I\Delta t}FV$$

$$\text{EC} = \frac{Iu\Delta t}{\left[(\text{COD})_i - (\text{COD})_f\right]V}$$

where $(\text{COD})_i$ and $(\text{COD})_f$ are initial and final chemical oxygen demand at time $t = 0$ and time Δt, respectively, mg/L; I is current, A; u is voltage, V; F is Faraday constant, 96,487 C/mol; V is volume of the solution, L; and 8 is a dimensional factor for unit consistency.

Results and Discussion

Effect of Anode and Structure of Coated SnO_2

Figure 1 shows nano-scale particles on the coated surface of Ti/SnO_2 electrodes. 21–25 nm particles were uniformly distributed, providing active sites for electrochemical oxidation of PAHs. EDAX of electrode surface confirms the presence of tin and oxygen at wt% of 61.28 and 35.91, respectively, (Fig. 2). Oxidation of H_2O on the electrode surface forms hydroxyl radicals ($OH^{\bullet}$) which oxidize the crystal lattice of SnO_2 to form higher oxides (SnO_{2+x}). Due to high reactivity toward most organic compounds, $OH^{\bullet}$ readily reacts with organic substrates near the anode surface to form CO_2 and H_2O. Electro-oxidation of organics is represented by Eqs. 1–4 below (Bouya et al. 2012). RO represents incomplete oxidation of organics and formation of intermediates. Complete oxidation of organics occurs more easily (Eq. 4).

Fig. 1 FESEM surface images of the Ti/SnO_2 surface

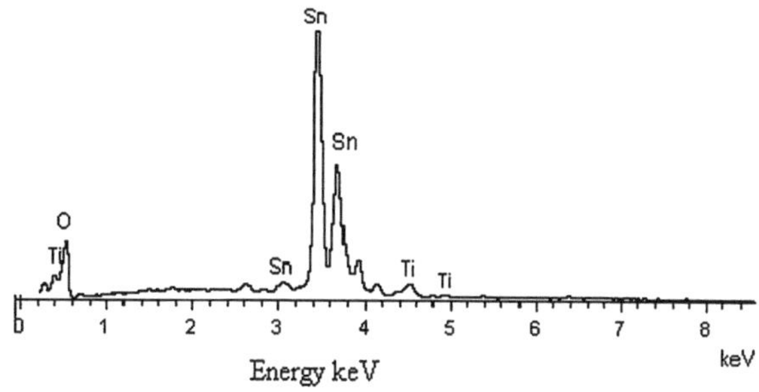

Fig. 2 EDAX of the Ti/SnO_2 electrode

$$SnO_2 + H_2O \rightarrow SnO_2(OH^\bullet)_x + xH^+ + xe^- \quad (1)$$

$$SnO_2(OH^\bullet)_x \rightarrow SnO_{2+x} + xH^+ + xe^- \quad (2)$$

$$R + SnO_{2+x} \rightarrow SnO_2 + RO \quad (3)$$

$$R + SnO_2(OH^\bullet)_x \rightarrow CO_2 + SnO_2 + xH_2O + xe^- \quad (4)$$

Effect of pH and Electrolyte

Tables 2 and 3 show PAH removal at initial pH of 3, 6, and 9 without and with electrolyte, respectively. Solution pH was adjusted using 1 M H_2SO_4 acid and NaOH. Naphthalene (2 benzene rings) was efficiently removed at all conditions tested. Complete oxidation of PAHs up to pyrene was observed in the presence of NaCl. The use of NaCl also enhanced degradation of the remaining PAHs. High degradation of PAHs with NaCl as electrolyte is due to conversion of chloride ions (Cl^-) to molecular chlorine (Cl_2) at the anode surface (Eq. 5), which reacts with OH^- ions to generate hypochlorite ions (ClO^-) (Eq. 6) and subsequently oxidizes the PAHs to CO_2. Bouya et al. (2012) obtained similar results for electrochemical degradation of cypermethrin pesticide on a SnO_2 anode using NaCl as electrolyte.

$$2Cl^- \rightarrow Cl_2 + 2e^- E^o + 1.36\,V \quad (5)$$

$$Cl_2 + 2OH^- \rightarrow Cl^- + ClO^- + H_2O \quad (6)$$

Among the three initial pHs (3, 6, and 0) tested, pH 6 was found to be the most favorable for oxidation of PAHs in aqueous solution. The results show that about 87 % removal of PAHs was obtained at pH 6 without using electrolyte and 70 and 80 % were obtained at pHs 3 and 9 respectively. Tran et al. (2010) also recorded 81 and 84 % PAHs removal near neutral pH values of 6 and 7, respectively. The presence of electrolyte increased the PAHs removal to 87, 93 and 87 % at initial pH of 3, 6, and 9, respectively.

Table 2 Concentration and removal % without using electrolyte

PAHs	RT (min)	Concentration (μg/l)				Removal (%)		
		Initial conc.	Initial pH			Initial pH		
			3	6	9	3	6	9
Nap	7.731	10.69	0	0	0	100	100	100
Acy	9.666	1.72	0	0	0	100	100	100
Ace	9.871	9.90	0	0	0	100	100	100
Flu	10.51	8.04	0.79	0	0	90.17	100	100
Phe	11.69	8.60	0.86	0.40	0.54	90.00	95.34	93.72
Ant	11.76	7.34	1.20	0.48	1.1	83.65	93.46	85.01
Fla	13.15	7.67	1.50	0.71	1.12	80.44	90.74	85.39
Pyr	13.44	7.37	1.40	0.65	1.01	81.00	91.18	86.29
BaA	15.01	6.30	2.62	1.01	1.51	58.41	83.96	76.03
Chr	15.07	6.56	3.59	1.40	2.40	45.27	78.65	63.41
B*b*F	16.65	6.48	3.21	1.41	1.96	50.46	78.24	69.75
B*k*F	16.69	5.41	3.58	1.45	2.45	33.82	73.19	54.71
B*a*P	17.19	5.18	1.61	0.50	1.50	68.91	90.34	71.04
Ind	19.50	5.24	4.05	2.26	2.72	22.70	56.87	48.09
DbA	19.56	5.03	3.28	1.36	2.33	34.79	72.96	53.67
BPer	20.19	5.51	3.48	1.39	2.47	36.84	74.77	55.17
Total removal (%)						70.88	87.33	80.27

Table 3 Concentration and removal % using electrolyte

PAHs	RT (min)	Concentration (μg/l)				Removal (%)		
		Initial conc.	Initial pH			Initial pH		
			3	6	9	3	6	9
Nap	7.731	10.69	0	0	0	100	100	100
Acy	9.666	1.72	0	0	0	100	100	100
Ace	9.871	9.90	0	0	0	100	100	100
Flu	10.51	8.04	0	0	0	100	100	100
Phe	11.69	8.60	0	0	0	100	100	100
Ant	11.76	7.34	0	0	0	100	100	100
Fla	13.15	7.67	0	0	0	100	100	100
Pyr	13.44	7.37	0	0	0	100	100	100
BaA	15.01	6.30	1.43	1.31	1.41	77.30	79.20	77.61
Chr	15.07	6.56	1.86	1.4	1.61	71.64	78.65	75.45
B*b*F	16.65	6.48	1.66	1.36	1.49	74.38	79.01	77.00
B*k*F	16.69	5.41	1.89	1.43	1.64	65.06	73.56	69.68
B*a*P	17.19	5.18	1.47	0	1.36	71.62	100	73.74
Ind	19.50	5.24	1.66	1.77	2.8	68.32	66.22	46.56
DbA	19.56	5.03	2.1	0	1.83	58.25	100	63.61
BPer	20.19	5.51	1.82	0	1.65	66.96	100	70.05
Total removal (%)						87.02	93.21	87.11

Table 4 Instantaneous Current Efficiency and Energy Consumption per mg of COD

	With Electrolyte			Without Electrolyte		
	COD Removal (%)	ICE	EC*	COD Removal (%)	ICE	EC*
pH 3	70.20	0.27	84.8	58.86	0.23	188.0
pH 6	80.31	0.31	74.3	67.37	0.26	189.5
pH 9	71.20	0.28	83.2	60.40	0.24	183.5

Unit of EC*: kWh/m^3

Instantaneous Current Efficiency (ICE)

Table 4 shows the instantaneous current efficiency and EC per mg of COD removed. Higher current efficiency was observed when NaCl (electrolyte) was used. Due to high COD removal at pH 6, current efficiency was also high compared to other pH values. The EC was lower at pH 6 when electrolyte was used, but was similar at all three pH values when no electrolyte was used.

Conclusion

Ti/SnO_2 anode produced substantially better results in the presence of NaCl electrolyte. Over 93 % PAHs removal was achieved. Efficiency of electrode was highest at pH 6 and low current density of 0.33 mA/cm^2. Instantaneous current efficiency was also highest at pH 6.

References

Bashir, M. J. K., Isa, M. H., Awang, Z. B., Aziz, H. A., Kutty, S. R. M., Mohajeri, S., et al. (2009). Landfill leachate treatment by electrochemical oxidation. *Waste Management, 29*(9), 2534–2541.

Bouya, H., Errami, M., Salghi, R., Bazzi, L. H., Zarrouk, A., Al-Deyab, S. S., et al. (2012). Electrochemical degradation of cypermethrin pesticide on a SnO_2 anode. *International Journal of Electrochemical Science, 7*, 3453–3465.

Ferrarese, E., Andreottola, G., & Opera, I. A. (2008). Remediation of PAH-contaminated sediments by chemical oxidation. *Journal of Hazardous Materials, 152*, 128–139.

Mohajeri, S., Aziz, H. A., Isa, M. H., Zahed, M. A., Bashir, M. J. K., & Adlan, M. N. (2010). Application of the central composite design for condition optimization for semi-aerobic landfill leachate treatment using electrochemical oxidation. *Water Science and Technology, 61*(5), 1257–1266.

Montilla, F., Morallon, E., & Vazquez, J. L. (2005). Evaluation of the electrocatalytic activity of antimony-doped tin dioxide anodes toward the oxidation of phenol in aqueous solutions. *Journal of the Electrochemical Society, 152*, B421–B427.

Pakpahan, E. N., Isa, M. H., Kutty, S. R. M., Chantara, S., Wiriya, W., & Faye, I. (2012a). Comparison of polycyclic aromatic hydrocarbons emission from thermal treatment of petroleum sludge cake in the presence of different additives. *Journal of Scientific Industrial Research, 17*, 430–436.

Pakpahan, E. N., Isa, M. H., Kutty, S. R. M., Chantara, S., & Wiriya, W. (2012b). Polycyclic aromatic hydrocarbon removal from petroleum sludge cake using thermal treatment with additives. *Environmental Technology, 34*(3), 407–416.

Pérez, S., Guillamón, M., & Barceló, D. (2001). Quantitative analysis of polycyclic aromatic hydrocarbons in sewage sludge from wastewater treatment plants. *Journal of Chromatography A, 938*, 57–65.

PHFS. (2009). Public health fact sheet. Polycyclic aromatic hydrocarbons (PAHs): health effects. Public health SA. www.health.sa.gov.au/pehs/environhealth-index.htm.

Santos, M. R. G., Goulart, M. O. F., Tonholo, J., & Zanta, C. L. P. S. (2006). The application of electrochemical technology to the remediation of oily wastewater. *Chemosphere, 64*, 393–399.

Sun, Z.-Q., Lu, H.-Y., Ren, X.-B., Huang, W.-M., Dong, Y.-J., & Lin, H.-B. (2009). Performances of Ti/SnO2-Sb2O5 anodes prepared by brush coating thermal decomposition method. *Acta Physico-Chimica Sinica, 25*(7), 1385–1390.

Tran, L. H., Drogui, P., Mercier, G., & Jean-Francois, B. (2010). Comparison between Fenton oxidation process and electrochemical oxidation for PAH removal from an amphoteric surfactant solution. *Journal of Applied Electrochemistry, 40*, 1493–1510.

Yaqub, A., Ajab, H., Isa, M. H., Jusoh, H., Junaid, M., & Farooq, R. (2012). Effect of ultrasound and electrode material on electrochemical treatment of industrial wastewater. *Journal of New Material Electrochemical Systems, 15*(4), 289–292.

Sonocatalytic Degradation of Acid Red Dye in Water Using Fe-Doped TiO_2 Deposited on HY Catalyst

A. H. Alwash, N. A. Jamalluddin, N. Ismail and A. Z. Abdullah

Abstract This study addresses the properties affecting the activity and stability of Fe/Ti-NaY heterogeneous catalysts with different Fe and Ti loadings in ultrasound-irradiated decolorization of Amaranth dye in aqueous solution. The catalysts were prepared using an ion-exchange method followed by an impregnation method. Different characterization techniques, i.e., XRD, AFM, UV–Vis, and SEM/EDAX, were performed to characterize the catalysts. TiO_2 was successfully encapsulated within the internal pore of the zeolite resulted in a blue shift toward shorter wave length. On the other hand, Fe deposition on the external surface caused a red shift and reduced the surface roughness. Loading of Ti and Fe slightly increased the hydrophilicity of the heterogeneous catalyst. Increasing amount of Fe above the optimum value could also cause the scavenging of $^{\bullet}OH$ radicals and higher consumption of hydrogen peroxide. External H_2O_2 could be used to accelerate the sonocatalytic reaction.

Introduction

Ultrasonic (US) waves have the ability to decompose water molecules into extremely reactive hydrogen ($H^{\bullet}$) and hydroxyl ($^{\bullet}OH$) radicals by the heat that is generated during the cavity implosion. The synergetic effect between ultrasound

A. H. Alwash
Department of Chemistry, College of Science, Al-Nahrain University, Al-Jadirya, Baghdad, Iraq

A. H. Alwash · N. A. Jamalluddin · A. Z. Abdullah (✉)
School of Chemical Engineering, Universiti Sains Malaysia, 14300 Nibong Tebal, Penang, Malaysia
e-mail: chzuhairi@eng.usm.my

N. Ismail
School of Industrial Technology, Universiti Sains Malaysia, 11800 Nibong Tebal, Penang, Malaysia

R. Pogaku et al. (eds.), *Developments in Sustainable Chemical and Bioprocess Technology*, DOI: 10.1007/978-1-4614-6208-8_13,

irradiation and heterogeneous catalyst provides advantages with respect to the desired formation of radical species. Titanium dioxide (TiO_2) is one of the most efficient photocatalysts, but it has poor catalytic activity when used alone due to the fast charge carrier recombination in addition to the recycling difficulties (Ökte and Yilmaz 2009a). Therefore, there is a strong need to overcome these drawbacks by doping suitable metals to increase its activity (Mirkhani et al. 2009), while the use of porous supports such as zeolite for the TiO_2 particles could result in a catalyst with higher surface area. Different advantages have been obtained from the loading of TiO_2 on zeolite, i.e., increasing the concentration of TiO_2 without the problem of aggregation, especially at higher concentration, uncomplicated separation or filtration process, the adsorption of intermediate products that are produced after the reaction, the ability to concentrate the pollutants near TiO_2 particles and better recyclability of the catalyst (Ökte and Yilmaz 2009b).

Certain metals especially Fe can be incorporated into the heterogeneous catalysts to be used in the ultrasound-accelerated process to enhance the generation of $^{\bullet}OH$ radicals during the reaction. Although homogeneous Fe catalyst (Fenton catalyst) can also be used in an US process, it is not a preferred approach. This is due to some drawbacks with respect to the high consumption of hydrogen peroxide, narrow range of applicable pH value, and difficulty in the separation of iron ions from the treated solution (Chen et al. 2010). To eliminate these drawbacks, heterogeneous catalysts can be developed by immobilizing Fe(III) ions on suitable supports. Jamalluddin and Abdullah (2011) reported the doping of Fe(III) into different supports such as TiO_2 and zeolite Y to produce active heterogeneous catalysts for sonocatalytic degradation of organic water pollutants. The removal of dyes was found to be significantly improved after doping of Fe ions into TiO_2 and zeolite Y as compared to the activity achieved with the pure support materials. However, the amount of Fe(III) ions incorporated into both TiO_2 and zeolite can significantly affect the characteristics and the stability of the catalysts.

Methodology

Catalyst Preparation

HY zeolite was first converted into its sodium form by means of an ion-exchange method (Tayade et al. 2007). TiO_2^- modified NaY zeolite was prepared using a similar procedure as reported by Liu et al. (1992). Different concentrations (5.6–28.2 mM) of potassium titanooxalate solution were first prepared in 100 ml water. The catalysts are denoted as Ti_1-NaY, Ti_2-NaY, Ti_3-NaY, and Ti_4-NaY according to the increasing loading of titanium in the catalyst. After calcination at 550 °C for 5 h, they were ground into fine powder and kept in a dry cabinet. The NaY material with the highest loading of titanium (Ti_4-NaY) was then chosen as a support for the loading of Fe ions ranging from 0.2 to 1.0 wt.% using impregnation

Table 1 The structure and properties of Amaranth

Properties	Info
Molecular structure	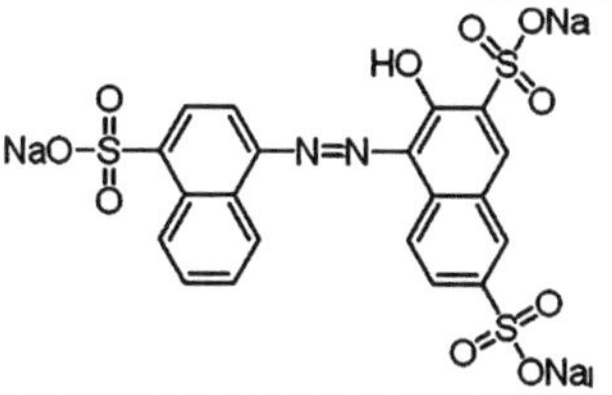
Other names	Acid red 27, FD&C Red 2, Food Red 9
Molecular weight	604.47 g/mol
Stability	Soluble in water
C.I.	16,185
λ_{max}	521 nm

method. The Fe/Ti_4-NaY catalyst was prepared using a wet impregnation method. The final catalysts are named according to the different Fe loadings that were loaded onto Ti_4-NaY such as 0.2 % Fe/Ti_4-NaY, 0.4 % Fe/Ti_4-NaY, 0.6 % Fe/Ti_4-NaY, 0.8 % Fe/Ti_4-NaY, and 1.0 % Fe/Ti_4-NaY. The characteristics of the catalysts produced are thoroughly discussed elsewhere (Al-Wash et al. 2012).

Sonocatalytic Process

All experiments were carried out in 100-ml cylindrical vessels, and they were placed in an US water bath with an output power fixed at 50 W and at an US frequency of 40 kHz. Before the sonocatalytic reaction was started, the solution was stirrer for 30 min at room temperature to maintain a good dispersion of catalyst with the dye solution and to allow any possible adsorption of the dye on the catalysts. The reaction was generally carried out with 10 ppm for the initial Amaranth dye concentration, a pH value of 2.5 and 2.0 g/l of the catalyst loading. The catalytic activity was examined by calculating the decolorization efficiency of Amaranth dye (properties given in Table 1) using a UV–Vis spectrophotometer, and the measurement was made at a wavelength of 521 nm.

Results and Discussion

Sonocatalytic decolorization of Amaranth was investigated in the presence of US irradiation and various catalysts. The decolorization efficiency of Amaranth was found to increase with increasing titanium oxide amount in the catalyst, reaching its maximum value of 50 % after 120 min of reaction. This increase was ascribed to the presence of more active sites within the internal pores of the zeolite to consequently increase the catalytic activity. This was clearly the advantage of the sonocatalytic process as compared to a photo-assisted process. The actual amount of the Ti species that was incorporated into the zeolite through the ion-exchange method was determined using an inductively coupled plasma (ICP) spectrophotometer for the Ti_4-NaY (that showed the highest decolorization efficiency), and it was found to be 2.97 mg/g. Therefore, the Ti_4-NaY was subsequently used in this study for the doping with Fe ions to further improve the sonocatalytic activity.

The effect of Fe(III) loading in Ti_4-NaY between 0.2 and 1.0 wt.% was subsequently investigated for the decolorization of Amaranth. The activity of the catalysts as shown in Fig. 1 was found to increase with increasing Fe doping and reached its highest level of 75 % in 120 min with the catalyst containing 0.8 wt.% of Fe. Increasing sonocatalytic activity from 50 % for Ti_4-NaY to 75 % at 0.8 wt.% Fe/Ti_4-NaY could be ascribed to the increasing number of Fe(III) active sites on the surface of the catalyst in addition to the existence of nanoparticles TiO_2 inside the internal pores of the zeolite. The reduction in the energy gap of the catalyst after the loading of TiO_2 in the internal pores led to enhancement the sonocatalytic activity through a well-documented mechanism (Jamalluddin and Abdullah 2011). Fe(III) can further improve the activity due to a Fenton-like mechanism as elaborated in an earlier report (Zhong et al. 2011). Besides, with an increase in the Fe ions loading in the catalyst, corresponding increase in the ability of the catalyst to inhibit the e^-–h^+ recombination would be achieved (Jamalluddin and Abdullah 2011). All these factors led to improvement in the sonocatalytic activity as observed in this study.

Panda et al. (2011) through their investigation on the effect of Fe loading onto silica surface concluded that higher catalytic activity at high Fe loading was attributed to an increase in the active sites in the catalyst. Consequently, this led an enhancement in the production of active radicals. However, the decrease in the decolorization efficiency to 45 % for 1.0 % Fe/Ti_4-NaY catalyst in this study could be ascribed to the reduced accessibility of dye molecules to the encapsulated TiO_2. Choi et al. (1994) reported that the average distance between the trapping sites could decrease with increasing loading of impregnated Fe. The experimental evidence in

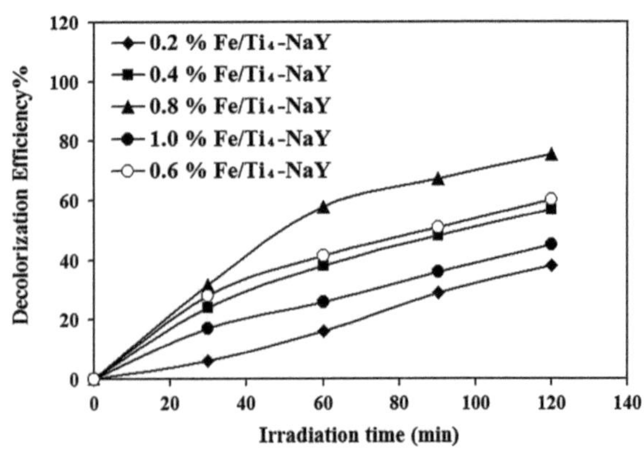

Fig. 1 Effect of Fe loading on decolorization efficiency of Amaranth (Conditions: Initial dye concentration 10 ppm, natural pH, and catalyst loading 1.5 g/l)

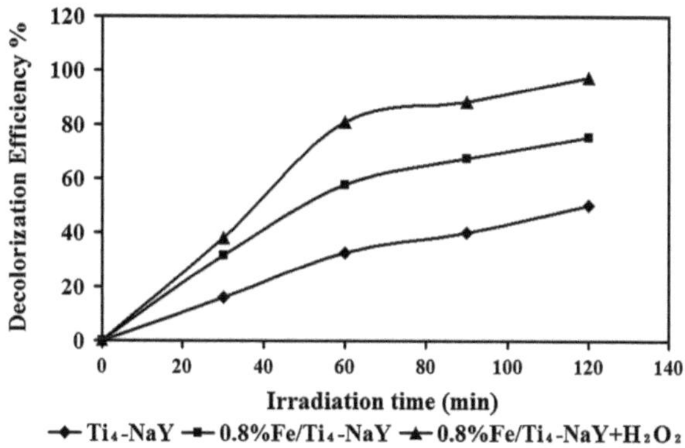

Fig. 2 Decolorization efficiency of Amaranth before and after the addition of hydrogen peroxide

which Fe deposition mainly occurred at the external surface could support this argument. In addition, increasing amount of Fe above the optimum value could also cause the scavenging of $^{\bullet}OH$ radicals and higher consumption of hydrogen peroxide (H_2O_2) (Jamalluddin and Abdullah 2011). The H_2O_2 in this study was produced through the sonolysis of H_2O, and it dissociated into $H^{\bullet}$ and $OH^{\bullet}$ radicals. These hydrogen atoms would react with the dissolved O_2 from the atmosphere to form the $^{\bullet}OOH$ which would then recombine to form H_2O_2 and O_2 molecules.

On the other hand, below the optimum Fe loading, i.e., at 0.2, 0.4, 0.6 wt.%, the highest decolorization activity achieved in 2 h or reaction was 38, 57, and 60 %, respectively. The activity of the catalyst was found to increase with increasing concentration of the metal. This behavior was ascribed to increasing the number of the trapping electron sites. Thus, the low activity of the catalyst below the optimum loading was ascribed to the insufficient amount of Fe(III) ions in the catalyst so that the amount of $HO^{\bullet}$ radicals generated was not high enough to rapidly degrade the Amaranth dye molecules.

In order to further enhance the decolorization efficiency of Amaranth to increase the production of $^{\bullet}OH$ radicals, an external supply of H_2O_2 was added to the reaction system. The decolorization efficiency of Amaranth was found to increase to 98 % after the addition of H_2O_2 at a fixed dosage of 20 mmol/100 ml H_2O_2, and the results are as shown in Fig. 2. The enhancement in the decolorization efficiency was attributed to the generation of more $^{\bullet}OH$ radicals from the disassociation of hydrogen peroxide into $H^{\bullet}$ and $OH^{\bullet}$ radicals in addition to the role of the heterogeneous sonocatalyst. As such, external H_2O_2 could be used to accelerate the sonocatalytic reaction.

Conclusions

A new heterogeneous catalyst was successfully synthesized by loading zeolite Y with two active metals, i.e., Ti and Fe(III), using an ion-exchange and an impregnation method, respectively. The catalyst showed high activity in Amaranth dye degradation, and increasing activity was demonstrated with increasing Fe loading to 0.8 wt.%. Further increase was demonstrated by the system when 20 mmol/100 ml H_2O_2 was added. Increasing amount of Fe above the optimum value could also cause the scavenging of $^{\bullet}OH$ radicals and higher consumption of hydrogen peroxide. Thus, this catalyst was highly potential in achieving high decolorization efficiency of Amaranth while showing a small amount of leaching into the solution.

References

Al-Wash, A. H., Abdullah, A. Z., & Ismail, N. (2012). Zeolite Y encapsulated with Fe-TiO_2 for ultrasound-assisted degradation of amaranth dye in water. *Journal of Hazardous Materials*, 233–234, 184–193.

Chen, F., Li, Y., Cai, W., & Zhang, J. (2010). Preparation and sono-Fenton performance of 4A-zeolite supported α-Fe_2O_3. *Journal of Hazardous Materials, 177*, 743–749.

Choi, W., Termin, A., & Hoffmann, M. R. (1994). The role of metal ion dopants in quantum-sized TiO_2: Correlation between photoreactivity and charge carrier recombination dynamics. *Journal of Physical Chemistry, 98*, 13669–13679.

Jamalluddin, N. A., & Abdullah, A. Z. (2011). Effect of pH and catalyst dosage on the leaching of Fe from Fe(III) doped zeolite Y used for sonocatalytic degradation of Acid Red B. *Research Journal of Chemistry and Environment, 15*, 860–865.

Liu, X., Iu, K.-K., & Thomas, J. K. (1992). Encapsulation of TiO_2 in zeolite Y. *Chemical Physics Letters, 195*, 163–168.

Mirkhani, V., Tangestaninejad, S., Moghadam, M., Habibi, M., & Rostami-Vartooni, A. (2009). Photocatalytic degradation of azo dyes catalyzed by Ag doped TiO_2 Photocatalyst. *Journal of Iranian Chemical Society, 6*, 578–587.

Ökte, A. N., & Yilmaz, Ö. (2009a). La and Ce loaded TiO_2-ZSM-5 catalysts: Comparative characterization and photocatalytic activity investigations. *Microporous and Mesoporous Materials, 126*, 245–252.

Ökte, A. N., & Yilmaz, Ö. (2009b). Characteristics of lanthanum loaded TiO_2-ZSM-5 photocatalysts: Decolorization and degradation processes of Methyl Orange. *Applied Catalysis, A: General, 354*, 132–142.

Panda, N., Sahoo, H., & Mohapatra, S. (2011). Decolourization of methyl orange using Fenton-like mesoporous Fe_2O_3–SiO_2 composite. *Journal of Hazardous Materials, 185*, 359–365.

Tayade, R. J., Kulkarni, R. G., & Jasra, R. V. (2007). Enhanced photocatalytic activity of TiO_2-Coated NaY and HY zeolites for the degradation of Methylene Blue in water. *Industrial and Engineering Chemistry Research, 46*, 369–376.

Zhong, X., Royer, S., Zhang, H., Huang, Q., Xiang, L., Valange, S., et al. (2011). Mesoporous silica iron-doped as stable and efficient heterogeneous catalyst for the degradation of C.I. Acid Orange 7 using sono-photo-Fenton process. *Separation and Purification Technology, 80*, 163–171.

Investigation of the Effect of Phosphoric Acid (H_3PO_4) on Rate of Oil and Solid Separation in Palm Oil Mill Clarifier

F. P. S. Wong and S. Y. Yong

Abstract In recent years, there is a growing demand for palm oil which is driven by the increase in human population and biodiesel production. In the palm oil milling process, a large quantity of water is consumed for the oil extraction process and consequently a large amount of wastewater is generated which is termed as palm oil mill effluent (POME). POME is a liquid waste having a high content of biochemical oxygen demand (BOD) that deteriorates the quality of river water and pollutes the environment. Despite the development of various treatment systems to recover the water, a reduction in the production of this wastewater is still the best choice. This research seeks to improve the rate of separation of oil and solid in the clarifier mixture (CM) to improve the rate of oil recovery, which will lead to a reduction in wastewater generation and faster water recovery. Experiments were carried out to investigate the effect of dilution at 0, 80:20, 70:30, and 60:40 CM–water ratios, operating temperature at 70, 80, and 90 °C, and phosphoric acid application of 0, 0.1, 0.3, and 0.5 % on the rate of separation. The results revealed that the best rate of oil flotation could be obtained at a temperature of 90 °C with the treatment of 0.3 % H_3PO_4 in the undiluted CM. The addition of H_3PO_4 could neutralize the negative charges of oil by the introduction of positively charged hydrogen ions to enhance the coalescences of oil droplets. It could also induce the degumming process that could remove the emulsifying components of palm oil such as phosphatide. This could lead to a higher rate of solid and oil separation because when the emulsion was broken, oil density was reduced and the viscosity of the CM was reduced.

Keywords POME · Clarifier · Rate · Separation · Phosphoric acid · Oil · Solid

F. P. S. Wong (✉) · S. Y. Yong
Chemical Engineering Department, School of Engineering and Science,
Curtin University, CDT 250, 98009 Sarawak, Malaysia
e-mail: florence.wong@curtin.edu.my

S. Y. Yong
e-mail: ysyee89@hotmail.com

R. Pogaku et al. (eds.), *Developments in Sustainable Chemical and Bioprocess Technology*, DOI: 10.1007/978-1-4614-6208-8_14,

Introduction

The oil palm industry is one of the highest pollution-generating agro-industries in Malaysia (Sumathi 2004). The largest source of pollution in this industry is the huge volume of palm oil mill effluent (POME). From the production in year 2011, the amount of POME generated was 36,021,943 tonnes as reported by Malaysian Palm Oil Board (2011). The residue oil and solids are the major components of POME that cause high biochemical oxygen demand (BOD) level of 25,000 mg/liter (Ahmad et al. 2003). This hazardous characteristic of POME has motivated a wide range of research on the treatment systems and technologies.

However, prevention is always better than cure. The reduction in the generation of waste is still a better approach to mitigate the harmful impact of POME. The existing measure adopted by the majority of palm oil mills (POM) for reduction of POME production is the use of 3-phase decanters. The effluent discharged contains 95–96 % water, 0.6–0.7 % (4,000 mg/l) oil and grease, and 4–5 % total solids (Ahmad et al. 2003).

POME with such characteristics requires hydraulic retention time (HRT) of about 40–60 days with the existing pond treatment system as practiced by 85 % of POM in Malaysia (Rupani et al. 2010). The long HRT is not environmental friendly because it deprives the environment of a huge amount of portable water for such long period of time. Therefore, further reduction in residue oil and solid is desirable to lower BOD level and to shorten the duration for treatment. However, the amount of residue oil in POME is so well emulsified in the suspended solid that it is not easily recovered both chemically and mechanically.

There are a range of studies done on the improvement of oil and solid separation from POME as a result of the implementation of discharge regulations by Department of Environment in Malaysia. Some conventional methods of treatment include gravitational methods such as API separators, corrugated plate interceptors (CPI), chemical treatment that uses sulfuric acid, aluminum sulfate and others as emulsion breaker, and dissolved air flotation (DAF). However, the efficiency of these methods in separating oil and solids from the oily wastewater is often unsatisfactory particularly when the oil droplets are finely dispersed in dilute concentration and there are inconsistent variations in the composition of the wastewater (Alade et al. 2011). More recent studies were done using adsorption techniques with various adsorbents such as Chitosan adsorbent and rubber powder which achieve about 80 % of oil removal. However, this method has only been done on lab scale. Its performance at pilot and commercial scales is yet to be demonstrated (Ahmad et al. 2005a, b). The application of membrane separation for the treatment of oily wastewater is demanded due to its high efficiency to handle a low concentration of oil. However, its operation can often be hindered by membrane fouling, and its economical advantage has not been that robust because of the high capital and maintenance cost (Alade et al. 2011).

This research aims at investigating the rate of separation of oil and solid in the clarifier mixture (CM) (40 % oil, 3 % solid, and 57 % water) to improve the rate of oil recovery by the addition of food-grade H_3PO_4 at the existing operating dilution and temperature of 90 °C. Experiments were also carried out to investigate the effect of dilution and operating temperatures on the rate of separation.

According to Deffense (2012), 0.1–0.3 % of food-grade H_3PO_4 can be used in the palm oil refineries for degumming purposes. The process is usually carried out at elevated temperature around 90 °C, as described in the literature (Mag and Reid 1980) and (Campbell et al. 1983). Studies done by Wong et al. (2005) had discovered that the rate of oil flotation from refined palm oil/cooking oil–water mixture is higher than that of crude palm oil–water mixture at constant temperature and oil concentration. The possible reason could be the degumming process that removes the phosphatide which is a good emulsifying agent present in the crude palm oil. The reduction in emulsifying activities and viscosity of the mixture will lead to a faster rate of coalescence of oil droplets and flotation. Sereewatthanawut et al. (2011) also described the removal of phospholipids and other antioxidants during refining process of another vegetable oil, rice bran. The reduction in the amount of different components will lead to a reduction in the density of the oil, causing a higher rate in oil flotation. Since the content of crude palm oil mill clarifiers is made of a mixture (40 % oil, 3 % solid and 57 % water) where a large part of it is well emulsified. It is therefore reasonable to try using H_3PO_4 to improve oil recovery in the clarifiers of POM which usually operate at 90 °C also.

Moreover, the positively charged hydrogen ions of H_3PO_4 will neutralize the negative charges of emulsified oil which are generated during mechanical mixing and pumping. This neutralization will enhance the breaking of emulsions, leading to coalescences of oil droplets and flotation (Kemmer 1988). If H_3PO_4 is used in the clarifiers of POM, it may partially or completely save the degumming stage in the refineries. This also will lead to a double benefit of an increase in oil extraction rate and a reduction in the volume of wastewater generated in the crude POM.

Methodology

The samples for CM were obtained at the inlet to the clarifiers in the palm oil mill of Alum Bumi Sdn. Bhd, Bintulu, Sarawak. Experiments were conducted to determine the separation efficiency of oil and solid in the CM under different treatment conditions. The oil separation rate with respect to heat treatment, different concentrations of CM in water, and H_3PO_4 treatment was studied. The samples for the tests were made up of different concentration ratios of CM to water of 60:40, 70:30, 80:20, and 100:0.

The first experiment was done using undiluted and diluted CM to determine the rate of oil flotation by observing the curve upon the change in interfacial height under temperature treatment of 70 °C. Before conducting the experiment, preheating of the CM in water bath was done. Next, four 100-ml glass cylinders were filled with the CM diluted with water at different ratios of 100:0, 80:20, 70:30, and

60:40. The glass cylinders were immersed in water bath for 60 min. A stop watch was used to observe the change in interfacial height of oil with time, and the result was recorded every minute. The experiment was repeated under temperature treatment of 80 and 90 °C.

The second experiment was conducted to determine the rate of oil flotation in CM with treatment of different concentrations of H_3PO_4. In this experiment, different composition ratios of CM to water such as 100:0, 80:20, 70:30, and 60:40 were used. Concentrations of 0.1, 0.3, and 0.5 % of H_3PO_4 were mixed together with each type of CM in four 100-ml glass cylinders which were then immersed in a water bath at a fixed temperature of 90 °C.

Results and Discussion

Treatment with a higher temperature clearly shows a positive effect on the rate of oil flotation from the CM as shown in Fig. 1a below. The highest initial rate of oil flotation from the CM can be obtained by heat treatment with 90 °C without dilution as shown in Fig. 1b. This is in agreement with Wong et al. (2005) that the kinetic order of oil floatation depends on the concentration of oil inside the oil–water emulsion. The undiluted CM containing higher amounts of palm oil and the droplets showed close behavior. This makes the coalescence of droplets faster as the increased temperature reduces the viscosity and speeds up the collision and attraction of the droplets. From Fig. 1b, it seems that at a higher concentration of CM, the oil flotation rate was not significantly affected by the increase in temperature. However, with a lower concentration of CM, the temperature increase produced a greater effect in the increase in oil flotation rate. This shows that the concentration of oil is an important factor that determines the effect of heat treatment for enhancing oil flotation rate.

The data in Table 1 which are also shown in Fig. 2a and b indicate that treatment with 0.3 % H_3PO_4 on the undiluted CM at 90 °C enables the fastest rate of oil flotation. The rate of oil flotation in undiluted CM treated with 0.3 % H_3PO_4 at 70 °C

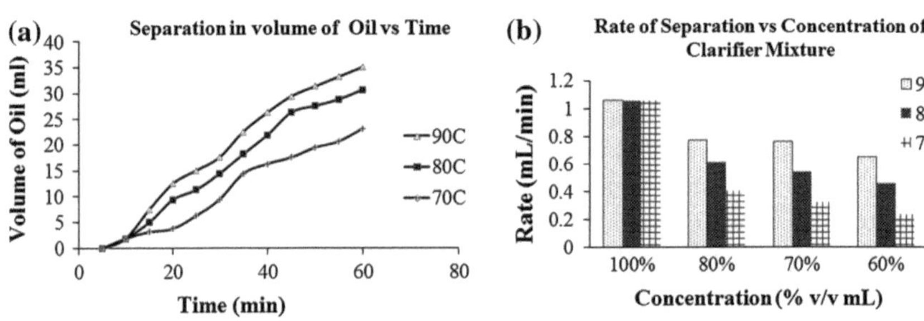

Fig. 1 **a** Volume of oil (ml) from 80 % CM at different temperatures. **b** Rate of oil flotation in CM at different temperatures

Table 1 Rate of oil flotation (ml/min) in CM treated with H_3PO_4 and heat

H_3PO_4 (%)	100 % CM, 90 °C	100 % CM, 80 °C	100 % CM, 70 °C	60 % CM, 90 °C	70 % CM, 90 °C	80 % CM, 90 °C	100 % CM, 90 °C
0	1.061	1.058	1.055	0.649	0.761	0.772	1.061
0.1	1.355	1.245	1.129				
0.3	1.363	1.262	1.132	1.001	1.107	1.236	1.363
0.5	1.224	1.215	1.084				
		Increase in rate		54.24 %	45.47 %	60.10 %	28.46 %

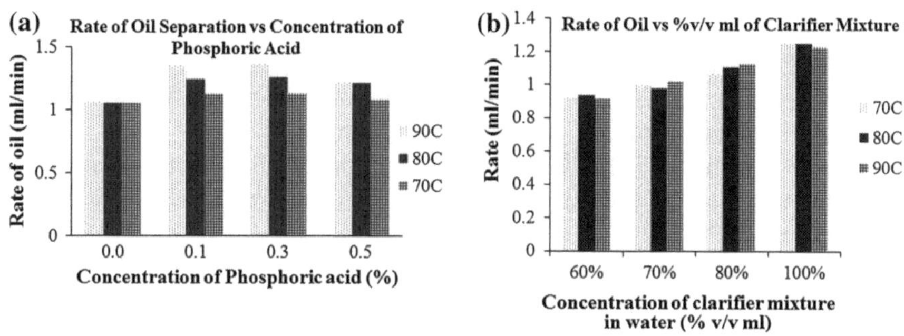

Fig. 2 **a** Rate of oil flotation for CM (100 %) treated with H_3PO_4 and heat. **b** Rate of oil flotation for different CM treated with 0.3 % H_3PO_4 and heat

was greater than the rate in the same sample without treatment of H_3PO_4 at 90 °C. This demonstrates that treatment with H_3PO_4 was an effective measure in improving oil extraction rate in clarifiers of POM. The degumming effect (Mag and Reid 1980) and (Campbell et al. 1983) and the demulsifying effect (Kemmer 1988) of the H_3PO_4 could be the reasons which led to the decrease in the oil density and viscosity of the mixture. The increase in oil extraction rate will reduce oil and grease content in the sludge discharged to POME ponds. According to (Hanaki et al. 1981), the long-chain fatty acid present in the oil will inhibit the anaerobic microbial degradation of the organic matter in POME. Therefore, the reduction in oil and grease content in POME will speed up the biodegradation of its organic matter, leading to a faster recovery of water from POME.

The percentage increase in rate shown in Table 1 suggests that the oil concentration in the sample treated with H_3PO_4 is again a factor that influences the rate of oil flotation. Studies done by Wong et al. (2005) also showed that hydrochloric acid tends to make the flotation rate more dependent on the oil concentration in the oil–water emulsion. These phenomena can be attributed to the dynamic equilibrium process of neutralization of positive ions from the acid and the negative ions of oil droplets that results in the coalescence of oil droplets.

Figure 2a indicates the effect of the concentration of H_3PO_4 on the effect of temperature for the rate of oil flotation. 0.1 to 0.3 % of H_3PO_4 at 900 °C for the undiluted CM will result in the highest oil flotation rate. Without the treatment of H_3PO_4, an increase in temperature does not lead to any significant increase in the rate of oil flotation in the undiluted CM.

Conclusion and Recommendation

The results in this study offer a promising route in the separation and recovery of oil from the clarifiers of POM. The result could be summarized as follows:

- The highest rate of oil flotation was obtained from the undiluted CM treated with 0.3 % H_3PO_4 at 90 °C.
- An increase in temperature increased the rate of oil flotation, but less significantly in the undiluted CM than in diluted mixtures.
- Treatment with 0.3 % H_3PO_4 increases the rate of oil flotation at all tested temperatures and dilutions of CM, but most significantly with the 80:20 CM to water ratio at 90 °C.
- It can be concluded that it is more feasible to treat the CM with H_3PO_4 for a higher recovery rate of oil and a reduction of POME production than to treat POME with HCl as suggested by Wong et al. (2005) due to too low percentage of oil present in POME.

Acknowledgments We wish to thank Alum Bumi Palm Oil Mill Sdn. Bhd and Curtin University, Sarawak, Malaysia for their support in this project.

References

Ahmad, A. F., Ismail, S., & Bhatia, S. (2003). Water recycling from palm oil mill effluent (POME) using membrane technology. *Desalination, 157*, 87–95.

Ahmad, A. L., Bhatia, S., Ibrahim, N., & Sumathi, S. (2005a). Adsorption of residue oil from palm oil mill effluent using rubber powder. *Brazilian Journal of Chemical Engineering*, *22*(3).

Ahmad, A. L., Sumathi, S., & Hameed, B. H. (2005b). Residue oil and suspended solid removal using natural adsorbents chitosan, bentonite and activated carbon: A comparative study. *Chemical Engineering Journal, 108*(1–2), 179–185.

Alade, A. O., Jameel, A. T., Muyubi, S. A., Abdul Karim, M. I., & Alam, M. Z. (2011). Removal of oil and grease as emerging pollutants of concern (EPC) in wastewater stream. *IIUM Engineering Journal, 12*(4).

Campbell, S. J., Nakayama, N., & Unger, E. H. (1983). Chemical degumming of crude vegetable oils, United Oilseed Products Ltd, Canadian Patent 1 157 883.

Deffense N.D. (2012). Edible oil processing—chemical degumming. http://lipidlibrary.aocs.org/processing/chem-degum/index.htm. Accessed 8 Aug 2012.

Hanaki, K., Matsuo, T., & Nagas, M. (1981). Mechanism of inhibition caused by long chain fatty acids in anaerobic digestion. *Biotechnology and Bioengineering, 12*, 1591–1610.

Kemmer, F. N. (1988). *Emulsion breaking in the Nalco water handbook* (2nd edn, pp. 11.1–11.8). Singapore: McGraw-Hill.

Mag, T. K., & Reid, M. P. (1980). Continuous process for contacting of triglyceride oils with an acid. Canada Packers Limited, US Patent 4,240, 972.

Malaysian Palm Oil Board N.D. (2011). Economic and statistics—FFB processed by mill 2011. http://bepi.mpob.gov.my/index.php/statistics/sectoral-status/75-sectoral-status-2011.html. Accessed 8 Aug 2012.

Rupani, P. F., Singh, R. P., Ibrahim, M. H., & Esa, N. (2010). Review of current palm oil mill effluent (POME) treatment methods: Vermicomposting as a sustainable practice. *World Applied Sciences Journal, 11*(1), 70–81.

Sereewatthanawut, I., Baptista, I. I. R., Boam, A. T., Hodgson, A., & Livingston, A. G. (2011). Nanofiltration process for the nutritional enrichment and refining of rice bran oil. *Journal of Food Engineering, 102*(1), 16–24.

Sumathi, S. (2004). Removal of residue oil from palm oil mill effluent (POME) using chitosan. Master's thesis. Malaysia: School of Chemical Engineering, Universiti Sains Malaysia.

Wong, F., Nandong, J., Samyudia, F., & Tan, K. K. (2005). *An experimental investigation on the oil separation from oil-water emulsion of palm oil mills effluent (POME)*. Conference on Chemical and Bioprocess Engineering (ICCBPE 2005), 15th–17th Dec. Kota Kinabalu, Sabah, Malaysia.

Polyester Thin Film Composite Nanofiltration Membranes Prepared by Interfacial Polymerization Technique for Removal of Humic Acid

M. N. Abu Seman, N. A. Jalanni, C. K. M. Faizal and N. Hilal

Abstract Nanofiltration (NF) polyester thin film composite membranes have been prepared through interfacial polymerization using a microporous polyethersulfone membrane as support. The thin polyester layer formed on the top surface of the microporous support layer was produced by the reaction of 6 %w/v of triethanolamine (TEOA) in aqueous solution and a solution containing trimesoyl chloride (TMC) for different reaction times (15, 25, and 35 min). The performance of the membranes was then characterized using permeation experiments with 0.001–0.1 M of salt solutions (NaCl and Na_2SO_4) and 15 mg/l of humic acid solution as a model for natural organic matter (NOM), normally found in surface water. This study has shown that through the interfacial polymerization technique, the variation of reaction time at a constant monomer concentration of 6 % w/v can affect the properties of the membrane produced and indirectly influences the membrane performance. Increasing the reaction time resulted in decreasing water permeabilities. However, a higher removal of humic acid was observed.

M. N. Abu Seman (✉) · N. A. Jalanni · C. K. M. Faizal
Faculty of Chemical and Natural Resources Engineering, Universiti Malaysia Pahang, Lebuhraya Tun Razak, 26300 Gambang, Kuantan, Pahang, Malaysia
e-mail: mazrul@ump.edu.my

N. A. Jalanni
e-mail: are_yin882004@yahoo.com

C. K. M. Faizal
e-mail: mfaizal@ump.edu.my

N. Hilal
Centre for Water Advanced Technologies and Environmental Research (CWATER), Swansea University, Swansea SA2 8PP, Swansea, UK
e-mail: n.hilal@swansea.ac.uk

R. Pogaku et al. (eds.), *Developments in Sustainable Chemical and Bioprocess Technology*, DOI: 10.1007/978-1-4614-6208-8_15,

Introduction

Interfacial polymerization (IP) is a common technique used for preparation of nanofiltration (NF) and reverse osmosis (RO) composite membranes either in laboratory scale or in commercial scale. In general, the produced membranes exhibit high water permeation flux and high salt rejection (Ahmad et al. 2004). Most commercial NF and RO membranes are produced by the interfacial polymerization method that has a polyamide (PA) thin layer on top of the membrane support. The polyamide produced from reaction of amine monomer with acyl chloride in organic phase. One of the common amine monomers used is m-phenylenediamine (MPD), and other amine monomers for production of polyamide are included. Besides the polyamide membrane, a few researchers explored this technique to produce the other polymeric thin film polyester and polyesteramide membranes, where BPA was used as the primary monomer during interfacial polymerization (Abu Seman et al. 2010; Mohammad et al. 2003; Jayarani and Kulkarni 2000). Even their membranes could exhibit very good performances; however, in the last few years, BPA toxicity is becoming a big issue as it may strongly contribute to some negative effects to human health. For example, manufacturers used polycarbonate plastic as the material for baby bottles because it is rigid, shatter resistant, and clear in colour. Plastic baby bottles have received many attentions recently as research came to light showing many models leach BPA, a suspected endocrine disruptor (Beverly 2011). Concerning this issue, the same thing could happen to this type of polymeric membrane which uses BPA as the monomer. As an alternative, polyester membrane still could be produced by choosing the right monomer which is safe and inexpensive. Previous study shows that polyester thin film composite membranes produced by triethanolamine (TEOA) have the ability to be used in desalination at different environmental pHs. On top of that, TEOA is an active monomer which is environment-friendly, economical, and easy to be obtained (Tang et al. 2008). In this chapter, the effect of interfacial polymerization parameters (i.e. monomer concentration and reaction time) on the membrane performance was investigated. Membrane performances have been evaluated in terms of membrane permeability and solute removal (salt and organic molecules). In this study, NaCl and Na_2SO_4 were selected to represent salt solutions, while humic acid was used as a model of natural organic matter (NOM) which is normally found in surface water.

Methodology

Materials

The commercial flat-sheet UF polyethersulfone membrane was supplied by Amfor, Inc. (China), and the membrane details are shown in Table 1. Triethanolamine

Table 1 Membrane characteristics provided by the manufacturer

Membrane	UF PES50
Materia	Polyethersulfone
Nominal MWCO	50,000 Da
Water flux @ 25°C & 50PSI	260 LMH

(TEOA) purchased from R & M Marketing (Essex, UK) was used as an active monomer of aqueous phase. Trimesoyl chloride (TMC) used as an active monomer of organic phase was obtained from Alfa Aesar (UK). Sodium hydroxide (NaOH) was mixed with TEOA purchased from Merck (Germany). n-hexane was used as the organic phase solvent, which was obtained from Merck (Germany). Humic acid was purchased from Sigma Aldrich (Switzerland).

Interfacial Polymerization

Aqueous TEOA solution with concentration 6 % w/v was prepared by dissolving 6 g TEOA in 100 ml NaOH aqueous solution (NaOH 1 % w/v). On the other hand, organic TMC solution with concentration 0.15 % w/v prepared in organic phase form by dissolved 0.15 g TMC in 100 ml n-hexane. UF polyethersulfone (PES) supporting membrane was cut into disc form and immersed in aqueous TEOA solution (6 % w/v) for 30 min. Then, the excess TEOA solution on the membrane surface was drained at room temperature for about 2 min. After that, the TEOA-coated membrane was immersed in TMC–hexane for 15 min for interfacial polymerization. The resulting aromatic composite polyester membrane was dried overnight at room temperature. Three membrane samples were produced at three different reaction times (15, 25, and 35 min).

Membrane Performance Testing

Before all NF experiments, each membrane was first immersed in distilled water and then placed in the filtration cell (Amicon Model 8200) about one hour and pressurized at 400 kPa for at least 10 min using distilled water. Subsequently, pure water experiments were conducted at 100–400 kPa to obtain the water flux. For desalination membrane performance, two different types of salt solutions (NaCl and Na_2SO_4) were prepared at different concentrations (0.001–0.1 M) and the salt concentration was analysed using a conductivity metre. NOM solution model was represented by an aqueous solution of 15 mg/L humic acid at pH7, and Hitachi UV–VIS (U-1800) was used to measure the humic acid concentration in both permeate and retentate. The salt and organic removal can be calculated by the following equation:

$$R = 1 - \frac{C_p}{C_b} \quad (1)$$

where C_p and C_b are permeate and bulk concentration, respectively.

Results and Discussion

Membrane Permeability Study

Figure 1 shows the effect of operating pressure on the water flux for membrane fabricated at different interfacial polymerization times. Then, the membrane permeability, P_m was determined by the gradient of the straight line in Fig. 1 for each fabricated membrane. Table 2 shows the water permeabilities, P_m for the three membranes prepared. As shown in the second column of Table 2, the permeability decreased significantly as the reaction time was increased. The range of values obtained was well within the range of values reported previously for NF membranes available commercially, which is between 1.331 and 50.50 L/m^2.h.bar as reported by Bowen and Mohammad (1998). From the water permeability data, the membrane produced in our study could be categorized as tight NF membrane. From the result obtained, it can be observed that the reaction time greatly affects the water permeability. Ji and Mehta (2001) reported that the growth of thin films depends largely on the reactant concentration and reaction time. In this case as the reaction was increased, the thin film composite layer was postulated to be thicker and thus resulted in lower permeabilities. In this case, longer reaction time will induce a thicker thin film layer on top of the polyethersulfone support.

Fig. 1 Permeate flux versus transmembrane pressure of different membranes

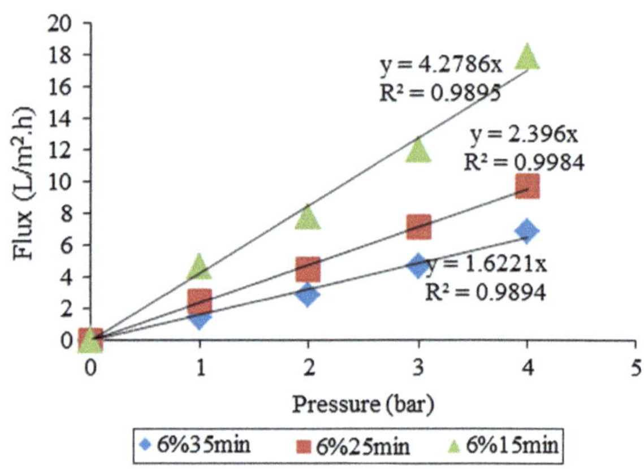

Table 2 Pure water permeability of membrane fabricated with 6 % w/v of TEOA at different interfacial polymerization reaction times

Membrane	Pure water permeability, P_m (L/m^2.h.bar)
6 % 15 min	4.28 ± 1.28
6 % 25 min	2.40 ± 0.60
6 % 35 min	1.62 ± 0.09

Salt Rejection Study

Figure 2 shows the effect of transmembrane pressure on NaCl rejection at different salt concentrations. It was observed that for any pressure value, membranes fabricated at the highest reaction time of 35 min exhibit the highest rejection of NaCl for both salt concentrations (0.1 and 0.001 M) and the highest NaCl removal could be achieved as high as 50 % at 400 kPa for salt concentration of 0.001 M. However, when the higher concentration of NaCl (0.1 M) was applied, the salt removal drastically decreased. The similar trend also observed for the other fabricated membranes in this study. The decrement of salt removal at higher salt concentration is due to Donnan exclusion effects where at higher ionic strength, the Donnan effect is no longer significant.

From the previous monovalence ions salt (NaCl) performance test, the membrane fabricated at 35 min shows an excellent salt removal at both concentrations, and this indicates that this membrane has high negatively charge density (comes from the ester functional group). In order to investigate the capability of this membrane to separate multivalence ions, another salt solution with di-valence ions (Na_2SO_4) was applied and compared with the monovalence ions (i.e. NaCl in this

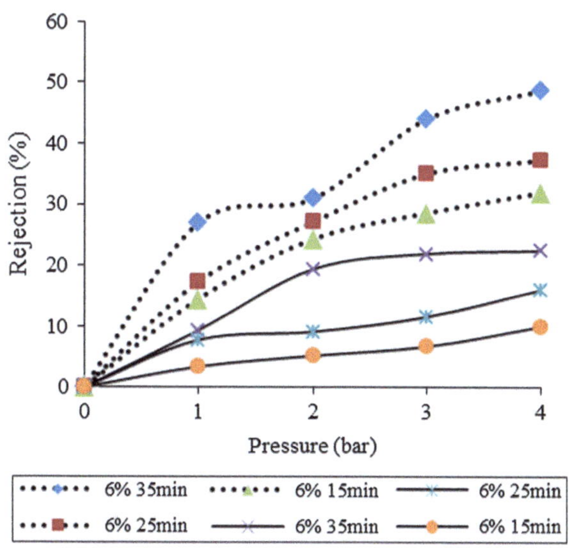

Fig. 2 Effect of transmembrane pressure on NaCl salt rejection at different salt concentrations (……. 0.001 M and _____0.1 M)

study). It was found that 0.01 M of Na_2SO_4 was highly rejected at a magnitude almost double that of NaCl removal. This behaviour could be explained by Donnan effect where SO_4^{2-} ion with higher valence than Cl^- ion relatively repelled far away from the negatively charged skin active layer. However, at higher concentration of 0.1 M, the different rejection factors between Na_2SO_4 and NaCl are not too much compared to solution at concentration of 0.01 M since the Donnan effect become less significant at higher ionic strength.

NOM Rejection

Besides the membrane performance in desalination (salt removal), the removal of NOM has also been investigated in this study. These three membranes were tested with 15 mg/L humic acid (pH7) at constant pressure of 400 kPa and 300 rpm of stirring rate. Again, membrane fabricated at 35 min shows a very good performance with rejection of humic acid up to 94 %, while the other two membranes exhibit less than 90 % of humic acid removal. This is due to the highest negatively charge of membrane active layer repel/reject the negatively charge of humic acid at pH7.

Conclusions

Based on the findings from this study, the variation of reaction time can affect the properties of the membrane produced. Increasing the reaction time resulted in decreasing water permeabilities due to the formation of a thicker layer of thin film on the top surface of the membrane support. Results show that membrane fabricated with 6 % w/v of TEOA at 35 min of reaction time has high performance in terms of removal monovalence/di-valence ions and removal of natural organic matter (i.e. humic acid in this study). This is due to highly negatively charge density of membrane surface that was produced after interfacial polymerization. Membranes produced by TEOA as monomer has a very high performance, and this indicates that BPA is no longer needed for producing polyester thin film composite NF membrane. However, further investigation would be required to fully characterize this type of membrane in terms of membrane chemical properties (hydrophilicity/hydrophobicity), membrane physical (SEM, AFM, or by using model). By this, all the membrane characteristics could be correlated with the overall membrane performance.

Acknowledgments The research work was financial supported through short-term Grant RDU 110325. Appreciation is also expressed to the Ministry of Higher Education of Malaysia for supporting this study (MyBrain15). The authors are thankful to all technical staffs, Chemical Engineering Research Laboratory for providing the necessary facilities, constant guidance, and encouragement.

References

Abu Seman, M. N., Khayet, M., & Hilal, N. (2010). Nanofiltration thin-film composite polyester polyethersulfone-based membranes prepared by interfacial polymerization. *Journal of Membrane Science, 348*, 109–116.

Ahmad, A. L., Ooi, B. S., Mohammad, A. W., & Choudhury, J. P. (2004). Development of a highly hydrophilic nanofiltration membrane for desalination and water treatment. *Desalination, 168*, 215–221.

Beverly, S. R. (2011). Bisphenol A: an endocrine disruptor with widespread exposure and multiple effects. *Journal of Steroid Biochemistry and Molecular Biology, 127*, 27–34.

Bowen, W. R., & Mohammad, A. W. (1998). Characterization and prediction of nanofiltration membrane performance—A general assessment. *IChemE Part A, 76*, 885–893.

Jayarani, M. M., & Kulkarni, S. S. (2000). Thin-film composite poly(esteramide)-based membranes. *Desalination, 130*, 17–30.

Ji, J., & Mehta, M. (2001). Mathematical model for the formation of thin-film composite hollow fiber and tubular membranes by interfacial polymerization. *Journal of Membrane Science, 192*, 41–54.

Mohammad, A. W., Hilal, N., & Abu Seman, M. N. (2003). A study on producing composite nanofiltration membranes with optimized properties. *Desalination, 158*, 73–78.

Tang, B., Huo, Z., & Wu, P. (2008). Study on a novel polyester composite nanofiltration membrane by interfacial polymerization of triethanolamine (TEOA) and trimesoyl chloride (TMC): I. Preparation, characterization and nanofiltration properties test of membrane. *Journal of Membrane Science, 320*, 198–205.

Adsorption of Lead Ions from Aqueous Solution onto Activated Carbon

M. A. Adrah and P. Ravindra

Abstract The adsorption of lead ions from aqueous solutions was performed using commercial activated carbon. The adsorption equilibrium data for lead ions found to follow both Langmuir and Freundlich isotherm models. The kinetic process of lead ions adsorption onto activated carbon was found to fit the pseudo-second-order kinetic model. The thermodynamic parameters of the adsorption demonstrate that the adsorption process was spontaneous and exothermics.

Introduction

An increasing interest to the activated carbon adsorption from aqueous solutions was initiated by the pollution of air and water due to fast-growing industrialization and the high usage of different kinds of chemicals in almost every facet of human endeavor (Bansal and Goyal 2005).

Recently, a greater attention has been given over the content of lead in drinking and natural water, due to extreme lead poisoning in humans causes severe damage to the kidneys, liver, brain, and nervous system while a long-term exposure may induce sterility, abortion, and neonatal death (Gercel and Gercel 2007). Adsorption onto activated carbon one of the various techniques was found to be a promising technique to effectively remove trace amounts of lead from solutions (Sreejalekshmi et al. 2009). High surface area, microporous character, and chemical nature of the activated carbons have made them potential adsorbents for the removal of heavy metals from wastewaters (Mohammadi et al. 2010).

M. A. Adrah · P. Ravindra (✉)
Chemical Engineering Programme, School of Engineering and Information Technology, Universiti Malaysia Sabah, Jalan UMS, 88400 Kota Kinabalu, Sabah, Malaysia
e-mail: ravindra@ums.edu.my

R. Pogaku et al. (eds.), *Developments in Sustainable Chemical and Bioprocess Technology*, DOI: 10.1007/978-1-4614-6208-8_16,

The aim of the study is to investigate the isotherms, kinetics, and thermodynamics of adsorption of lead ions from aqueous solution by using commercial activated carbon.

Methodology

Adsorbent

Commercial activated carbons obtained from a local supplier were used as adsorbent. The properties of the activated carbon are as follows:

Form	Powder
Color	Black
pH	6.0–9.0 at 40 g/L at 25 °C
Relative density	0.250–0.600 g/cm^3

Aqueous Solution

The standard solution of lead ions (1000 mg/L) was used in the preparation of aqueous solution. The 10 mg/L of lead ions aqueous solution was prepared by dissolving 0.01 L of the standard solutions in 1 L volumetric flask with distilled water.

Batch Adsorption Studies

Batch adsorption studies were carried out by shaking a known dosage of the commercial activated carbon with 50 mL of the aqueous solution of lead ions in a different conical flask at 100 rpm. The mixtures were separated by filtration using Whatman membrane filters, and the filtrate was analyzed.

The concentration of lead ions before and after adsorption was determined using an atomic absorption spectrophotometer (Z-5000 Polarized Zeeman). The amount of lead ions adsorbed per unit mass of adsorbents was calculated using the equation:

$$q = \frac{v(C_0 - C)}{m}$$

where C_0 and C are the initial and equilibrium concentration of metal ion solution (mg/L) respectively, m is the amount of adsorbents (g), and v is the volume of solution (L).

Adsorption Isotherms

For modeling lead ions adsorption from aqueous solution, Langmuir and Freundlich isotherms were used:

Langmuir isotherm $$\frac{C}{q} = \frac{1}{K_L q_m} + \frac{1}{q_m} C$$

Freundlich isotherm $$\log q = \log K_F + \frac{1}{n} \log C$$

where q is the amount of metal ions adsorbed per specific amount of adsorbent (mg/g), C is equilibrium concentration (mg/L), q_m is amount of metal ions required to form a monolayer (mg/g), K_L is Langmuir equilibrium constant, and K_F and n are Freundlich equilibrium constants.

Adsorption Kinetics

To describe the adsorption mechanism and adsorption characteristics, pseudo-first-order and pseudo-second-order kinetic models are selected. Both models can be expressed in a linear form as:

Pseudo-first-order $$\ln(q_e - q_t) = \ln q_e - k_1 t$$

Pseudo-second-order $$\frac{t}{q_t} = \frac{1}{k_2 q_e^2} + \frac{t}{q_e}$$

where k_1 is the adsorption rate constant for the first order adsorption, k_2 is the second order reaction constant, q_t is the amount of heavy metal adsorbed at time t (mg/g) and q_e is the amount of heavy metal adsorbed at saturation (mg/g).

Adsorption Thermodynamics

The changes in Gibbs free energy (ΔG), enthalpy (ΔH) and entropy (ΔS) for adsorption process were determined using the following equations:

$$\Delta G = -RT \ln K_c$$

$$K_c = \frac{C_{Ac}}{C_e}$$

$$\Delta G = \Delta H - T\Delta S$$

where K_c is the equilibrium constant, C_{Ac} is the equilibrium concentrations (mg/L) of metal ion on adsorbent, C_e is equilibrium concentrations (mg/L) of metal ion in the solution, T is temperature in kelvin and R is gas constant (8.314×10^{-3} kJ/mol.K).

Enthalpy (ΔH) and entropy (ΔS) were obtained from the slope and intercept of Vant Hoff plots of $\log K_c$ versus $\frac{1}{T}$.

$$\log K_c = \frac{\Delta S}{2.303R} - \frac{\Delta H}{2.303RT}.$$

Results and Discussion

Adsorption Isotherms

The experimental data were fitted to the Langmuir and Freundlich isotherms. The plot of $\frac{C}{q}$ versus C, and $\log q$ versus $\log C$ for various adsorbent dosages was found to be linear in Figs. 1 and 2, respectively.

The R^2 values shown in Table 1 indicated that the Langmuir and Freundlich isotherms were able to adequately describe the relationship between the amount of lead ions adsorbed by the commercial activated carbon and its equilibrium concentration in the solution.

Fig. 1 Langmuir isotherm curve

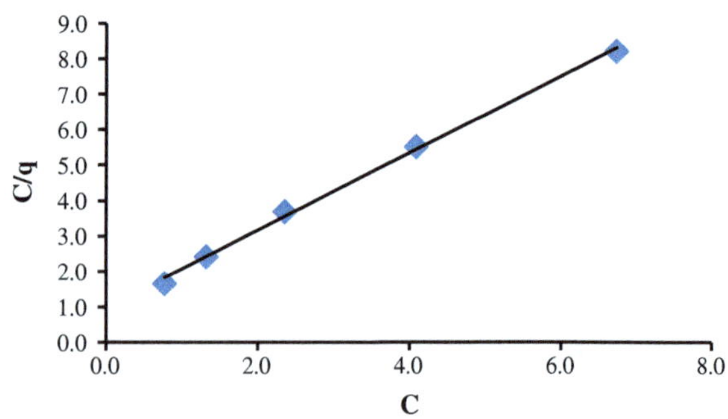

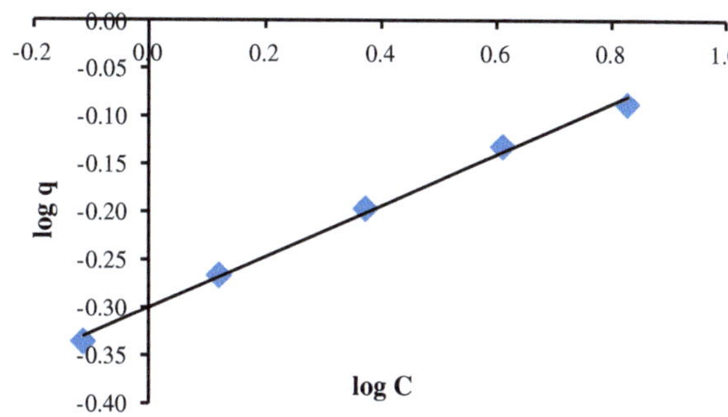

Fig. 2 Freundlich isotherm curve

Table 1 Langmuir and Freundlich constants for the adsorption of lead ions onto commercial activated carbon

Isotherm type	Linearity equation	R^2
Langmuir	$\frac{C}{q} = 0.9884 + 1.0894\ C$	0.9977
Freundlich	$\log q = -0.3004 + 0.2665 \log C$	0.9962

Adsorption Kinetics

The experimental data were fitted to the pseudo-first-order and pseudo-second-order kinetic models. The plot of $\log(q_e - q_t)$ against t, and $\frac{t}{q_t}$ against t was found to be linear in Figs. 3 and 4 respectively.

The R^2 values shown in Table 2 indicated that the pseudo-second-order kinetic model is applicable to the lead adsorption on commercial activated carbon.

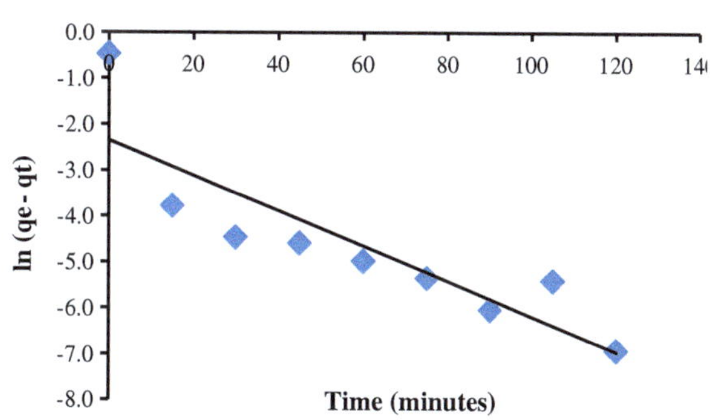

Fig. 3 Pseudo-first-order kinetic curve

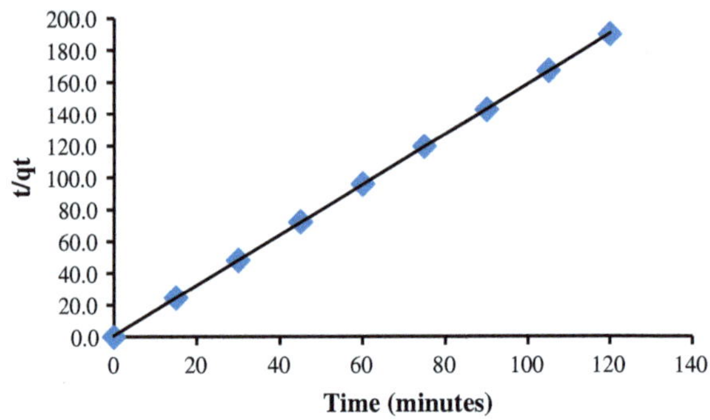

Fig. 4 Pseudo-second-order kinetic curve

Table 2 Pseudo-first-order and pseudo-second-order constants for the adsorption of lead ions onto commercial activated carbon

Kinetic model	Linearity equation	R^2
Pseudo-first-order	$\ln(q_e - q_t) = -2.3487 + 0.0384t$	0.7516
Pseudo-second-order	$\frac{t}{q_t} = 0.6966 + 1.5811t$	1.0000

Adsorption Thermodynamics

The thermodynamics parameters namely Gibbs free energy $(\Delta \boldsymbol{G})$, enthalpy $(\Delta \boldsymbol{H})$ and entropy $(\Delta \boldsymbol{S})$ were determined, and the values obtained are presented in Table 3.

The negative values of $(\Delta \boldsymbol{H})$ show that the adsorption of lead ions on the commercial activated carbon is exothermic. The negative values of $(\Delta \boldsymbol{G})$ indicate the feasibility and spontaneous nature of adsorption of lead ions on the adsorbent. The entropic change occurring from adsorption is thought to be negligible since $(\Delta \boldsymbol{S})$ has been estimated to be very small.

Table 3 Equilibrium constants and thermodynamic parameters for the adsorption of lead ions onto commercial activated carbon

Temperature (°C)	K_c	ΔG(kJ/mol)	ΔS (kJ/mol.K)	ΔH (kJ/mol)
303.15	3.234	−2.958	−0.002	−5.457
313.15	2.959	−2.825	−0.002	
323.15	2.794	−2.761	−0.002	
333.15	2.656	−2.706	−0.001	

Conclusions

The results of the research clearly show that commercial activated carbon can successfully remove lead ions from aqueous solution. The equilibrium adsorption data for lead ions follow both Langmuir and Freundlich isotherms. The kinetic process of lead ions adsorption onto commercial activated carbon was found to fit the pseudo-second-order kinetic model. The negative values of ($\Delta \boldsymbol{H}$) showed the exothermic nature of the process. The negative values of ($\Delta \boldsymbol{G}$) indicate the feasibility and spontaneous of adsorption of lead ions on the commercial activated carbon.

Acknowledgments We, authors would like to thank the Universiti Malaysia Sabah, our beloved family, and friends for their support in completing the research.

References

Bansal, R. C., & Goyal, M. (2005). *Activated carbon adsorption*. Boca Raton: CRC Press.

Gercel, O., & Gercel, H. F. (2007). Adsorption of lead(II) ions from aqueous solutions by activated carbon prepared from biomass plant material of *Euphorbia rigida*. *Chemical Engineering Journal, 132*, 289–297.

Mohammadi, S. Z., Karimi, M. A., Afzali, D., & Mansouri, F. (2010). Removal of Pb(II) from aqueous solutions using activated carbon from Sea-buckthorn stones by chemical activation. *Desalination, 262*, 86–93.

Sreejalekshmi, K. G., Krishnan, K. A., & Anirudhan, T. S. (2009). Adsorption of Pb(II) and Pb(II)-citric acid on sawdust activated carbon: Kinetic and equilibrium isotherm studies. *Journal of Hazardous Materials, 161*, 1506–1513.

Part II
Chemical and Bioprocesses

Review: Pre-treatments and Fermentation of Seaweed for Bioethanol Production

R. F. Mansa, H. Mansuit, K. F. Fong, C. S. Sipaut, F. Y. Chye and S. M. Yasir

Abstract This article reviews the current studies on the production of bioethanol from seaweed with a focus on the process pre-treatments and variety of microorganisms used in the process. Pre-treatment selection is essential to maximize the amount of reduced sugar for the fermentation to produce bioethanol. Specific microbial strains are matched to their ability to utilize sugar sources. Some studies focus mainly on general processing with variable microbial strains to gauge their abilities in fermentation. A summary of the current studies was carried out, and it is evident that two or more yield increasing techniques can coexist within a single process. The integration of the findings may be the key to make seaweed fermentation more efficient and affordable to serve as a sustainable and renewable energy source.

Introduction

The world is facing a crisis of depleting fossil fuel energy sources. In order to replace the dependency on fossil fuel, bioethanol has attracted attention as an alternative renewable fuel. The world has been confronted with an energy crisis due to depletion of finite resources of fossil fuel (Meinita and Hong 2012). Now,

R. F. Mansa (✉) · H. Mansuit · K. F. Fong · C. S. Sipaut
Energy and Materials Research Group, School of Engineering and Information Technology, Universiti Malaysia Sabah, Sabah, Kota Kinabalu, Malaysia
e-mail: rfmansa@ums.edu.my

F. Y. Chye
School of Food Science and Nutrition, Universiti Malaysia Sabah, Kota Kinabalu, Sabah, Malaysia

S. M. Yasir
Seaweed Research Unit, Universiti Malaysia Sabah, Jalan UMS, Kota Kinabalu 88400 Sabah, Malaysia

R. Pogaku et al. (eds.), *Developments in Sustainable Chemical and Bioprocess Technology*, DOI: 10.1007/978-1-4614-6208-8_17,

there are various types of renewable energies that are available; however, one of the most important energy sources in near future is biomass. Biofuel is a starch/cellulose biomass fermentation product that is a renewable energy source which can be used as an alternative for petroleum fuels. According to John et al. (2011), globally, ethanol is expected to be the most widely used biofuel.

The first-generation biofuel was based on edible crops such as corn, beet, and potatoes. It produced sufficient yields of ethanol. However, it has raised morality and ethics issues as millions of people around the world are currently suffering from malnutrition and hunger (Goh and Lee 2009). It also raises the issue on the increasing proportion of land use for biomass crops. Yet the "food versus fuel" debate is not applicable to macroalgae or seaweed (Adams et al. 2009) due to it being a non-terrestrial plant. For the purpose of consistency, macroalgae will be referred to as seaweed in this review.

Seaweed are crops that are self-sustainable, easily mass-produced and has minimal land requirement. According to John et al. (2011), seaweed has the ability to grow at a fast rate and yield huge amounts of biomass. Besides that, it can be cultivated in the ocean under the sun with no need for nitrogen-based fertilizers (Buck and Buchholz 2004). In addition, seaweed contains no lignin; hence, it is a suitable biomass material that can be converted to ethanol without any expensive lignin removal process (Park et al. 2012).

In this review, various types of seaweed was reported to have been put through a series of processes to extract higher amounts of fermentable sugars and later undergo fermentation using different types of microbes such as yeasts under aerobic conditions.

Seaweed Pre-treatments Before Fermentation

The main process in ethanol production from seaweed is the conversion of reduced sugar into ethanol via fermentation. Although seaweed has high amounts of carbohydrates, but the initial content of reduced sugars found in seaweed is insignificant. The carbohydrates can be broken down to simple sugars via saccharification. The common saccharification pre-treatment methods for the fermentation of seaweed are as follows: acid hydrolysis and enzymatic hydrolysis. However, most of the recent finding prefers acid hydrolysis pre-treatment method, since the enzymatic method is costly and promotes complexity of the saccharification process (Huimin et al. 2007). Acid hydrolysis was preferred because of its low-cost, short hydrolysis time, and simple operation when compared to other saccharification methods. However, acid hydrolysis also produces by-products that can inhibit ethanol-producing yeast and bacteria such as furans (Larsson et al. 1999; Klinke et al. 2004).

It is interesting to note that alkaline pre-treatment was not found to be used for the pre-treatment of seaweed fermentation. Yet it is commonly used for agar or carrageenan extraction especially from red seaweed species. Istini et al. (1994)

proved that alkali pre-treatment was successful in the extraction of agar or carrageenan from seaweed. The authors used 5 % NaOH at 70 and 80 °C for 2 h to extract agar from *Gracilaria chorda*, *Gracilaria fisheri*, *Gracilaria verrucosa,* and *Gracilaria Iemanei ormis*. In carrageenan extraction, 6 % KOH was used for the treatment of seaweed species of *Kapphaphycus alvarezii* and *Eucheuma spinosum*.

Table 1 presents a non-exhaustive review of seaweed pre-treatments prior to fermentation to produce bioethanol. The data and parameters are extracted from numerous findings and journals according to the seaweed subspecies and also to compare different materials and outcomes of using other pre-treatment techniques. Yoon et al. (2012) used a combination of gamma irradiation and dilute sulfuric acid treatment on *Undaria* sp. to increase the final content of reducing sugar (0.235 g/L). While Wi et al. (2009) employed *E. coli to* perform enzymatic hydrolysis on *Gelidium amansii* in a buffered solution to achieve 70 % glucose yield in six hours. Jeong et al. (2011) used a dual stage acid hydrolysis with 4 % sulfuric acid at 190 °C on *Gelidium amansii*, to yield 68.58 % from 74.6 % carbohydrate content of seaweed. It was found that the major trend in pre-treatment was to use dilute sulfuric acid hydrolysis at high temperature. Improvements that can be added to this method are gamma irradiation, dual stage, biological action and usage of buffered solution.

Fermentation of Seaweed

In fermentation of seaweed, microbes consume fermentable sugars from pre-treated seaweed to produce bioethanol in an anaerobic condition. *Saccharomyces cerevisiae* is commonly used in anaerobic fermentation. However, Lee et al. (2011) reported that a small amount of oxygen can be used to induce facultative respiration among limited species of microbes such as *Pichia stipitis*. This microbe consumes reduced sugars and produce carbon dioxide and ethanol. The fermentation process will go through a lag phase, exponential phase, a stationary phase and then a death phase. In the stationary phase, the microbe reproduction slows down due to elevated concentrations of ethanol (usually 10–13 %) which is an inhibitory product. The fermentation stops when substrate is depleted or the microbes die.

Various seaweed species have different characteristics and require specific process parameters. It is important to understand the seaweed characteristics prior to bioconversion to bioethanol. Table 2 presents a review of fermentation on various seaweed species. Meinita et al. (2012) used brewer yeast or *Saccharomyces cerevisiae* directly for bioethanol fermentation. The authors obtained 84 % galactose from 30.5 g/L reducing sugar after acid hydrolysis. The ethanol yield achieved after 24 h fermentation at 30 °C from 84 % galactose was 1.31 g/L ethanol. Candra et al. (2011) used 10 % v/v *Saccharomyces cerevisiae* for the fermentation of 16 mg reducing sugar/ml (0.016 g reducing sugar/g dry

Table 1 Summary of pre-treatments

Seaweed species r	Initial content	Materials, pre-treatment method, neutralization method	Final content	Reference(s)
Undaria sp.	Alginate, Laminarin, Mannitol	Cobalt-60 gamma irradiator, gamma irradiation, N/A	0.048 g/L of reducing sugar	Yoon et al. (2012)
		H_2SO_4, Acid Hydrolysis, Using $CaCO_3$	0.017 g/L of reducing sugar	
		Cobalt-60 gamma irradiator & 1 % H_2SO_4, combined gamma irradiation and acid hydrolysis, $CaCO_3$	0.235 g/L of reducing sugar	
Gelidium amansii	90 % Glucan	4 % H_2SO_4, 210 seconds, 190 °C, Acid hydrolysis/N/A	38.09 % glucose yield	Jeong et al. (2012)
	53.4 % Carbohydrate	*E. coli*, 6 h, with sodium acetate as buffer solution, enzymatic hydrolysis with sodium chlorite, NH_4OH	70 % glucose yield	Wi et al. (2009)
	74.6 % Carbohydrate	H_2SO_4, 190 °C 4 %, acid hydrolysis/dual stage, N/A	68.58 % glucose yield	Jeong et al. (2011)
Ulva pertusa	Cellulose, starch, β-1,3-glucan	Meicelase 50 °C, 120 h, enzymatic hydrolysis, using buffer solution	43.0 % glucose	Yanagisawa et al. (2011)
Alaria crassifolia			67.2 % glucose	

Table 2 Summary of seaweed fermentation results

Seaweed species	Initial Content,Method,pH	Culture strain	Type of sugars utilized	Ethanol Yield (v/v) or (g/L)	Reference(s)
Laminaria japonica	30.54 % Mannitol 6.98 % Glucose, SSF, 5.5 using HCl	Recombinant *E. coli* KO11 and *Saccharomyces cerevisiae*	Mannitol and glucose	25.8 g ethanol/L after 116 h by using *E. coli* KO11	Kim et al. (2011)
Gelidium amansii	65.1 g/L galactose and 7.1 g/L glucose, Anaerobic	Modified *B. custersii* (KCTC18154P)	Glucose and galactose	Continuous reactor: 27.6 g/L in 39 h Batch reactor: 11.8 g/L in 56 h	Park et al. (2012)
Kappaphycus alvarezii	51.02 % of carrageenan, pH 6.4–6.8	*Saccharomyces cerevisiae* NCIM 3523	Galactose from carrageenan	2.06 %	Khambhaty et al. (2012)
Kappaphycus alvarezii	25.6 g/L of galactose, Anaerobic, pH 5	*Saccharomyces cerevisiae*	Galactose	1.31 g/L	Meinita et al. (2012)
Kappaphycus alvarezii	Anaerobic pH 5	*Saccharomyces cerevisiae*	Glucose	4.60 %	Candra et al. (2011)
Laminaria japonica	0.278 g/g of glucose, Anaerobic, pH 6.5	*Saccharomyces cerevisiae*	Glucose	41.2 %	Ge et al. (2011)

seaweed). They obtained 4.6 % ethanol yield after 168 h fermentation at 28–30 °C. Ge et al. (2011), fermented seaweed (*Laminaria japonica*) waste hydrolysate using *Saccharomyces cerevisiae* at 30 °C for 36 h and claimed that the conversion of ethanol was 0.143 L from 1 kg of seaweed hydrolysate. The authors concluded that the fermentation of glucose after saccharification can be totally fermented by the yeast of *Saccharomyces cerevisiae*.

This review showed that *Saccharomyces cerevisiae* is the most commonly used for ethanol production. There are also other microbes such as modified *B.custersii* (KCTC18154P) and *E. coli* KO11. These microbes are able to utilize more than one type of sugar. Production of bioethanol was proven to be viable using the microbes as presented in the table. Bioethanol production from seaweed (*L. japonica*) was demonstrated by Kim et al. (2011) using ethanogenic recombinant *Escherichia coli* KO11 with ethanol production of 0.4 g ethanol/g of carbohydrate. Park et al. (2012) used modified *B. custersii* (KCTC18154P) and achieved higher yield of ethanol in a continuous reactor of 30 °C and 39 h which was 38.2 % from 72.2 g/L of glucose and galactose content.

Higher alcohol yield was achieved by using two different organisms or two-step process; however, it would be more efficient when using a single organism that can utilize multiple substrates simultaneously (Park et al. (2012). The criterion for an effective bio-agent is the matching of the type of culture strain to the spectrum of sugars that it utilizes to produce bioethanol. Higher versatility and affinity to various types of sugars, generally glucose, galactose, and mannitol contributes to higher concentration of alcohol.

Conclusions

This review had reported mainly on acid hydrolysis and enzymatic pre-treatments for single microbe fermentation methods. It was found that the pre-treatment methods were unable to convert all of the complex sugars in seaweed into reducible sugars. Yet it was proven that bioethanol production from pre-treated seaweed via bioconversion is a viable process. It is essential to increase the amount of reducible sugars in order to increase ethanol yield. Hence, it may be interesting to investigate the potential of using alkaline hydrolysis pre-treatment for fermentation. Following that pairing of the reducible sugar content and the bio-agent would ensure higher amounts of sugars were utilized during bioconversion.

Acknowledgments Authors would like to extend their appreciation and gratitude for the funding from Seaweed Research Unit of University Malaysia Sabah under the EPP#3 NKEA Malaysian Government Project.

References

Adams, J. M., Gallagher, J. A., & Donnison, I. A. (2009). Fermentation study on *Saccharina latissima* for bioethanol production considering variable pre-treatments. *Journal of applied Phycology, 21*, 569–574.

Buck, B. H., & Buchholz, C. M. (2004). The offshore-ring: A new system design for the open ocean aquaculture of macroalgae. *Journal of Applied Phycology, 16*, 355–368.

Candra, K. P., Sarwono, Sarinah. (2011). Study on bioethanol production using red seaweed *Eucheuma cottonii* from Bontang sea water. *Journal of Coastal Development 15*(1), 45–50.

Ge, L., Wang, P., & Mou, H. (2011). Study of saccharification techniques of seaweed wastes for the transformation of ethanol. *Journal of Renewable Energy, 36*, 84–89.

Goh, C. S., & Lee, K. T. (2009). A visionary and conceptual macroalgae-based third-generation bioethanol (TGB) biorefinery in Sabah, Malaysia as an underlay for renewable and sustainable development. *Renewable Sustainable Energy Revolution, 14*, 842–848.

Huimin, Q. I., Daxin, L. I., Zhang, J. J., Liu, L., & Zhang, Q. B. (2007). Study on extraction of agaropectin from *Gelidium amansii* and its anticoagulant activity. *Chinese Journal of Oceanology and Limnology, 26*(2), 186–189.

Istini, S., Ohno, M., & Kusunose, H. (1994). Methods of analysis agar, carrageenan and alginate in seaweed. *Bulletin of Marine Science and Fisheries—Kochi University, 14*, 49–55.

Jeong, T. S., Choi, C. H., Lee, J. Y., & Oh, K. K. (2011). Two-stage acid saccharification of fractionated *Gelidium amansii* minimizing the sugar decomposition. *Journal of Bioresource Technology, 102*, 10529–10534.

Jeong, T. S., Choi, C. H., Lee, J. Y., & Oh, K. K. (2012). Behaviors of glucose decomposition during acid-catalyzed hydrothermal hydrolysis of pretreated *Gelidium amansii. Journal of Bioresource Technology, 116*, 435–440.

John, R. P., Anisha, G. S., Nampoothiri, K. M., & Pandey, A. (2011). Micro and macroalgal biomass: A renewable source for bioethanol. *Bioresource Technology, 102*, 186–193.

Khambhaty, Y., Mody, K., Gandhi, R. M., Thampy, S., Maiti, P., Brahmbhatt, H., et al. (2012). *Kappaphycus alvarezii* as a source of bioethanol. *Journal of Bioresource Technology, 103*, 180–185.

Kim, N., Li, H., Jung, K., Chang, H. N., & Lee, P. C. (2011). Ethanol production from marine algal hydrolysates using *Escherichia coli* KO11. *Bioresource Technology, 102*, 7466–7469.

Klinke, H. B., Thomsen, A. B., & Ahring, B. K. (2004). Inhibition of ethanol producing yeast and bacteria by degradation products produced during pre-treatment of biomass. *Applied Microbiology and Biotechnology, 66*, 10–26.

Larsson, S., Palmqvist, E., Hagerdal, B. H., Tengborg, C., Stenberg, K., Zacchi, G., et al. (1999). The generation of fermentation inhibitors during dilute acid hydrolysis of softwood. *Enzyme and Microbiology Technology, 24*, 151–159.

Lee, S. M., & Lee, J. H. (2011). The isolation and characterization of simultaneous saccharification and fermentation microorganisms for *Laminaria japonica* utilization. *Journal of Bioresource Technology, 102*, 5962–5967.

Meinita, M. D. N., & Hong, Y. (2012). Detoxification of acidic catalyzed hydrolysate of *Kappaphycus alvarezii* (cottonii). *Bioprocess and Biosystems Engineering, 35*, 93–98.

Meinita, M. D. N., Kang, J., Jeong, G., Koo, H. M., Park, S. M., & Hong, Y. (2012). Bioethanol production from the acid hydrolysate of the carrageenophyte *Kappaphycus alvarezii* (cottonii). *Journal of Applied Phycology, 24*, 857–862.

Park, J., Hong, J., Jang, H. C., Oh, S. G., Kim, S., Yoon, J., et al. (2012). Use of *Gelidium amansii* as a promising resource for bioethanol: A practical approach for continuous dilute-acid hydrolysis and fermentation. *Bioresource Technology, 108*, 83–88.

Wi, S. G., Kim, H. J., Mahadevan, S. A., Yang, D. J., & Bae, H. J. (2009). The potential value of seaweed Ceylon moss (*Gelidium amansii*) as an alternative bioenergy resource. *Journal of Bioresource Technology, 100*, 6658–6660.

Yanagisawa, M., Nakamura, K., Ariga, O., & Nakasaki, K. (2011). Production of high concentrations of bioethanol from seaweeds that contain easily hydrolyzable polysaccharides. *Process Biochemistry, 46*, 2111–2116.

Yoon, M., Choi, J., Lee, J. W., & Park, D. H. (2012). Improvement of saccharification process for bioethanol production from *Undaria* sp. by gamma irradiation. *Radiation Physics and Chemistry, 81*, 999–1002.

Assessment on Bioavailability of Flavonoids from *Orthosiphon Stamineus* During Spray Drying

S. F. Pang, M. M. Yusoff and J. Gimbun

Abstract This article presents findings of a microencapsulation study of flavonoids obtained from *Orthosiphon stamineus* leaves. The plant extract was obtained using ultrasonic-assisted extraction method. Protein and a polysaccharide derivative (maltodextrin) were used as the microencapsulation agent. Aqueous extract and dried powder were tested for their total flavonoids content by means of aluminum chloride colorimetric assay and detailed analysis of its polyphenol content by high-performance liquid chromatography. Morphology of the dried powder was analyzed by field emission scanning electron microscopy. Microencapsulation was found to use the least amount of protein (1.42 wt%) and has the highest total flavonoids retention (85.6 %) besides having the highest retention of the bioactive markers, sinensetin (81.8 %), and eupatorin (70.1 %). The formulation of 1:9 protein/maltodextrin ratio is also found to be more effective with 86.7 % total flavonoids retention compared other formulation, viz., 1:1 and 9:1.

Keywords Orthosiphon stamineus · Microencapsulation · Flavonoids · Spray drying · Bioavailability

S. F. Pang · J. Gimbun
Faculty of Chemical and Natural Resources Engineering, Universiti Malaysia Pahang, Gambang, Malaysia

M. M. Yusoff
Faculty of Industrial Sciences and Technology, Universiti Malaysia Pahang, Gambang, Malaysia

M. M. Yusoff
Central Laboratory, Universiti Malaysia Pahang, Gambang, Malaysia

J. Gimbun (✉)
Centre of Excellence for Advanced Research in Fluid Flow (CARIFF), Universiti Malaysia Pahang, 26300 Gambang, Pahang, Malaysia
e-mail: jolius@ump.edu.my

R. Pogaku et al. (eds.), *Developments in Sustainable Chemical and Bioprocess Technology*, DOI: 10.1007/978-1-4614-6208-8_18,

Introduction

Orthosiphon stamineus (vernacular name: "misai kucing") is traditionally used in Malaysia for the treatment of eruptive fever, edema, hepatitis, jaundice, hypertension, diabetes mellitus, gout, rheumatism, diuretic, anti-inflammatory, and influenza (Ho et al. 2010; Saravanan et al. 2006). Previous scientific studies revealed that extracts of *O. stamineus* contained useful bioactive compounds that exhibited excellent antibacterial, antifungal, antimicrobial, and antitumor properties (Saravanan et al. 2006). Furthermore, *O. stamineus* had been reported to possess anti-inflammatory (Masuda et al. 1992), antihypertensive (Ohashi et al. 2000), and hypoglycemia activity (Mariam et al. 1996).

A systematic method to prepare the water-soluble extract of *O. stamineus* is well understood (e.g., Akowuah et al. 2005). However, a solid pharmaceutical dosage in the form of tablets is also desirable for convenience of administration besides having longer shelf life and ease of handling. A method to produce high-quality solid powder products from *O. stamineus* extracts must be established as the product of conventional method via hot spray is prone to thermal degradation. Thus, assessment of bio availability of solid powder is the ultimate aim of this study.

Bioactive compounds such as flavonoids, vitamins, protein, and antioxidant are known to suffer from thermal degradation when exposed to high temperature over a long period, for example, during the powder-making process. Anandharamakrishnan et al. (2008) reported more than 60 % protein denaturation in a tall pilot-scale spray dryer at temperatures above 100 °C, commonly used conditions for spray drying operation. Thermal degradation of other bioactive compounds, such as vitamin E, vitamin A, and antioxidants, has also been reported by Xie et al. (2010). Thermal degradation is undesirable, because the degraded product is of low nutritional value and consequently hampers the intention to produce a nutraceutical product. Spray drying is a common method of producing powder. One of the objectives of this work was to minimize degradation of bioactive compounds from *O. stamineus* extract during spray drying using the microencapsulation technique. Furthermore, limited literature concerning microencapsulation of flavonoids from *O. stamineus* extract is presently available in the literature.

Materials and Methods

Chemicals

Sodium nitrite, methanol, and sodium hydroxide were obtained from Merck (Malaysia). HPLC grade methanol and quercetin were obtained from Fisher Malaysia. HPLC grade tetrahydrofuran, dimethylsulfoxide, aluminum hexachloride, eupatorin, and sinensetin were obtained from Sigma Aldrich (Malaysia). A lactose-free whey protein isolates (WPI) powder was obtained from Ultimate

Nutrition (Fleetwood, Lancashire, UK) with 99 % denatured whey proteins. Maltodextrin of 10 DE was obtained from San Soon Seng Food Industries Sdn. Bhd. (Malaysia).

Plant Material

The leaves were collected in Gambang, Pahang, Malaysia from a white-flowered *O. stamineus* plants. Freshly collected leaves were washed with deionized water and dried at 37 °C for 3 days before powdering. Prior to use the powder was kept in an air-tight plastic bag at room temperature to prevent moisture absorption.

Ultrasonic-Assisted Extraction

The powdered plant material was weighed (8 g) and mixed with aqueous methanol (100 ml; 50 %) in a 250 ml sealed Erlenmeyer flask. Extraction was carried out in an ultrasonic bath (CREST P1800D, US) at 45 kHz for 90 min and temperature setting at 50 °C. The supernatant was then separated from the residue by vacuum filtration using Whatman no. 1 filter paper. The solvent was evaporated *in vacuo* at 40 °C.

HPLC Analysis

HPLC analysis was performed according to Akowuah et al. (2005) using Agilent Technologies 1,200 system (Agilent Technologies, USA) equipped with an automatic injector, a column oven, and a UV detector. A Zorbax Eclipse plus C18 (Agilent Technologies, USA) column (5 μm, 250 × 4.6 mm i.d.) was used for the chromatographic separation. The temperature was maintained at 25 °C, with injection volume of 10 μl and flow rate of 1 ml/min. All markers were separated under isocratic condition with methanol–water–tetrahydrofuran (45:50:5 v/v) as mobile phase. The peaks were detected at 340 nm and verified by comparison with standards [eupatorin (Eup) and sinensetin (Sin)]. HPLC quantification was carried out using external standards method and the results reported as mg marker/ml sample.

Total Flavonoid Content

Total flavonoid content (TFC) was measured by the aluminum chloride colorimetric assay (Sameh et al. 2008). Sample or standard (quercetin) aliquots (0.2 ml) was added to a 15 ml centrifuge tube containing deionize water (4.8 ml). $NaNO_2$ (0.3 ml, 5 %) was then added and mixed using a vortex mixer for 5 min. Subsequently, $AlCl_3$ (0.3 ml, 10 %) was added followed by the addition of NaOH solution (2 ml, 1 M), and the total volume adjusted with deionized water (10 ml). The solution was mixed well, and absorbance measured against a blank at 510 nm. Total flavonoid content of sample solutions are expressed as milligram quercetin equivalents per milliliter of sample solution by comparison with the quercetin calibration curve.

Moisture Content

In order to calculate the exact amount of sample needed for comparison in TFC and HPLC analysis, the moisture content of spray-dried powders was determined using a moisture analyzer (AND MS-70, Japan).

Field Emission Scanning Electron Microscope

Morphologies of the spray-dried particles were evaluated by Field Emission Scanning Electron Microscope (FESEM) (JEOL JSM-7800F, Japan). Samples were mounted on specimen stubs with double-sided adhesive carbon tapes. The specimen was coated with platinum and was examined at 1–3 kV with a magnification ranging from 500 to 20,000x.

Spray Drying

Solution of various solid concentrations containing maltodextrin, WPI and *O. stamineus* ex-tract were spray-dried using a laboratory-scale spray drier (Lab Plant SD06A, UK). The atomizer had an inside diameter of 0.5 mm. Air flow rate of about 4 m/s was set constant throughout the experiment. The inlet temperature was set at 180 °C and maintained at ±1 °C by the proportional–integral–derivative controller. Feed was metered into the dryer by means of a peristaltic pump at 400 ml/h. In order to ensure a fair comparison, similar setups were employed for all experiments. After passing through an electric heater, the inlet drying air flowed concurrently with the feed spray in the main drying chamber. The main chamber was made of thick transparent glass and had an inside diameter of 21.5 cm, a total height of 50 cm, and the bottom of the chamber is dished shaped. The distance between the tip of the atomizer and the axis of the side exit tube was about 47 cm. A cyclone air/powder separator was used to recover powder particles. Dried powder samples were collected from a Schott bottle attached at the bottom of the cyclone.

Results and Discussion

Moisture Content

Average moisture contents (g water/g total %) of the spray-dried powder samples are shown in Table 1. The results revealed that increasing solid content reduces the moisture content of the final product for all feed concentrations. This is due to reduction in total water content in the sample as the solid concentration increases despite exposure to the same amount of heat. Similar trends were also observed in other published reports (Anandharamarakrishnan et al. 2008). The driest product resulted from initial feed concentration of 13.47 %.

Influence of Total Solid Content

In this study, effects of total solid content and degradation of polyphenol were assessed by comparing the microencapsulation efficiency of WPI at various concentrations ranging from 4 to 13 %. The initial solution and the dried powder were examined for their flavonoids content by HPLC and proximate analysis using aluminum chloride colorimetric assay. Total solid content of the initial solution and moisture content of the dried powder were determined to ensure a proper comparison by dissolving the same solid content. All formulations of microencapsulates tested in this work show remarkably good retention (ranging from 68.6 to 85.6 %) of total flavonoids in *O. stamineus* extract. Total flavonoids content analysis show that the sample with lowest solid concentration (4.25 %) had better total flavonoids retention (85.6 %) as depicted in Fig. 1. Similar findings were obtained from HPLC analysis, the highest sinensetin (81.8 %) and eupatorin (70.1 %) retention detected corresponded with the lowest solid concentration. This is due to higher protein denaturation, solid concentration increase corresponds to loss of solubility (Anandharamakrishnan et al. 2008). In such a case, microencapsulated sinensetin

Table 1 Total solid content

Sample	Total solid content (wt%)	Moisture content (wt%)
Extract	2.83	
TS 4 %	4.25	8.43
TS 7 %	7.22	8.32
TS 10 %	10.39	8.07
TS 13 %	13.47	7.21
W1M9	7.50	6.63
W1M1	7.47	7.32
W9M1	7.45	8.44

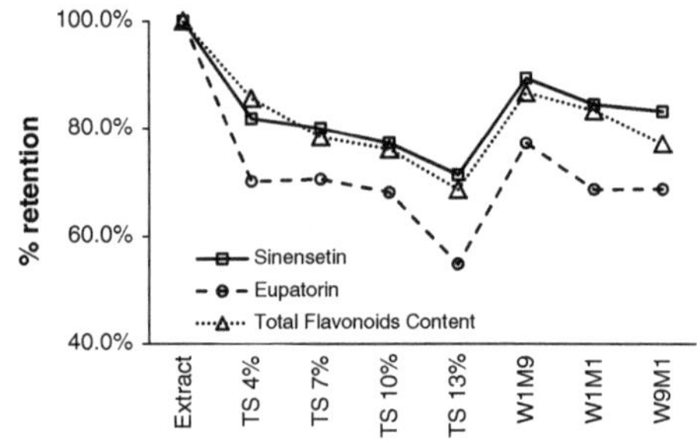

Fig. 1 Total flavonoids content, sinensetin and eupatorin in extract and spray-dried powder. *TS* total solid content, W1M9 = WPI/maltodextrin ratio 1:9, W1M1 = WPI/maltodextrin ratio 1:1, W9M1 = ratio WPI/maltodextrin 9:1

and eupatorin will not be soluble for analysis and lower flavonoids concentration corresponding with increase in solid concentration.

Influence of WPI to Maltodextrin Ratio

Proteins are known to be effective microencapsulating agents at very low concentrations compared with existing materials such as maltodextrin (Jafari et al. 2008). Earlier Rosenberg and Sheu (1996) have successfully improved volatiles retention via microencapsulation using a formulation comprising WPI and lactose. In this study, various formulations of microencapsulation agents were prepared, viz., WPI to maltodextrin ratio of 1:9, 1:1, and 9:1. The results shown in Fig. 1 suggest a better retention of total flavonoids when a lower concentration of WPI is employed, i.e., WPI to maltodextrin ratio of 1:9, which showed 86.7 % retention. Higher concentrations of WPI reduced the flavonoids retention that can be attributed to a higher percentage of protein denaturation, hence, loss of solubility. Results from HPLC analysis for eupatorin and sinensetin also showed higher retention for 1:9 WPI/maltodextrin formulation.

Particle Morphology

Spray-dried samples prepared from WPI and maltodextrin were observed for granular structure using FESEM (Fig. 2). Results clearly showed significant differences in size and shape. At a lower WPI concentration (1.42 %), the particle appears to have more wrinkled surface, contrary to the ones at higher a concentration (10.65 %) which appeared more spherical. Formation of dented surfaces of spray-dried particles has been attributed to the shrinkage of the particles during the drying process (Rosenberg et al. 1985). There is a possibility that bubble inflation may be more prevalent at a lower WPI concentration (1.42 %) as there was more water available to evaporate and also the surface may be soft enough to allow the particle to expand, and therefore, more dented surface formation resulted in the

Fig. 2 FESEM image of dried powder of various microencapsulation formulations

final stage of the drying process. Particle morphology is also affected by the amount of maltodextrin content in the mixture. Particles with higher maltodextrin content (WPI/maltodextrin, 1:9) appeared to have sharper wrinkles than those with a higher WPI content (Fig. 2). This indicates that maltodextrin is more susceptible to shrinkage during the drying stages than WPI. Regardless of their morphology, the particle surface irregularity did not affect the microencapsulation of flavonoids in *O. stamineus* extract.

Conclusion

Microencapsulation of flavonoids from *O. stamineus* by spray drying in wall systems consisting of WPI or a mixture of WPI and maltodextrin resulted in a high retention of flavonoids content. The highest retention was achieved by employing the least amount of WPI (1.42 wt%), which resulted in 85.6 % total flavonoids retention. This corresponds to highest retention of sinensetin and eupatorin at 81.8 and 70.1 %, respectively. The findings suggest that a mixture of 1:9 protein to maltodextrin ratio is more effective with 86.7 % total flavonoids retention compared with the 1:1 and 9:1 formulations. Results obtained from this work demonstrated that a whey proteins and maltodextrin mixture can be an effective microencapsulating agent for flavonoids derived from *O. stamineus*.

Acknowledgments Funding from the Ministry of Higher Education Malaysia (RACE RDU 121308) is gratefully acknowledged.

References

Akowuah, G. A., Ismail, Z., Norhayati, I., & Sadikun, A. (2005). The effects of different extraction solvents of varying polarities on polyphenols of *Orthosiphon stamineus* and evaluation of the free radical-scavenging activity. *Food Chemistry, 93*, 311–317.

Anandharamakrishnan, C., Rielly, C. D., & Stapley, A. G. F. (2008). Loss of solubility of α-lactalbumin and β-lactoglobulin during the spray drying of whey proteins. *LWT, 41*, 270–277.

Ho, C. H., Ismail, N., Shaida-Fariza, S., & Ahmad, R. (2010). *In vitro* antibacterial and antioxidant activities of *Orthosiphon stamineus* Benth extracts against food-borne bacteria. *Food Chemistry, 122*, 1168–1172.

Jafari, S. M., Assadpoor, E., He, Y., & Bhandari, B. (2008). Encapsulation efficiency of food flavours and oils during spray drying. *Drying Technology, 26*(7), 816–835.

Mariam, A., Asmawi, M. Z., & Sadikun, A. (1996). Hypoglycaemic activity of the aqueous extract of *Orthosiphon stamineus. Fitoterapia, 67*(5), 465–468.

Masuda, T., Masuda, K., Shiragami, S., Jitoe, A., & Nakatani, N. (1992). Orthosiphol A and B, novel diterpenoid inhibitors of TPA (12-O-tetradecanoylphorbol–13–acetate)–induced inflammation, from orthosiphon stamineus. *Tetrahedron, 48*, 6787–6792.

Ohashi, K., Bohgak, T., & Shibuya, H. (2000). Anti-hypertensive substances in the leaves Kumis Kucing (*Orthosiphon stamineus*) in Java Island. *YakugakuZasshi, 120*, 474–482.

Rosenberg, M., & SHEU, T. Y. (1996). Microencapsulation of volatiles by spray-drying in whey protein-based wall systems. *International Dairy Journal, 6*, 273–284.

Rosenberg, M., Kopelman, I. J., & Talmon, Y. (1985). A scanning electron microscopy study of microencapsulation. *Journal of Food Science, 50*, 139–144.

Sameh, F., Abou, Z., & GAMAL, M. E. (2008). Increase in flavonoids content in red onion peel by mechanical shredding. *Journal of Medicinal Plants Research, 2*(9), 258–260.

Saravanan, D., Hossain, M. A., Zhari, S., Gam, L. H., & Ismail, Z. (2006). The use of principal component analysis and self-organizing map to monitor inhibition of calcium oxalate crystal growth by *Orthosiphon stamineus* extract. *Chemometrics and Intelligent Laboratory System, 81*, 21–28.

Xie, Y. L., Zhou, H. M., Liang, X. H., He, B. S., & Han, X. X. (2010). Study on the morphology, particle size and thermal properties of vitamin A microencapsulated by starch octenylsuccinate. *Agricultural Sciences in China, 9*(7), 1058–1064.

A Chimney of Low Height to Diameter Ratio for Solar Crops Dryer

S. Kumaresan, M. M. Rahman, C. M. Chu and H. K. Phang

Abstract Sabah, Malaysia is rich with solar energy where the daily mean daylight is the for is available for 4–8 h. The sunlight can be used effectively in a suitable solar crops drying system. Solar crops dryer with mechanical system is an efficient and suitable option for crops drying. Mechanical system can be replaced with a modified chimney and a conceptual design is proposed. Previously, in the laboratory, a number of experiments had been conducted by using conventional chimney and modified chimney with four different heat loads varied from 1 to 2.5 kW and also for different chimney heights varying from 0.3 to 1.2 m as well as three different types of chimney model with face areas of 0.56, 1.00, and 2.25 m^2. It had been found that the modified chimney significantly reduced the effect of flow reversal or cold inflow and restored the loss in flow rate. The current study using computational fluid dynamics (CFD) managed to show draft enhancement by the modified chimney and revealed the pressure prior to exit to be higher than ambient compared with the lower pressure in normal chimney, indicating vulnerability of normal chimneys to flow reversal but positive prevention in a modified chimney.

Keywords Flow reversal · Cold inflow · Wire mesh · Solar chimney · Natural draft · Enhancement

C. M. Chu (✉) · H. K. Phang
Chemical Engineering Programme, School of Engineering and Information Technology, Universiti Malaysia Sabah, Jalan UMS 88400 Kota Kinabalu, Sabah, Malaysia
e-mail: chrischu@ums.edu.my

S. Kumaresan
Materials and Mineral Research Unit, School of Engineering and Information Technology, Universiti Malaysia Sabah, Jalan UMS 88400 Kota Kinabalu, Sabah, Malaysia
e-mail: shiva@ums.edu.my

M. M. Rahman
Mechanical Engineering Programme, School of Engineering and Information Technology, Universiti Malaysia Sabah, Jalan UMS 88400 Kota Kinabalu, Sabah, Malaysia
e-mail: mizanur@ums.edu.my

R. Pogaku et al. (eds.), *Developments in Sustainable Chemical and Bioprocess Technology*, DOI: 10.1007/978-1-4614-6208-8_19,

Introduction

The efficient and economical uses of renewable energy reduce the dependency on energy supply from finite conventional energy sources. Solar energy can be used directly (photovoltaic cell, solar thermal collectors) or indirectly (passive cooling technique) to convert into usable source of energy. Solar radiation has potential value towards developing solar crops dryer in Sabah, Malaysia. According to the Malaysian Metrological Department, the solar radiation data were recorded from 13 observation stations in Malaysia. Malaysia was found to have mean daily solar radiation for low land areas ranging from 16.10 MJ/m^2 in Kuching to 22.70 MJ/m^2 in Kota Kinabalu in August, 2012; and 16.50 MJ/m^2 in Pulau Langkawi and Kuching to 20.30 MJ/m^2 in Kota Kinabalu in July 2012 (Malaysian Meteorological Department 2012). Sabah had the highest mean daily solar radiation in these two months. The daily average sun shining hours in Malaysia ranges from 4 to 8 h (Christopher 2012). Therefore, it is an essential issue to utilize solar energy in Sabah, Malaysia in different sector such as electricity production, house ventilation, drying agricultural product and clothes. Solar dryer of either forced or natural ventilation types for crops and agricultural products are more cost-effective and faster in drying process than sun drying (Oguntola et al. 2010; Shaik 2009; Ameen and Saiful 2004).

Jensen and Kristensen (2001) conducted test on a solar crop dryer which had transparent area 4.77 m^2 and drying bed area 1.35 m^2. The solar crop dryer had a 12 V DC fan with capacity 300 m^3/hr at 40 Pa. Tarigan and Tekasakul (2005) carried out experiment on solar dryer with biomass burner and heat storage back-up heater in Thailand. The capacity of the burner was designed to provide 400 W/m^2 of energy in the system. Forsona et al. (2007) designed a crop dryer at solar irradiance of 400 W/m^2 where the ambient temperature was 25.1 °C and the relative humidity was 77.8 %. In this system, a chimney was used to create sufficient draft and to confirm natural circulation of air inside the drying chamber (Forsona et al. 2007). Santos et al. (2005) designed a solar crop dryer having an air flow rate of 2.10 m^3/min. A solar dryer was also constructed in Bhutan to dry agricultural crops. The solar dryer had a blower to create air velocity of 0.75 m/s and air mass flow rate of 3120 kg/hr (Fuller et al. 2005). A natural convection solar maize dryer was also designed by Folaranmi (2008) at an irradiation of 400 W/m^2 and the saturated hot air passed through a chimney to avoid contamination of dried product with condensate water.

In this study, a new type of chimney for natural draft ventilation can be constructed at low height and large diameter. This type of chimney has been tested to suffer much less cold inflow than normal conventional chimneys. The new chimney design is fitted with wire mesh of pore size appropriate for the application and this has a bouncing effect to the cold air at its exit thus preventing inflows as demonstrated by smoke tests on laboratory chimneys (Chu et al. 2012). The smaller pore size probably reduced the boundary layer making the flow reversal more difficult and also creates a resistance so that the chimney preferentially draws

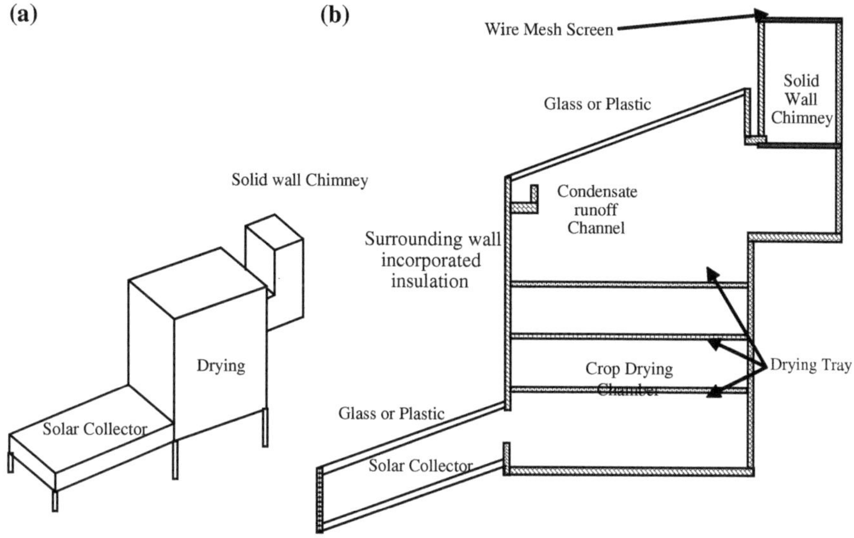

Fig. 1 Conceptual solar crop dryer. **a** Design layout. **b** Sectional side view

cold air or gas from below. By building them at lower heights and larger diameters, the throughput is often higher than chimneys of equal volume, and at less cost than conventional chimneys. The efficiency of such chimneys can be improved by up to 90 % depending on the conditions of the setup. The new type of low height and large diameter chimneys reduce capital cost of construction and hazard of maintenance. Compared with turbo ventilators, it has no moving parts and no noise. Dust may collect on the screen over time but since it is downstream, the dust does not affect the quality of the dried agricultural produce. It can easily be cleaned like air-conditioner air filters. The design and schematic diagram of a concept solar crop dryer is shown in Fig. 1.

Methodology

The primary aim of this study is to investigate the draft due to chimney effect in the solar crops dryer by using computational fluid dynamics (CFD) software (PHOENICS 2011). Two types of chimney, conventional and modified, are used for investigation. Both chimneys have no wire mesh at the bottom but modified chimney only has a single mesh at the top of the chimney. Using a typical scale dryer, the heating chamber (solar collector) and drying chamber each has dimensions of approximately $2 \times 2 \times 2$ m, while the chimney has the dimensions of $2 \times 2 \times 3$ m. The ambient conditions are atmospheric pressure of 101,325 Pa, ambient temperature of 30 °C, and the turbulence model of Chen and Kim (1987) was implemented.

Results and Discussion

The experimental results showed that modified chimney has significant effect on flow natural convection flow losses. The modified chimney restores draft losses by reducing effect of cold inflow or flow reversal in the system (Chu et al. 2012). The simulation results are shown in Fig. 2. The predicted average z-velocity at the exit of the solar crops dryer is shown to be at 0.167 ms^{-1} for modified chimney and 0.163 ms^{-1} for conventional chimney, where the higher draft of modified chimney demonstrates that CFD has simulated flow reversal which is detrimental to draft. However, an important difference can be seen between the two types of chimney, which is the pressure just prior to exit of the chimney, showing positive with respect to ambient for the modified and negative for the conventional. This signals that a conventional chimney is vulnerable to flow reversal, while the modified chimney is not. For larger face dimensions than 2×2 m^2, it has been found that flow reversals and therefore loss of draft are substantial in conventional chimneys.

The simulation program results are in agreement with the data collected during experiments in the laboratory for conventional chimney and modified chimney in three different models having cross-sectional areas of 0.56, 1.00, and 2.25 m^2 and the chimney heights of 0.3, 0.6, 0.9, and 1.2 m and heat loads ranging approximately from 1.00 to 2.5 kW (Chu et al. 2012) where the enhancement was 50–90 %. The enhancement of draft is probably because of the wire mesh at the top of chimney significantly reduces the effect of cold air inflow and thereby improves the performance of the chimney. The chimney performance dominates the air ventilation rate of the solar crops dryer. According to Abdullah et al. (2006), the forced draft solar drying system shows better performance than normal solar drying system. A mechanical system is used to create draft in the forced draft

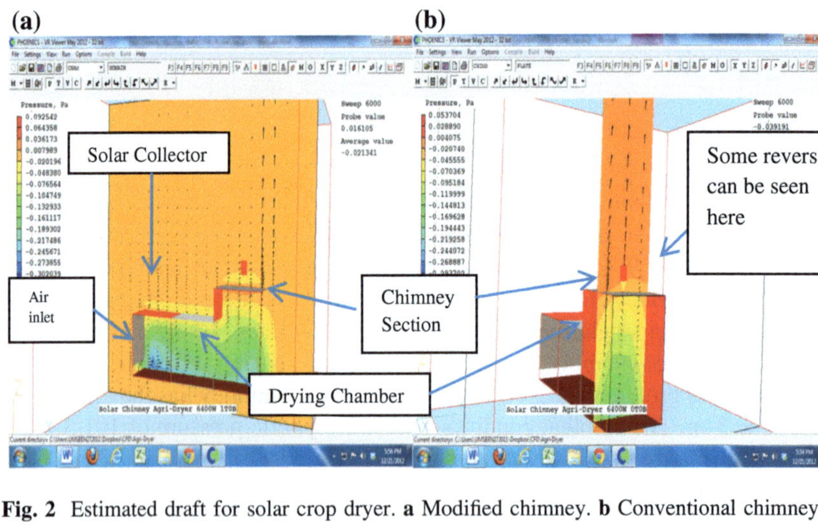

Fig. 2 Estimated draft for solar crop dryer. **a** Modified chimney. **b** Conventional chimney

solar drying system. If the mechanical system can be replaced with a non-mechanical system, then the system will be more cost-effective and environmental friendly.

Conclusion

Chimney with solar crops dryer will be a suitable option compared with normal drying and forced draft drying processes for house hold, hotels, hospital, etc. The modified chimney has been found by both experiment and simulation to significantly increase ventilation of solar crops dryer that can be used as a replacement of mechanical system used in the forced ventilation solar crops dryer. Modified chimney reduces the effect of cold inflow or flow reversal at exit of the solar dryer that restore the losses of draft during operation and enhance the performance of solar crops dryer.

References

Abdullah, M. O., Mikie, F. A., & Lam, C. Y. (2006). Drying performance and thermal transient study with solar radiation supplemented by forced-ventilation. *International Journal of Thermal Science, 45*, 1027–1034.

Ameen, A., & Saiful, B. (2004). Investigation into the effectiveness of heat pump assisted clothes dryer for humid tropics. *Energy Conversion and Management, 45*, 1397–1405.

Chen, Y.S. & Kim, S.W. (1987). Computation of turbulent flows using an extended k-e turbulence closure model, NASA CR-179204.

Christopher, T. B. S. (2012). Electricity from solar energy in Malaysia: Clean, renewable, and abundant energy source, so what's the problem? http://christopherteh.com/blog/2012/05/solar-malaysia/ Posted on May 4th, 2012.

Chu, C., Rahman, M. M., & Kumaresan, S. (2012a). Effect of cold inflow on chimney height of natural draft cooling towers. *Nuclear Engineering and Design, 249*, 125–131.

Chu, C. M., Chu, R. K. H., & Rahman, M. M. (2012b). Experimental study of cold inflow and its effect on draft of a chimney. *Advanced Computational Methods and Experiments in Heat Transfer XII, WIT Transactions on Engineering Sciences, 75*, 73–82.

Folaranmi, J. (2008). Design construction and testing of solar maize dryer. *Leonardo Electronic Journal of Practices and Technologies, 13*, 122–130.

Forsona, F. K., Nazhab, M. A. A., Akuffoa, F. O., & Rajakaruna, H. (2007). Design of mixed-mode natural convection solar crop dryers: Application of principles and rules of thumb. *Renewable Energy, 32*, 2306–2319.

Fuller, R.J., Lhendup, T., & Aye, Lu. (2005). Technical and financial evaluation of a solar dryer in Bhutan solar 2005. ANZSES Annual Conference, Dunedin NZ, 28–30 Nov 2005.

Jensen, S. O. & Kristensen E. F. (2001). Test of a solar crop dryer. Danish Technological Institute and Danish Institute of Agricultural Science, www.solenergi.dk/SEC/rapporter/sec-r-6.pdf.

Malaysian Meteorological Department. (2012). Monthly Weather Bulletin-2012. www.met.gov.my.

Oguntola, J. A., Collins, N. N., & Olayinka, A. (2010). Design and construction of a domestic passive solar food dryer. *Leonardo Journal of Sciences, 16*, 71–82.

PHOENICS. (2011). Parabolic hyperbolic or elliptic numerical integration code series, computational fluid dynamics software, CHAM Ltd., http://www.cham.co.uk.

Santos, B. M., Queiroz, M. R., & Borges, T. P. F. (2005). A solar collector design procedure for crop drying. *Brazilian Journal of Chemical Engineering, 22*(2), 277–284.

Shaik, Z. A. (2009). Design and development of a solar cloth dryer. Bachelor of Technology in Mechanical Engineering, National Institute of Technology Rourkela. http://ethesis.nitrkl.ac.in/1126/1/project.pdf.

Tarigan, E., & Tekasakul, P. (2005). A mixed-mode natural convection solar dryer with biomass burner and heat storage back-up heater solar 2005. ANZSES Annual Conference, Dunedin NZ, 28–30 Nov 2005.

Thermal Denaturation Kinetics of Whey Protein and Maltodextrin Mixture

N. M. Yusoff, J. Gimbun and M. M. Yusoff

Abstract This chapter presents thermal denaturation studies of whey protein and maltodextrin mixture. Different compositions of whey protein–maltodextrin mixture were exposed to various temperature and holding time. The main protein contents, i.e., β-lactoglobulin (β-lg) and α-lactalbumin (α-lac) before and after heat exposures were analyzed using ultra-performance liquid chromatography. Significant decreases in protein contents were found as temperature and holding time increases. Addition of maltodextrin to whey protein solution reduces the protein denaturation significantly. Formulation of 9:1 WPI to maltodextrin ratio was shown to reduce α-lac denaturation by up to 15.1 % and even higher reduction for β-lg which is up to 74.6 % after 20 min at 70 °C. Results from UPLC analysis of the residual native protein suggests that the thermal sensitivity of β-lg was higher compared with α-lac. Over 97 % of β-lg was denatured after 20 min as opposed to about 60 % for α-lac at 80 °C. Protein denaturation was also observed to increase proportionally with temperature. Results from this work may be useful to preserve the protein content in milk during processing involving thermal exposure.

Keywords Protein denaturation · Maltodextrin · Whey protein · Mixture · UPLC

N. M. Yusoff · J. Gimbun (✉)
Faculty of Chemical and Natural Resources Engineering, Universiti Malaysia Pahang, Gambang, Malaysia
e-mail: jolius@ump.edu.my

J. Gimbun
Centre of Excellence for Advanced Research in Fluid Flow (CARIFF), Universiti Malaysia Pahang, Gambang, Malaysia

M. M. Yusoff
Faculty of Industrial Sciences and Technology, Universiti Malaysia Pahang, Gambang, Malaysia

M. M. Yusoff
Central Laboratory, Universiti Malaysia Pahang, Universiti Malaysia Pahang, 26300 Gambang, Pahang, Malaysia

R. Pogaku et al. (eds.), *Developments in Sustainable Chemical and Bioprocess Technology*, DOI: 10.1007/978-1-4614-6208-8_20,

Introduction

Whey protein isolate (WPI) is a dietary supplement created by filtering milk protein. WPI is commonly used in infant formula and widely used as a supplement in sports due to its higher protein content. WPI is known to exhibit thermal denaturation especially when exposed to high-temperature operations such as drying processes in the spray dryer. Numerous studies have been conducted to elucidate denaturation kinetics of the major whey protein, e.g., Parris et al. (1991), Cairolli et al. (1994), Apenten (1995), Curda et al. (1997), Dalgeish et al. (1997), Law et al. (1997), Law and Leaver (1999), Qi et al. (1995), Relkin (1994), Baeza and Pilosof (2002), Sochava et al. (1985), Havea et al. (2000, Indyk et al. (2008) and Goetz and Koehler (2005). Although the kinetic of pure protein denaturation is well understood, however, no research has been performed for protein mixed with other substances such as maltodextrin, and therefore, this is the aim of this study. All previous studies only examined the extent of denaturation of major whey protein and may not imply similar denaturation behavior when WPI–maltodextrin mixture is employed. The main objective of this study was to determine the extent of denaturation of major whey protein (α-lac and *β*-lg) mixed with starch derivative.

Materials and Methods

Chemicals

The WPIs powder (WPI, 80 % protein, 4.7 % moisture, 6 % fat, 6.6 % lactose, 2.7 % minerals) was obtained from Ultimate Nutrition (Lancashire, UK). Meanwhile, individual bovine milk proteins, including α-lac, *β*-lg, and BSA with 85, 80, and 96 % purity, respectively, were obtained from Sigma (St. Louis, MO, USA) as calibration standard. HPLC grade trifluoroacetic acid (TFA) and acetonitrile (MeCN) were obtained from Merck (Germany) for the UPLC analysis. Maltodextrin of 10 DE was obtained from San Soon Seng Food Industries Sdn. Bhd. (Malaysia).

Sample Preparation

WPI solutions (5 % w/v) were prepared at room temperature by dissolving WPI in ultrapure water. Samples of mixed maltodextrin and WPI mixture (of either 1:1 or 9:1) were prepared with a total protein content of 5 % w/w. Protein was mixed homogenously with maltodextrin using a magnetic stirrer at room temperature.

Kinetic Study Design and Heat Treatment

Triplicate samples (2.3 ml) of the mixture were placed into clean and dry stainless steel tubes, then heated at 70 and 80 °C using a water bath at times ranging from 3 to 20 min. Heated samples were cooled immediately in iced water and centrifuged at 12,000 rpm for 30 min at 20 °C to remove the insoluble matter. Clear supernatant was collected and stored at 4 °C until further analysis. All samples were analyzed within 48 h. Unheated and filtered 5 % w/v native WPI solutions were used as control samples.

UPLC Analysis

WPI solutions were diluted to 1:20 dilution factors with ultrapure water and then filtered with PES syringe filter (0.22 μm, Millipore Corp., Bedford, MA). Separations were carried out by ultra-performance liquid chromatography (UPLC) using Acquity BEH 300 C4 column. Filtered samples were injected (10 μl) into a reversed phase column (2.1 × 100 mm; 1.7 μm). The column was equilibrated in 65 % Solvent A (0.1 % v/v TFA in ultrapure water), and after sample injection, a 1-min isocratic period was applied followed by a series of linear gradients to 100 % of solvent B (0.1 % v/v TFA in MeCN). Elution gradient was set as follows :1–1.46 min, 38–42 % B, 1.46–3.13 min, 42–46 % B, 3.13–4.38 min, 46–90 % B, 4.38–4.79 min, 90 % B, 5–6.04 min, 90–35 % B, 6.04–7.09 min, 35–38 % B. The flow rate was set at 0.4 mL/min, and column temperature was maintained at room temperature. The peak areas were determined by absorbance at a wavelength of 205 nm. Residual native protein was calculated as % area relative to unheated protein area (mAU*ml).

Results and Discussion

Comparison of UPLC chromatogram was made between the standard and sample of WPI to quantify WPI composition in each sample. The major peaks corresponding to α-lactalbumin (α-lac) and β-lactoglobulin (β-lg) appeared at elution times within 3 and 4 min, respectively. The peak for bovine serum albumin (BSA) appeared at 2.7 min. The UPLC method developed for the first time during this study is capable of a rapid and accurate qualitative and quantitative analysis of protein content in WPI. The analysis time was approximately 6 min, which is over three times faster than those previously reported, e.g., Anandharamakrishnan et al. (2008) at 30 min and Cheison et al. (2012) at 18 min. The degree denaturation of each WPI sample was calculated according to the following equation (Fureby et al. 2001):

$$\text{Degree of denaturation protein} = \left(\frac{1 - \text{soluble protein}}{\text{soluble protein in unheated sample}}\right) \times 100$$

Figure 1 shows the percentage of denaturation of α-lac with and without maltodextrin at different ratios. The results suggest that a small amount of maltodextrin may help to reduce α-lac denaturation but not when higher concentration is added (i.e., 1:1). Formulation of 9:1 WPI to maltodextrin ratio shows a small reduction in α-lac denaturation to 16 % after 20 min at 70 °C as oppose to the pure WPI (18.2 %). Although further addition of maltodextrin from 9:1 to 1:1 WPI to maltodextrin ratio is not an improvement as the α-lac denaturation increases to 40 %. This may be attributed to alteration of rheological characteristic of WPI when more maltodextrin was added, thus promoting higher denaturation (Trivedi et al. 2008). Similar trends were observed at a higher temperature (80 °C). A larger reduction from 60.4 to 45.3 % was achieved after 20 min when 9:1 WPI to maltodextrin ratio was used as opposed to pure WPI.

β-lg denaturation was up to 90 % after 20 min without maltodextrin at 70 °C as shown in Fig. 1. In the case of β-lg, addition of maltodextrin helps to reduce protein denaturation. The formulation of WPI to maltodextrin 9:1 reduces β-lg denaturation remarkably from 90 % to 15.4 %, while the 1:1 formulation achieved 48.2 % after 20 min at 70 °C. Effect of maltodextrin addition to β-lg denaturation was more apparent, because it was more susceptible to thermal degradation. For the case of β-lg at 70 °C, maltodextrin addition can significantly reduce the denaturation by up to 41.8 and 74.6 % for 1:1 and 9:1 mixture, respectively. The effect of maltodextrin addition to β-lg denaturation is clearly distinguishable at 70 °C; however, the difference was not significant at a higher temperature (80 °C)

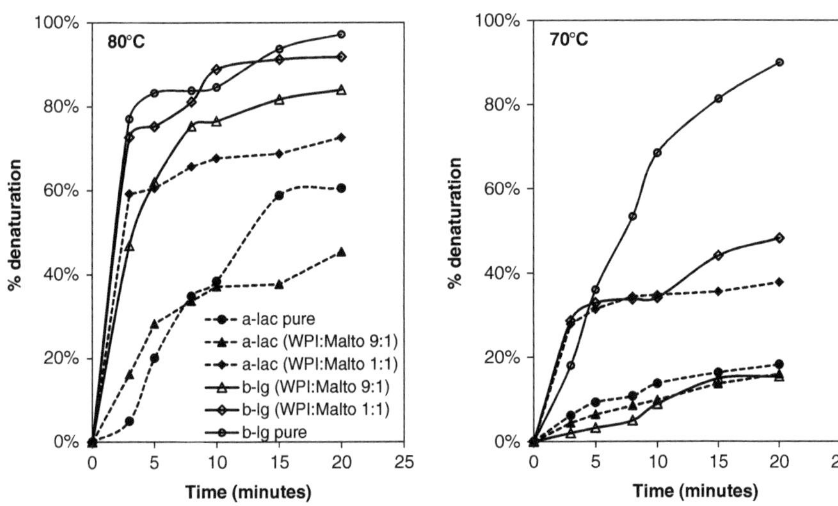

Fig. 1 Percentage denaturation of α-lactalbumin and β-lactoglobulin

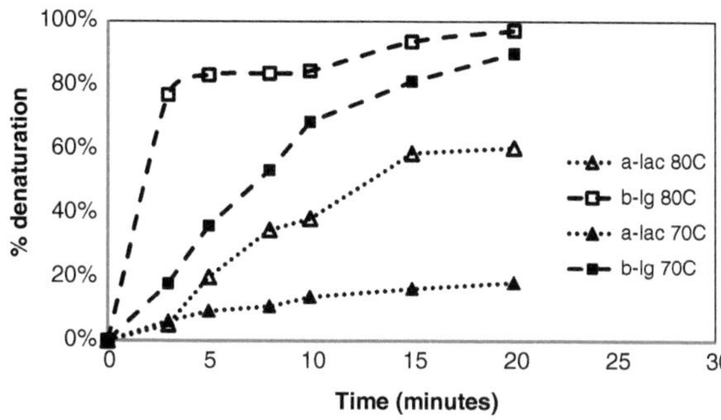

Fig. 2 Comparison percentage denaturation of α-lactalbumin and β-lactoglobulin at different temperatures

as shown in Fig. 1. Nevertheless, reduction in β-lg denaturation after 20 min is still visible with up to 5.4 % for 1:1 and 13.3 % for 9:1, respectively.

The results from this study also reveal that the thermal sensitivity of β-lg is higher compared with α-lac as shown in Fig. 2. Denaturation of β-lg reaches 90 % after 20 min at 70 °C and further increase to 97.2 % at 80 °C. A very steep increase in β-lg denaturation is also observed at 80 °C with over 80 % of protein denaturation after just 3 min. Only a minimal denaturation of α-lac (18.2 %) was observed at 70 °C after 20 min, but the value increases to 60.4 % at 80 °C. Nevertheless, denaturation of α-lac is over 37 % lower than that of β-lg in both cases. This is attributed to the heat stability of α-lac as opposed to β-lac. Earlier studies (e.g., Indyk et al. 2008; Law et al. 1997) has shown the order of protein susceptibility on thermal denaturation was immunoglobulins > serum albumin > β-lactoglobulin > α-lactalbumin. Thermal denaturation of bovine β-lg involves two stage processes, which is unfolding of the polypeptide chain followed by self-association with other protein via sulfhydryl-disulfide interchange reactions (Sawyer et al. 2002). Bovine α-lac tends to be more stable than β-lg due to the differences in its environment rather than molecular structure. Denaturation of bovine α-lac is related to the complex formation with β-lg. This is due to lack of a free sulfhydryl group in α-lac which prevents it from undergoing sulfhydryl-disulfide interchange reactions in the absence of a protein containing a free sulfhydryl group such as β-lg (Huppertz et al. 2004).

Conclusion

Maltodextrin addition to protein solution is potentially good at reducing protein denaturation due to thermal exposure. Formulation of 9:1 WPI to maltodextrin ratio was shown to reduce α-lac denaturation by up to 15.1 % and even higher reduction for β-lg which is up to 74.6 % after 20 min at 70 °C. Results from UPLC analysis of the residual native protein suggested that the thermal sensitivity of β-lg was higher compared with α-lac. Over 97 % of β-lg was denatured after 20 min as opposed to about 60 % for α-lac at 80 °C. Protein denaturation was also observed to increase proportionally with temperature. These results may be useful in considering preservation strategies of protein content in milk during processing involving thermal exposure.

Acknowledgments Nuraini Mohd Yusoff acknowledges funding from Universiti Malaysia Pahang in the form of GRS110353 and from Malaysian Toray Science Foundation (MTSF) in the form of RDU 121501.

References

Anandharamakrishnan, C., Rielly, C. D., & Stapley, A. G. F. (2008). Loss of solubility of α-lactalbumin and β-lactoglobulin during the spray drying of whey proteins. *LWT-Food Science and Technology, 41*, 270–277.

Apenten, R. K. O. (1995). A three-state heat-denaturation of bovine α-lactalbumin. *Food Chemistry, 52*, 131–133.

Baeza, R. I., & Pilosof, A. M. R. (2002). Calorimetric studies of thermal denaturation of β-lactoglobulin in the presence of polysaccharides. *LWT-Food Science and Technology, 35*, 393–399.

Cairolli, S., Iametti, S., & Bonomi, F. (1994). Reversible and irreversible modifications of β-lactoglobulin upon exposures to heat. *Journal of Protein Chemistry, 13*, 347–354.

Cheison, S. C., Bor, E. K., Faraj, A. K., & Kulozik, U. (2012). Selective hydrolysis of α-lactalbumin by acid protease a offers potential for β-lactoglobulin purification in whey proteins. *LWT-Food Science and Technology, 49*, 117–122.

Curda, L., Belhacova, M., Uhrova, J., & Fukal, L. (1997). Assessment of heat induced denaturation of whey proteins. *Journal of Chromatography A, 772*, 231–234.

Dalgeish, D. G., Senaratne, V., & Francois, S. (1997). Interaction between α-lactalbumin and β-lactoglobulin in the early stages of heat denaturation. *Journal of Agricultural and Food Chemistry, 45*, 3459–3464.

Fureby, A. M., Elofson, U., & Bergenstahl, B. (2001). Surface composition of spray-dried milk protein-stabilised emulsions in relation to pre-heat treatment of proteins. *Colloids and surfaces B, Biointerfaces, 21*, 47–58.

Goetz, J., & Koehler, P. (2005). Study of thermal denaturation of selected of whey and egg by low resolution of NMR. *LWT-Food science and technology, 38*, 501–512.

Havea, P., Singh, H., & Creamer, L. (2000). Formulation of new protein structures in heated mixtures of BSA and α-lactalbumin. *Journal of Agricultural and Food Chemistry, 48*, 1548–1556.

Huppertz, T., Fox, P. F., & Kelly, A. L. (2004). High pressure-induced denaturation of α-lactalbumin and β-lactoglobulin in bovine milk and whey: a possible mechanism. *Journal of Dairy Research, 71*, 489–495.

Indyk, H. E., Willliams, J. W., & Patel, H. A. (2008). Analysis of denaturation of bovine IgG by heat and high pressure using an optical biosensor. *International Dairy Journal, 18*, 359–366.

Law, A. J. R., Leaver, J., & Xavier, M. F. (1997). Effect of protein concentration on rates of thermal denaturation of whey proteins in milk. *Journal of Agricultural and Food Chemistry, 45*, 4255–4261.

Law, A. J. R., & Leaver, J. (1999). Factors that affecting the heat denaturations of whey proteins in cow milk. *International Dairy Journal, 9*, 407–408.

Parris, N., Purcell, J., & Ptashkin, S. (1991). Thermal denaturation of whey proteins in skim milk. *Journal of Agricultural and Food Chemistry, 39*, 2167–2170.

Qi, X. L., Bronlow, S., & Holt, C. (1995). Thermal denaturation of β-lactoglobulin; effect of protein concentration at pH 6.75 and 8.05. *Biochimica et Biophysica Acta, 1428*, 43–49.

Relkin, P. (1994). Differential scanning calorimetry: A useful tool for studying protein denaturation. *Thermochimica Acta, 246*, 371–386.

Sawyer, L., Barlow, P., Boland, M. J., Creamer, L. K., Denton, H., Edward, P. J. B., et al. (2002). Milk protein structure—what does it tell the dairy industry? *International Dairy Journal, 12*, 299–310.

Sochava, I. V., Belospolskaya, T. V., & Smirnova, O. I. (1985). DSC study of reversible and irreversible thermal denaturation of concentrated globular protein solutions. *Biophysical Chemistry, 22*, 323–336.

Trivedi, D., Bennet, R. J., Hemar, Y., Reid, D. C. W., Lee, S. K., & Illingworth, D. (2008). Effect of different starches on rheological and microstructural properties of (1) model processed cheese. *International Journal of Food Science and Technology, 43*, 2191–2196.

Production of α-Amylase by Immobilized *Bacillus Subtilis* in Polymeric PolyHIPE Matrix

D. N. Jimat, C. Harwood and G. Akay

Abstract The microcellular polymer known as polyHIPE polymer (PHP), with modified physico-chemical characteristics, was developed as a cell matrix for the immobilization of the starch-degrading bacterium, *Bacillus subtilis*. Suspension of *B. subtilis* spores was inoculated into a synthesized PHP matrix which is chemically modified, and this PHP matrix was named sulphonated PHPs (SPHPs). These inoculated spores were then activated by supplying continuously well-aerated culture medium (LB medium) and placed in a 37 °C constant temperature room for 24-h incubation. The growth and enzyme productivity data were evaluated and compared. Three different pore and interconnect sizes of SPHPs were evaluated: 42.0 ± 0.61, 36.0 ± 0.50 and 30.0 ± 0.64 μm. The collected samples obtained from the 24-h cultivation were used to determine α-amylase productivity and the loss of cells from the matrices. The morphology, viability and proliferation of the immobilized cells on PHP matrices were observed by scanning electron microscopy (SEM). The SPHP with a pore size of 36.0 μm performed better with respect to the production of α-amylase and the penetration of cells through the whole matrix compared to other SPHPs. The data showed that the total productivity of α-amylase enzyme produced by immobilized cells (on the basis of SPHP volume) was 7.6-fold higher than the batch cell culture. However, if productivity was determined on the basis of used total volume of nutrient medium, that of the immobilized cells was relatively low compared to batch cell culture.

D. N. Jimat (✉)
Biotechnology Engineering Department, Kulliyyah of Engineering, International Islamic University, P.O. Box 10 50728 Kuala Lumpur, Malaysia
e-mail: jnoraini@iium.edu.my

C. Harwood
Cell and Molecular Biosciences, Medical School, Newcastle University, Newcastle upon Tyne, UK

G. Akay
Process Intensification and Miniaturization Centre, School Of Chemical Engineering and Advanced Materials, Newcastle University, Newcastle upon Tyne, UK

R. Pogaku et al. (eds.), *Developments in Sustainable Chemical and Bioprocess Technology*, DOI: 10.1007/978-1-4614-6208-8_21,

Introduction

Enzymes are biological products that can act as catalysts. Owing to their specific, effective and versatile features, they tend to exhibit much higher reaction rates as compared to reactions catalysed by chemical catalysts. One of the world's largest industrial enzyme producers is Novozymes (Denmark), which is involved in the development, production and distribution of microbial enzymes. These enzymes are produced from genetically modified microorganisms, indicating that microbial fermentation is the major route for the production of industrial enzymes. There is a need to explore novel processes that could improve the production systems to meet the anticipated increased demand for industrial enzymes.

The application of immobilized cell technology can improve bioreactor productivity by maintaining high cell concentrations as well as operational stability. Additionally, immobilized cell technology can reduce processing time and downstream processing as it can act as a biomass separator. A number of previous studies have attempted to investigate the influence of immobilization on cell growth and metabolic activity (Duran-Paramo et al. 2000; Argirakos et al. 1992; Dobreva et al. 1998; Konsoula and Liakopoulou-Kyriakides 2006). The selection of the materials used for the matrices is of the utmost importance because it determines their characteristics. Currently, most of the support matrices used for cell immobilization is derived from natural sources such as alginate, carrageenan, gelatine and chitosan. These natural matrices lack mechanical rigidity and stability and are therefore prone to damage during long processing conditions. This is due to the impact of rapidly growing colonies on the outer surface of the matrix and subsequently leads to the release of cells into the medium. Additionally, the concentration of cells in the interior of these matrices is inconsistent because some treatments allow significant number of cells to escape. This means that more cells are observed growing on the outer surface rather than migrating deeper into the matrix. As a result, the efficiency of immobilization is compromised, affecting the final output of the bioreactor. The main factor that causes these problems is the limited mass transfer rates for nutrients, products or waste to and from the immobilized matrix. Thus, there is a need to improve the characteristics of the immobilized matrices used as cell supports. The matrix of the immobilized cell support must be porous enough to provide easy access to nutrients and waste but retain most or all of the bacteria. In this study, a polymeric material known as polyHIPE polymer (PHP) was used to immobilize the starch-degrading bacterium, *Bacillus subtilis*. This polyHIPE matrix is synthesized through the high internal phase (HIPE) route and its physico-chemical properties can be modified as desired. This type of polymeric matrix has a highly interconnected porous structure and is mechanically durable in the presence of harsh conditions such as high temperature, pH and toxic chemicals. Therefore, it can be sterilized and used for long operating conditions without significant physical or chemical degradation.

The PHP matrix used in the developed system was designed to improve the production of α-amylase by *B. subtilis*. The confined microenvironment was designed to promote the growth and metabolic activity of this bacterium. Moreover, the applied forced flow seeding technique yielded a more uniform distribution of cells within the polymeric support, which also helps to improve nutrient transport. This approach is essential in the area of bioprocess development specifically for microbial fermentations. The results of these studies may provide a platform for developing microreactors with sufficient mass transfer capacities, to compete with traditional fermenters.

Methodology

Materials

Phenol, styrene, DVB (divinylbenzene), sorbitanmonooleate (Span 80), potassium persulfate, vinylpyridine, sulphuric acid and isopropanol were purchased from Sigma.

Microorganisms and Growth Medium

B. subtilis 168(pKTH10) strain was obtained from Cell and Molecular Biosciences, Medical School, Newcastle University. The spore suspensions of *B. subtilis* 168 (pKTH10) were induced in Schaeffer's sporulation medium as described in Harwood et al. (1990). After overnight incubation, the cell cultures were centrifuged at 10,000 g for 10 min. The pelleted cells were washed five times with sterilized distilled water to remove all cell debris. The supernatant was discarded, and the spores were suspended in sterilized distilled water and stored at −20 °C. Luria–Bertani medium containing 1 % (w/v) Bacto-tryptone, 0.5 % (w/v) Bacto-yeast extract and 1 % (w/v) NaCl was used to cultivate the spore suspension in both immobilization and batch culture.

Preparation of Sulphonated PolyHIPE Polymer Matrix

PHPs were prepared by polymerization of a HIPE emulsion as described by Akay et al. (2005b) the chemical modification of PHPs via sulphonation was carried out using methods developed by Akay et al. (2005a, b).

Preparation of Immobilized Cell Reactor

The PolyHIPE (PHP) discs were mounted into a sealed PTFE block and sterilized in an autoclave at 121 °C for 15 min. Then, this PHP disc was initially wetted with distilled water overnight and then with LB media for an hour before use. Spores of *B. subtilis* strain 168 (pKTH10) ($\sim 2 \times 10^8$) were suspended in 2 ml of LB broth and pre-germinated by incubating for 90 min at 37 °C. The pre-germinated spores were force seeded into the PHP by passing the suspension through the matrix at a flow rate of 0.55 ml/min using a syringe pump. The seeded PHP was left for an hour to allow the pre-germinated spore to attach to the surface of the PHP, and then culture medium (LB broth) was pumped continuously through the matrix at a flow rate of 1.0 ml/min. This experimental setup was placed in a 37 °C constant-temperature room. Samples were collected at various time intervals to determine the numbers of cells released from the PHP matrix and α-amylase activity. At the end of experiment, sterilized 0.1-M phosphate-buffered saline (PBS) was pumped through the matrix at the same flow rate to remove the remaining medium. As a control, the same amount of medium was also added into the PHPs discs, without inoculated spores and then incubated overnight at 37 °C. The matrix samples were fixed in 2 % glutaraldehyde/PBS and stored at 4 °C until required.

Determination of Pore and Interconnect Size by Scanning Electron Microscopy

Pore and interconnect sizes of each type of the PHPs were analysed using scanning electron microscopy (SEM, Cambridge s240). Each sample was fractured in order to obtain a fresh internal surface and was then mounted onto aluminium stubs at room temperature using conductive carbon cement (LEIT carbon or glued sticker). Prior to analysis, the samples were then left overnight followed by sputter coated with gold at 20 nm, using a Polaron E5100 sputter coater and stored in silica gel. The analysed micrograph images of each polymer were used to determine the diameter of 50 pores and interconnect using ImageJ software (NIH image).

Quantification of Released Cell Numbers and Cell Dry Weight

The number of cells released from the PHP matrix was determined by plating serial dilutions of the outflowing medium onto LB agar plates containing 1 % starch. The plates were incubated overnight at 37 °C, and the number of colonies was determined. Exposing the surfaces of these plates to iodine vapour revealed clear zones around the colonies able to hydrolyse starch. Cells were recovered from batch culture samples (with volume of 1 ml) by centrifugation (3,000 g, 10 min, 4 °C) and washed twice with 0.9 % NaCl. The resulting pellet was dried at 105 °C for 24 h to constant weight. In the case of cells in the polyHIPE matrix, they were washed by passing distilled water through the matrix to remove culture medium, and the matrix dried to constant weight (by weighing several times) and then weighed. The cell dry weight was determined by subtracting the initial weight of the polyHIPE and its final weight. Weight of blank polyHIPE disc (without cells) was also determined as a control.

Quantification of Protein

1 ml of protein assay reagent (Coomassie Plus (Bradford) Assay reagent, Thermo Scientific) was added to 1 ml of sample and mixed by using vortex. The mixture was incubated for 10 min at room temperature, and the absorbance was determined at 595 nm against distilled water blank. Each sample was treated in duplicate, and the average value was determined. The protein concentration of each sample was determined from the standard curve using the known concentrations of bovine serum albumin (BSA).

Quantification of α-amylase Assays

Phadebas®amylase tablets, a water-insoluble starch substrate, was used to determine α-amylase activity. One tablet of Phadebas was suspended in 5 ml of sterile distilled water. 40 μl of sample was added to 0.8 ml of Phadebas suspension, and the mixture was incubated with frequent mixing at 37 °C in a water bath for 15 min. The reaction was stopped by the addition of 200 μml of 0.5 M NaOH and vortexed immediately. The mixture was centrifuged at 10,000 g for 5 min, and the supernatant was transferred into a cuvette. The absorbance of the supernatant was measured at 620 nm against a LB broth blank subjected to the same assay conditions. Samples with very high α-amylase activity were diluted up to fivefold using distilled water. Each sample was duplicated, and the average was taken. One unit (U) of α-amylase activity is defined as the amount of enzyme catalyzing the hydrolysis of 1-μmol glucosidic linkage per minute at 37 °C.

Preparation of Matrix Samples for SEM Analysis

The samples were dehydrated in an ethanol series (each step for 15 min excepting 100 % ethanol treatment was for 1 h). The dehydrated sample in 100 % ethanol was critical-point dried with liquid CO_2. The sample was mounted on aluminium stubs and coated with gold using a Polaron E1500 sputter coater. The sample was examined under SEM (Cambridge s240). SEM analysis for samples without cells was also undertaken. In this case, the samples were directly coated with gold without dehydration and critical drying.

Results and Discussions

Architecture of PolyHIPE Polymer Matrices

In this study, PHPs were synthesized through the water-in-oil HIPE route in order to obtain a good structural stability polyHIPE. The resulting PHPs have high porosities and interconnectivities. In order to enhance its hydrophilicity and diffusivity, the resulting w/o HIPE was subjected to sulphonation using sulphuric acid, and the treated porous, hydrophilic polymeric support is called sulphonated PHP (SPHP), and this chemically modified PHP was then used to immobilize the industrial starch-degrading bacterium, *B. subtilis*. The microstructure of SPHP was examined by SEM, and representative images (Fig. 1) and their characteristics are shown in Table 1.

As can be seen in Fig. 1a, the pore sizes in sample SPHP-2 (t_D5/t_M5) were significantly larger than those for samples SPHP-4 and SPHP-6. The pore and interconnect sizes of samples SPHP-4 and SPHP-6 were well distributed and uniform, and this may be due to the additional 5-min dosing time, allowing for the uniform formation of interconnects throughout the structure. The increase in mixing time resulted in the emulsion reaching higher energy state, with a resultant decrease in droplet size. However, since the pore and interconnect sizes of SPHP-6 were only marginally smaller than those of SPHP-4, it appears that the additional 10 min of mixing time had little or no effect on the final structure of the PHP. This was likely to be due to the emulsion having reached a highly viscous and stabile state such that the further break-up of the droplets was not possible without increasing the vigour of the stirring.

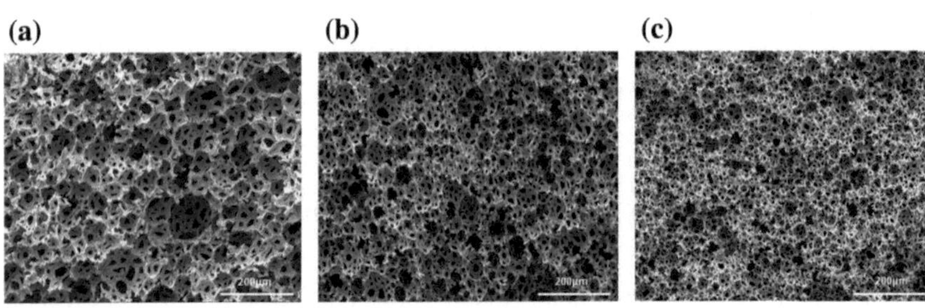

Fig. 1 Micrograph images illustrating the effects of dosing and mixing rates on the microphysical architecture of sulphonated polyHIPEs. **a** SPHP-2. **b** SPHP-4. **c** SPHP-6

Table 1 Characteristics of the polyHIPE polymer (PHP) used in this study

PHP code no.	Dosing time (t_D, min)	Mixing time (t_M, min)	Pore size (D, μm)	Interconnect size (d, μm)
SPHP-2	5	5	42.0 ± 0.61	22.0 ± 0.10
SPHP-4	10	10	36.0 ± 0.50	16.0 ± 0.89
SPHP-6	10	20	30.0 ± 0.64	11.0 ± 0.76

α-Amylase Production in Batch and Immobilized Cultures

In this study, a strain of *B. subtilis* 168 encoding pKTH10 was used for the high-level production of α-amylase. Recombinant plasmid pKTH10, based on the high-copy-number plasmid pUB110, encoded the structural gene for the *Bacillus amyloliquefaciens* α-amylase AmyQ (Palva 1982). As can be seen from Fig. 2, the cells grow rapidly in exponential phase for the first 8 h. Cell mass then reached a plateau for the next 12 h, before slightly declining for the last 4 h. The specific growth rate, μ (h^{-1}), and doubling time, td (h^{-1}), of the batch culture were determined from the growth curve; the values are 0.41 h^{-1} and 35 min, respectively. The concentration of α-amylase in the culture medium continued to increase from the exponential phase until the first 5 h of stationary phase.

In the case of immobilized cultures, the production of α-amylase by cells of *B. subtilis* immobilized in sulphonated polyHIPE matrices with different pore sizes (42.0 ± 0.61, 36.0 ± 0.50 and 30.0 ± 0.64 μm) was determined at various times during cultivation (Fig. 3). The data show that SPHP with a pore size 36.0 ± 0.50 μm (SPHP-4) was the most productive—it reached an optimum production rate (~0.5 U/ml) after ~12 h and maintained a similar rate throughout the experiment. This optimum enzyme production rate by SPHP-4 matrix was 20 % more compared to SPHP-6 matrix (~0.4 U/ml). In the case of matrix with a pore size 42.0 ± 0.61 μm (SPHP-2), the production of α-amylase from *B. subtilis*

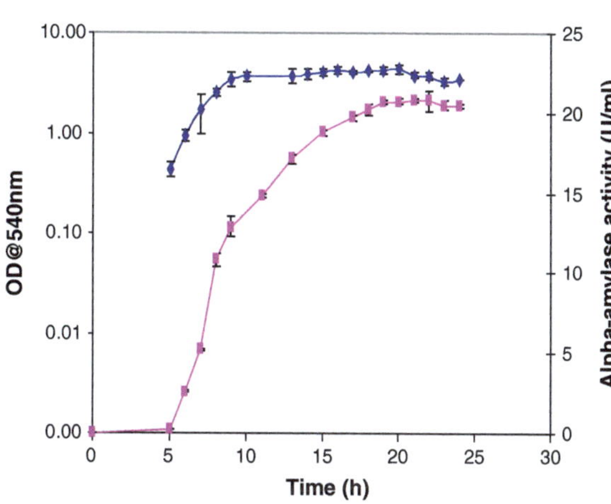

Fig. 2 Growth curve and kinetic of extracellular α-amylase production by a batch culture of *B. subtilis* 168 (pKTH10). *Filled diamond* optical density, *Filled square* α-amylase production. The data are the result of independent triplicate experiments

Fig. 3 The time course of the production of α-amylase and released cells in 10^6 cells/ml by *B. subtilis* immobilized in sulphonated polyHIPE with pore sizes of 42.0, 36.0 and 30.0 μm (for 24-h incubation with 1 ml/min of nutrient medium flow rate). The data are the result of independent triplicate experiments

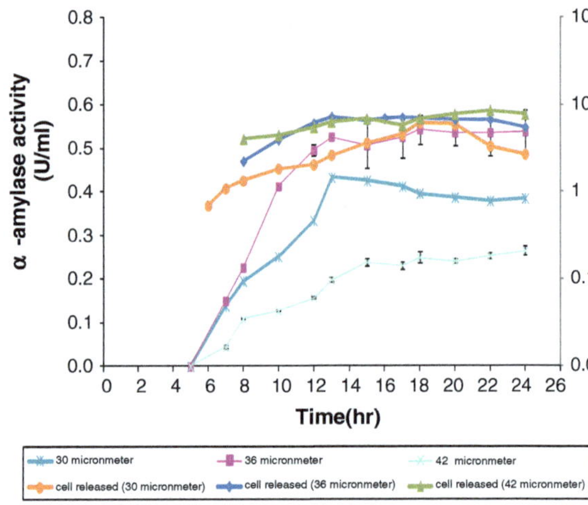

immobilized on the SPHP-2 matrix reached a maximum production rate (~0.2 U/ml) after 15 h and maintained at 0.25 U/ml towards the end of the experiment. The release of bacterial cells from the SPHP matrices was determined from samples taken every 2 h (Fig. 3). The cells were enumerated by counting the number of colonies on nutrient agar plates after growth overnight at 37 °C. The data show that the SPHP with the smallest pore size (30 μm) released the fewest number of cells. A comparison of performance of sulphonated polyHIPE matrices with batch cell culture is shown in Table 2. The productivity of α-amylase enzyme produced by immobilized cells (on the basis of SPHP volume) with pore size of 36 μm was 7.6-fold higher than the batch cell culture. However, the volumetric productivities of α-amylase per total nutrient medium of immobilized cells on the SPHP matrices were relatively low (<1) compared to that of the batch cell cultures.

Morphology of Immobilized B. subtilis *Cells*

After 24 h cultivation at 37 °C with continuous supply of well-aerated LB medium at a flow rate of 1 ml/min, the surfaces of all three sulphonated polyHIPEs were covered by a dense homogeneous layer of vegetative *B. subtilis* cells that were free of an extensive extracellular matrix (Fig 4d, h, l). Following inoculation and incubation, a large number of vegetative cells were observed at the surface of all sulphonated polymeric matrices: SPHP-2, SPHP-4 and SPHP-6 after 24 h incubation at 37 °C (Fig. 4a–c, e–g, i–k). The germinated spores had successfully migrated throughout the matrix of the polyHIPE supports with pore sizes of 42 μm (Fig. 4b) and 36 μm (Fig. 4f). However, in the case of SPHP-6, a thick layer of cells had accumulated on the surface (Fig. 4k) only migrating up to approximately

Table 2 Comparison of sulphonated PolyHIPE (SPHPs) with batch cultures

	Strain	Incubation time (h)	Per volume of polyHIPE disc (3 ml)			Per ml of used total medium (1,440 ml)	
			Volumetric productivity (U/ml/h)	Specific activity of α-amylase (U/mg protein)	Cell dry weight (mg/ml)	Volumetric productivity (mU/ml/h)	Specific activity of α-amylase (U/mg protein)
[a]Microreactor with polyHIPE matrices	***B. subtilis* 168 (pKTH10)**	24				14.26 ± 0.0001	17.14 ± 0.08
SPHP-2 ($\phi = 42$ μm)			2.89	3,470	0.18 ± 0.05		
SPHP-4 ($\phi = 36$ μm)			6.84	8,210	0.33 ± 0.03		
SPHP-6 ($\phi = 30$ μm)			5.12	6,140	0.23 ± 0.04		

	Per ml of used total medium (100 ml)		
	Volumetric productivity (U/ml/hr)	Specific activity of α-amylase (U/mg protein)	Cell dry weight (mg/ml)
[b]Shake flask	0.90	1,083	1.30 ± 0.0001

[a] PolyHIPE disc with volume of 3 ml
[b] Shake flask with 100 ml medium in 250-ml conical flask

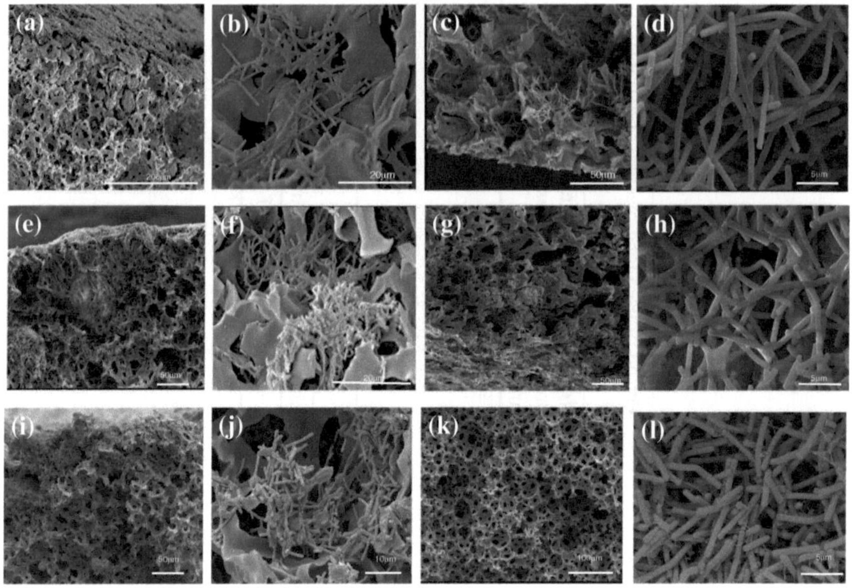

Fig. 4 Electron micrograph of cross section of the SPHP-2 (ϕ = 42 μm) matrix, showing the appearance of vegetative cells of *B. subtilis* immobilized on after 24-h cultivation. **a** Top edge, **b** further deep (x2000), **c** *bottom edge* and **d** *top* surface at high magnification. SPHP-4 (ϕ = 36 μm) matrix, showing the appearance of vegetative cells of *B. subtilis* immobilized on after 24-h cultivation. **e** *Top edge*, **f** in the middle of cross-section matrix x2000, **g** *bottom* edge and **h** *top* surface at high magnification. SPHP-6 (ϕ = 30 μm) matrix, showing the appearance of vegetative cells of *B. subtilis* immobilized on after 24-h cultivation. **i** *Top edge*, **j** in the middle of cross-section matrix (x2000), **k** deep further and **l** *top* surface at high magnification

3 mm into the support. The pore walls had an almost continuous layer of bacterial cells on the top surfaces, whereas smaller cell numbers were found further deep into the matrices.

Discussion

One of the design characteristics of matrices used for the immobilization bacterial cells is the need to optimize pore and interconnect size to ensure good mass transport into and out of matrix. In this study, sulphonated polyHIPE matrices with three different pore sizes were evaluated. The results indicated that pore size significantly affected the production of α-amylase by cells immobilized on sulphonated polyHIPE matrices (Fig. 4). SPHP matrices with a pore size of 36 μm (SPHP-4) yielded the highest total α-amylase enzyme and cell dry weight. The volumetric productivity obtained by SPHP-4 was 2.4-fold and 1.3-fold higher than that obtained by the SPHPs with pore sizes of 42 and 30 μm, respectively. As expected, there was a direct relationship between dry cell mass and α-amylase production and entrapping a high proportion of cell in the matrix a high priority. The low level of α-amylase production and low cell mass in the matrix with the smallest pore size (30 μm) are likely to be due to diffusional limitations that restrict the access of nutrients and oxygen to the cells and the release of the secreted enzyme to the medium. It might be also due to limited loading of cells onto this matrix.

Steady-state α-amylase production was achieved after 13 h, indicating that the matrix had reached the limit of its capacity to sustain further increases in biomass. Such a steady state represents the most efficient state for the production of α-amylase since net cell growth and nucleic acid syntheses are repressed while protein synthesis is still active (Kinoshita et al. 1968). This represents a shift from biomass increase to cell maintenance and product formation. Moreover, in *B. subtilis*, the secretion of macromolecular hydrolases like α-amylase is higher during stationary phase than exponential phase (Antelmann et al. 2001). Bacterial cells were released from all of the SPHP matrices (Fig. 3) although the SPHP matrix with the smallest pore size (30 μm) released the fewest numbers. The release of cells from the SPHP matrices with pore sizes of 36 and 42 μm was similar.

Previous studies have shown that polyHIPE provides a suitable matrix for the immobilization of both bacterial (Akay et al. 2005a, b; Erhan et al. 2004) and eukaryotic (Akay et al. 2004; Bokhari et al. 2005) cells. These studies have shown that *B. subtilis* can migrate up to 3 mm into such matrices over 30-day cultivation. Multicellular layers of osteoblast were seen on the top surface of polyHIPE, migrating to a maximum depth of 1.4 mm within the matrix. Previous studies used styrene–divinylbenzene polyHIPEs, which are hydrophobic and therefore exhibit reduced permeability to aqueous fluids. In the current study, we used sulphonated polyHIPEs, which are hydrophilic and show increased permeability to aqueous fluids. Sulphonated polyHIPEs (SPHPs) were used to immobilize *B. subtilis* spores, which were then germinated and cultured in situ. After 24-h continuous culture, a dense layer of *B. subtilis* cells was seen on the top surface of all the SPHPs matrices, irrespective of pore size. SEM analysis showed that the cells were elongated rods with little evidence of an extracellular matrix or cell lysis. This result indicates that the *B. subtilis* spores can be efficiently germinated and cultured throughout hydrophilic polymeric supports that provide a permissive microenvironment. A thick homogeneous layer of cells on the top surface of the polymeric supports is likely to be due to the presence of high nutrient and oxygen concentrations in this location. The reduction in cell concentration within the matrix probably reflects the fact that nutrient and oxygen concentrations are likely to decrease as the culture medium passes through the matrix. The production and excretion of extracellular substances, such as polysaccharides, are other parameters that affect the diffusive resistance in a cell mass (Karel et al. 1985). However, in this study, virtually no exopolysaccharide matrix was formed, and there was little or no evidence of cell debris resulting from cell lysis. This indicates that the microporous architecture of the polymeric matrix provided efficient dynamic flow conditions and presumably the continuous removal of dead and degraded cells and toxic materials. The only minor exception was the sulphonated polyHIPE, with pore size of 30 μm (SPHP-6), which was found to contain small amounts of cell debris. The results obtained have shown that the microarchitecture of the matrices is important for the growth, viability and penetration of cells, and was dependent on the pore size and the physico-chemical properties of the matrix. The volumetric productivity of α-amylase of the immobilized cells was significantly higher than that of the batch cell cultures if calculated on the basis of the volume occupied by the SPHP matrix. However, the volumetric productivities of α-amylase per used total nutrient medium of immobilized cells on the SPHP matrices were relatively low (<1) compared to that of the batch cell cultures. Consequently, the overall yield of α-amylase by immobilized cell cultures was relatively low compared to batch cell cultures.

Conclusion

The sulphonated polyHIPE matrix with a pore size of 36 μm was observed to give better cell growth, with cells penetrating throughout the matrix. The volumetric productivity of immobilized cells per volume occupied by SPHP matrix was 7.6-fold higher than the batch cell cultures, but its productivity was relatively low (<1) if determined on the basis of used total volume of medium.

Acknowledgments This work was funded by Ministry of Higher Education (MOHE), Malaysia. I would like to thank Tracey Davey and Pauline Carrick for their technical help.

References

Akay, G., Birch, M. A. and Bokhari, M. A. (2004). Microcellular polyHIPE polymer supports osteoblast growth and bone formation in vitro *Biomaterials*, *25*(18), 3991–4000.

Akay, G. (2005). Bioprocess and chemical process intensification. In S. Lee (Ed.), *Encyclopedia of chemical processing*. New York: Taylor & Francis Group, CRC Press.

Akay, G., Dogru, M., Calkan, B., & Calkan, O. F. (2005a). Flow induced phase inversion phenomenon in process intensification and micro-reactor technology. In Y. Wang & J. D. Holladay (Eds.), *Microreactor technology and process intensification (ACS Symposium Series, vol. 914)*.

Akay, G., Erhan, E., & Keskinler, B. (2005b). Bioprocess intensification in flow-through monolithic microbioreactors with immobilized bacteria. *Biotechnology and Bioengineering, 90*(2), 180–190.

Antelmann, H., Tjalsma, H., Voigt, B., Ohlmeier, S., Bron, S., & Van Dijl, J. M, et al. (2001). A proteomic view on genome-based signal peptide predictions. *Genome Research, 11*(9), 1484–1502.

Argirakos, G., Thayanithy, K., & Wase, D. A. J. (1992). Effect of immobilization on the production of alpha- amylase by an industrial strain of *Bacillus-Amyloliquefaciens*. *Journal of Chemical Technology and Biotechnology, 53*(1), 33–38.

Bokhari, M. A., Akay, G., Zhang, S., & Birch, M. A. (2005). The enhancement of osteoblast growth and differentiation in vitro on a peptide hydrogel -polyHIPE polymer hybrid material. *Journal of Biomaterials, 26*, 5198–5208.

Duran-Paramo, E., Garcia-Kirchner, O., Hervagault, J. F., Thomas, D., & Barbotin, J. N. (2000). α-Amylase production by free and immobilized Bacillus subtilis. *Applied Biochemistry and Biotechnology—Part A Enzyme Engineering and Biotechnology, 84–86*, 479–485.

Dobreva, E., Tonkova, A., Ivanova, V., Stefanova, M., Kabaivanova, L., & Spasova, D. (1998). Immobilization of Bacillus licheniformis cells, producers of thermostable α-amylase, on polymer membranes. *Journal of Industrial Microbiology and Biotechnology, 20*(3), 166–170.

Erhan, E., Yer, E., Akay, G., Keskinler, B., & Keskinler, D. (2004). Phenol degradation in a fixed-bed bioreactor using micro-cellular polymer-immobilized *Pseudomonas syringae*. *Journal of Chemical Technology and Biotechnology, 79*(2), 195–206.

Harwood, C. R. and Archibald, A. R. (1990) Growth, maintenance and general techniques', in Colin R. Harwood, S. M. C.(ed), *Molecular Biological Methods for Bacillus*. Wiley: Chichester : New York.

Karel, S. F., Libicki, S. B., & Robertson, C. R. (1985). Immobilization of whole cells: Engineering principles. *Chemical Engineering Science, 40*(8), 1321–1354.

Kinoshita, S., Okada, H., & Terui, G. (1968). On the nature of α-amylase forming system in *Bacillus subtilis. Fermentation Technology, 46*, 427–436.

Konsoula, Z., & Liakopoulou-Kyriakides, M. (2006). Thermostable alpha-amylase production by *Bacillus subtilis* entrapped in calcium alginate gel capsules. *Enzyme and Microbial Technology, 39*(4), 690–696.

Palva, I. (1982). Molecular cloning of a-amylase gene from *Bacillus amyloliquefaciens* and its expression in *B. subtilis*. *Gene, 19*(1), 81–87.

Production of Extracellular 5-Aminolevulinic Acid by *Rhodopseudomonas Palustris* in Solid-State Fermentation

P. Y. Ong, C. T. Lee, M. R. Sarmidi, H. M. Awad, L. S. Chua and F. Razali

Abstract Nowadays, 5-aminolevulinic acid (ALA) is widely used as non-toxic herbicide, as insecticide, and also as plant growth enhancer in agricultural application. However, their use is limited by high production cost via the route of chemical synthesis. Agricultural waste, which serves as inexpensive solid support medium, can be used to produce ALA through solid-state fermentation (SSF). In this study, production of ALA using empty fruit bunch (EFB) as solid support medium inoculated with *Rhodopseudomonas palustris* (*R. palustris*) NRRL-B4276 was conducted through SSF. Several environmental factors for the SSF process including the initial pH (2, 3, 4, 5, and 6), initial moisture content (40, 50, 60, 70, and 80 %) of the solid support medium, and incubation period (1–5 days) were investigated to study their effects for the production of ALA in SSF. The maximal yield of ALA was achieved at 38.22 ± 0.15 mg/kg substrate at the total viable cell count of $35 \pm 3 \times 10^6$ CFU/g in SSF using EFB as the solid support medium inoculated with *R. palustris*. The optimized operating parameters including the initial pH, initial moisture content of the solid support medium, and incubation period were found to be pH 3, 60 %, and 3 days (72 h), respectively, in the presence of 40 % (v/w) inoculum at 37 °C aerobically with light under static condition.

P. Y. Ong · C. T. Lee (✉) · M. R. Sarmidi · F. Razali
Department of Bioprocess Engineering, Faculty of Chemical Engineering, Universiti Teknologi Malaysia (UTM), 81310 Johor Bahru, Johor, Malaysia
e-mail: ctlee@utm.my

M. R. Sarmidi · H. M. Awad · L. S. Chua
Institute of Bioproduct Development, Universiti Teknologi Malaysia (UTM), 81310 Johor Bahru, Johor, Malaysia

H. M. Awad
Chemical of Natural and Microbial Product Department, National Research Center, El-Bohouth Street, Dokki 12622 Cairo, Egypt

R. Pogaku et al. (eds.), *Developments in Sustainable Chemical and Bioprocess Technology*, DOI: 10.1007/978-1-4614-6208-8_22,

Introduction

5-aminolevulinic acid (ALA) is a biosynthetic precursor metabolite for the production of tetrapyrrole compounds such as chlorophyll, vitamin B12, and heme (Choorit et al. 2011) that are essential elements for plant growth. To date, ALA is widely applied in diverse fields especially in the agricultural area as effective plant growth promoter under normal or stress condition, and also as non-toxic biodegradable herbicide and insecticide (Saikeur et al. 2009). Due to high demand of ALA for agricultural application, a wide diversity of production methods were studied and developed (Kang et al. 2011). The conventional production of ALA by chemical synthesis involves high cost, energy, and time-consuming. Therefore, in this study, the production of ALA using the single strain of *Rhodopseudomonas palustris* (*R. palustris*) inoculated on the solid support medium, i.e., the empty fruit bunch (EFB) under solid-state fermentation (SSF) process was studied. SSF is typically defined as a microbial cultivation process on solid substrate medium in the absence (or near absence) of water. Nowadays, SSF has raised a fresh attention and becoming a modern technology even though it was introduced since eighteenth century (Pandey et al. 2008). The SSF process is able to lead to higher productivities and yields, better quality of products, low capital and operating costs, and simple equipments (Mirón et al. 2010). SSF is expected to offer better stability for the growth of beneficial microorganism and the production of bioproducts such as ALA. The fermented solids can also be applied directly without tedious downstream separation processes to yield the bioproduct. Importantly, EFB has been extensively reported as one of the inexpensive local agricultural waste that could serve as the solid substrates and as the natural habitat for microbial growth and composting application (Mohammad et al. 2012). From the previous studies, ALA is extensively produced by recombinant *Escherichia coli*, algae, classical propionic acid bacteria, photosynthetic bacteria such as *Rhodobacter sphaeroides*, *R. palustris* and *Rhodobacter capsulatus* (Kang et al. 2011; Zhang et al. 2008). Among these, *R. palustris* has received much attention and considered as a beneficiary microorganism for agricultural application due to its attractive properties. It is not only serve as a plant growth promoter to increase the yield of fruits and crops, enhancing acid tolerance in plant, but also able to structurally break down the aromatic compounds such as lignin monomer (Kantha et al. 2010; Harwood 2009). However, there was no publication reporting the production of ALA by *R. palustris* through solid-state fermentation (SSF). This study aims to investigate the most suitable culture condition using locally available, inexpensive agrowaste, namely the EFB as the solid support medium to produce ALA by *R. palustris*.

Methodology

Materials

Shredded empty fruit branch (EFB) was kindly provided by Felda Palm Oil Industry (Kulai Penggeli, Johor, Malaysia). Grinded and sieved EFB was then dried using oven at 100 °C for 24 h and stored in freezer at −4 °C. All chemical substances were purchased from Merck (Merck, Darmstadt, Germany) and Sigma-Aldrich (Sigma, St. Louis, Mo, USA).

Preparation of Starter Culture and Solid-State Fermentation

R. palustris NRRL-B4276 was obtained from the National Center for Agricultural Utilization Research (ARS Culture Collection, NRRL). The working culture of *R. palustris* was cultivated and maintained in 50 % glycerol at −80 °C in deep freezer. The starter inoculum of *R. palustris* was cultured in 50 ml GM medium (g/l distilled water: 0.5 yeast extract, 3.7 monosodium glutamate, 0.8 $(NH_4)_2SO_4.7H_2O$, 0.5 K_2HPO_4, 0.5 KH_2PO_4, 0.2 $MgSO_4.7H_2O$) with an initial pH of 7 and incubated in orbital shaker incubator at 37 °C at 200 rpm for 24 h. Cells were harvested, rinsed twice with 20 ml of sterilized phosphate buffer saline (PBS) solution and resuspended with the same amount of PBS solution by centrifugation at 10,000× g for 15 min. The resuspended cell served as the starter inoculum and its total cell concentration was adjusted to $OD_{660nm} \approx 1.0$ and maintained at 4 °C prior to each experiment.

Five grams of sieved and air-dried EFB as the solid substrate medium was prepared in 250-ml Erlenmeyer flask and moisturized with modified GM as basal media to achieve a final relative humidity of 60 %. After sterilized at 121 °C for 15 min, the flasks were cooled and inoculated with 2 ml of the starter inoculum (40 % v/w) and incubated aerobically with a low light intensity of 1,000 Lux at 37 °C for 48 h under static condition. Low intensity of light is known to serve as an energy source for the growth of *R. palustris* as high density of light would impose negative effect on cell growth (Klug and Masuda 2008); in addition, most of the previous fermentation studies on *R. palustris* were conducted under the condition with light (Choorit et al. 2011; Saikeur et al. 2009).

Effect of Initial pH, Moisture Content, and Incubation Period

ALA production by SSF was optimized using classical factor-by-factor method. The initial pH of the basal media (2, 3, 4, 5, and 6) was adjusted using 2 M NaOH or 2 M HCl before being mixed into the solid substrate medium. The SSF experiments were conducted under the same conditions throughout the experiment to study the effect of initial pH on the production of ALA. The optimum initial pH of the solid substrate determined through this step was used for other subsequent experiments. To investigate the effect of the initial moisture content of the solid substrate on the ALA yield, different initial moisture contents were prepared for the EFB solid substrate medium (40, 50, 60, 70, and 80 %) and were adjusted using the GM basal media. The optimized initial pH and moisture content (%) were then fixed for the investigation of the effect of incubation period on the yield of ALA. Different incubation periods (1, 2, 3, 4, and 5 days) were applied to study the effect of incubation period on ALA production. All experiments were in duplicate and the mean values reported.

Extraction and Determination of 5-aminolevulinic Acid

ALA was extracted by vigorously mixing 50 ml of acetate buffer (pH 4.7) into each fermented sample and agitated at 200 rpm, room temperature for 30 min on the orbital shaker. The fermented slurry was then filtered through Whatman No.3 filter paper, and the resulted extract was centrifuged at 5,000 x g for 5 min. The clear supernatant was then stored at −4 °C for further end-product analysis within 24 h. One milliliter of clear supernatant was mixed with 35 μl of acetylacetone in a test tube. The mixed solution was then boiled in water bath for 15 min and cooled to room temperature (30 ± 2 °C). 700 μl of modified Ehrlich's reagent consisted of p-dimethylaminobenzaldehyde (1 g), glacial acetic acid (42 ml), and 70 % perchloric acid (8 ml) was then mixed to each sample. The concentration of ALA was measured by spectrophotometer at an absorbance wavelength of 556 nm, followed by 10 min of reaction at room temperature (Saikeur et al. 2009).

Determination of Cell Growth

The growth of *R. palustris* was determined by mixing thoroughly the fermented sample with 50 ml of sterile distilled water and shaken for 10 min on orbital shaker, followed by serial dilutions and poured plate using nutrient agar. The serial dilution nutrient agar plates were then incubated at 37 °C for 24 h. The viable colonies of *R. palustris* colonies in the fermented samples were then counted and expressed as colony-forming unit (CFU) per gram of fermented substrate.

Results and Discussion

Effect of Initial pH, Moisture Content, and Incubation Period on the Production of ALA

pH is one of the most important parameters for the cellular and metabolic activities and thereby affects the production of the metabolites in SSF. The yield of ALA gradually increased and achieved the optimal level of 33.82 mg/kg substrate at pH 3.0 (Fig. 1). This observation was significant as ALA was reported to be chemically stable at relatively low pH and *R. palustris* possess an acid-tolerant characteristic (Choorit et al. 2011). However, increment of pH from 3 to 6 resulted in the reduction in ALA production. It was widely reported that the metabolic activities of many microorganisms are sensitive to pH change (Ellaiah et al. 2004).

For the effect of the initial moisture content, a highest yield of ALA (35.57 mg/kg substrate) was achieved at the initial moisture content of 60 % as shown in Fig. 2. According to Rodriguez-Leon et al. (2008), the moisture content (%) of substrates plays a critical role in the SSF process to affect microbial activity and the production rate of metabolites (Rodriguez-Leon et al. 2008). A decrease in the

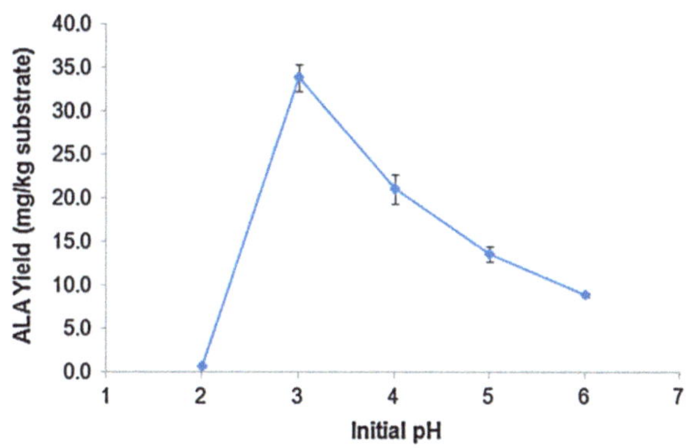

Fig. 1 Effect of initial pH on the production of ALA in SSF

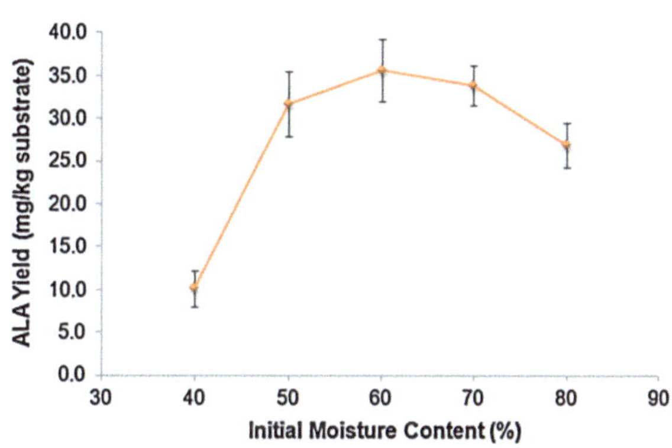

Fig. 2 Effect of initial moisture content on the production of ALA in SSF

Table 1 Kinetics of ALA production by *R. palustris* under SSF

Incubation period (days)	Viable cell count (CFU/g)			Final pH	ALA yield (mg/kg)
	10^2	10^4	10^6		
1	TNTC	TNTC	50 ± 2	4.42	17.80 ± 4.04
2	TNTC	TNTC	36 ± 4	3.97	37.63 ± 0.57
3	TNTC	TNTC	35 ± 3	4.20	38.22 ± 0.14
4	TNTC	TNTC	33 ± 3	4.45	33.33 ± 3.75
5	TNTC	TNTC	30 ± 5	4.44	24.75 ± 2.60

yield of ALA was observed when the initial moisture content was lower or higher than the optimum level. It is likely that at lower moisture content, the solubility and availability of nutrient to the culture were limited by heat exchange and oxygen transfer. Conversely, if the moisture content of the substrate medium in SSF was too high, it reduces the mass transfer process in terms of the diffusion rate of solute or gas and also the porosity within the solid matrixes in the medium during the SSF (Ellaiah et al. 2004; da Silva et al. 2010).

The production of ALA with the optimized parameters including pH 3, 60 % moisture content was set for the investigation of the last parameter, with the incubation period of up to 5 days. The maximal yield of ALA was obtained on day 3 (72 h), i.e., 38.22 ± 0.14 mg/kg substrate at a total viable cell count of $35 \pm 3 \times 10^6$ CFU/g as shown in Table 1. The yield of ALA decreased beyond this incubation period. It was likely that *R. palustris* produced ALA within the first 3 days (72 h) and beyond which there could be further transformation of ALA to other secondary metabolites. This process is typically blocked using an inhibitor such as levulinic acid (Saikeur et al. 2009). It was not used in this study as it was not the scope of the study.

Conclusions

The maximum production of ALA (38.22 ± 0.14 mg/kg substrate) with the total viable cell count of $35 \pm 3 \times 10^6$ CFU/g was achieved in an SSF system using EFB as the solid substrate medium and *R. palustris* as the culture. The optimized operating parameters including the initial pH, initial moisture content of the solid substrate medium, and incubation period were found to be pH 3, 60 %, and 3 days (72 h), respectively, in the presence of inoculum level of 40 % (v/w) at 37 °C aerobically with light under static condition. However, further investigation of other environmental parameters such as inoculum density, temperature, and substrate selection is highly recommended.

Acknowledgments This work is supported by Universiti Teknologi Malaysia (UTM) and the research facility in Institute of Bioproduct Development (IBD), UTM. This research work is funded by the Malaysian Ministry of Higher Education under the Fundamental Research Grant Schemes (FRGS) (Vote No: 4F079), UTM Research University Grant (GUP) (Vote No: Q.J130000.7125.00H73), and *MyBrain15-MyPhD* Program under the Ministry of Higher Education, Malaysia.

References

Choorit, W., Saikeur, A., Chodok, P., Prasertsan, P., & Kantachote, D. (2011). Production of biomass and extracellular 5-aminolevulinic acid by *Rhodopseudomonas palustris* KG31 under light and dark conditions using volatile fatty acid. *Journal of Bioscience and Bioengineering, 111*, 658–664.

da Silva, L. C. A., Honorato, T. L., Franco, T. T., & Rodrigues, S. (2010). Optimization of Chitosanase production by *Trichoderma koningii sp.* Under solid-state fermentation. *Food and Bioprocess Technology*, 1–9.

Ellaiah, P., Srinivasulu, B., & Adinarayana, K. (2004). Optimisation studies on neomycin production by a mutant strain of *Streptomyces marinensis* in solid state fermentation. *Process Biochemistry, 39*, 529–534.

Harwood, C. S. (2009). Degradation of aromatic compounds by purple nonsulfur bacteria. In C. N Hunter, F. Daldal, M. C. Thurnauer & J. T. Beatty (Eds.), *The purple phototrophic bacteria.* New York: Springer Science.

Kang, Z., Wang, Y., Gu, P., Wang, Q., & Qi, Q. (2011). Engineering *Escherichia coli* for efficient production of 5-aminolevulinic acid from glucose. *Metabolic Engineering, 13*, 492–498.

Kantha, T., Chaiyasut, C., Kantachote, D., Sukrong, S., & Muangprom, A. (2010). Selection of photosynthetic bacteria producing 5-aminolevulinic acid from soil of organic saline paddy fields from the Northeast region of Thailand. *African Journal of Microbiology Research, 4*(17), 1848–1855.

Klug, G., & Masuda, S. (2008). Regulation of genes by light. In C. N. Hunter, F., Daldal, M. Thurnauer & J. T. Beatty (Eds.), *The purple phototrophic bacteria.* Netherlands: Springer.

Mirón, J., Vázquez, J. A., González, P., & Murado, M. A. (2010). Enhancement glucose oxidase production by solid-state fermentation of Aspergillus Niger on polyurethane foams using mussel processing wastewaters. *Enzyme and Microbial Technology*, *46*, 21–27.

Mohammad, N., Alam, M. Z., Kabbashi, N. A., & Ahsan, A. (2012). Effective composting of oil palm industrial waste by filamentous fungi: A review. *Resources, Conservation and Recycling, 58*, 69–78.

Pandey, A., Soccol, C. R., & Larroche, C. (2008). Introduction. In A. Pandey, C. R. Soccol, & C. Larroche (Eds.), *Current developments in solid-state fermentation.* New York: Springer.

Rodriguez-Leon, J. A., Soccol, C. R., Pandey, A., & Rodriguez, D. E. (2008). In: A. Pandey, C. R. Soccol & C. Larroche (Eds.) Factors affecting solid-state fermentation current developments in solid-state fermentation. New York: Springer.

Saikeur, A., Choorit, W., Prasertsan, P., Kantachote, D., & Sasaki, K. (2009). Influence of precursors and inhibitor on the production of extracellular 5-aminolevulinic acid and biomass by *Rhodopseudomonas palustris* kG31. *Bioscience, Biotechnology, and Biochemistry, 73*, 987–992.

Zhang, W. F., Zhang, F., Raziuddin, R., Gong, H. J., Yang, Z. M., Lu, L., et al. (2008). Effects of 5-aminolevulinic acid on oilseed rape seedling growth under herbicide toxicity stress. *Journal of Plant Growth Regulation, 27*, 159–169.

Experimental Investigation of Combustion Behavior of Flash Pyrolysis Oil

N. Ibrahim, P. A. Jensen, K. Dam-Johansen, M. K. A. Hamid, R. M. Kasmani and R. R. Ali

Abstract The aim of this work is to investigate the combustion behavior of pyrolysis oils derived from wheat straw and pine wood. The technique of thermogravimetric analysis (TGA) was applied to study the thermal treatment of the pyrolysis oils under well-controlled temperature in an oxidative (O_2) and a non-oxidative (N_2) environment. The thermogravimetric curves indicated that the combustion process of bio-oil was divided into three stages. It was found that the removal of water and evaporation of light volatiles occur at room temperature up to 200 °C, the cracking of heavy fraction occurs at temperatures of between 200 and 500 °C and, finally, the char combustion occurs at temperatures above 500 °C. The thermal degradation of the both bio-oil samples showed an almost similar behavior in TGA experiment.

N. Ibrahim (✉) · R. M. Kasmani
Gas Engineering Department, Faculty of Petroleum and Renewable Energy Engineering, Universiti Teknologi Malaysia, 81310 Johor Bahru, Johor, Malaysia
e-mail: norazana@petroleum.utm.my

R. M. Kasmani
e-mail: rafiziana@petroleum.utm.my

P. A. Jensen · K. Dam-Johansen
Combustion and Harmful Emission Control (CHEC), Department of Chemical and Biochemical Engineering, Technical University of Denmark, DK-2800, Kgs. Lyngby, Denmark

M. K. A. Hamid
Process Systems Engineering Centre (PROSPECT), Faculty of Chemical Engineering, Universiti Teknologi Malaysia, 81310 Johor Bahru, Johor, Malaysia
e-mail: kamaruddin@cheme.utm.my

R. R. Ali
Polymer Engineering Department, Faculty of Chemical Engineering, Universiti Teknologi Malaysia, 81310 Johor Bahru, Johor, Malaysia
e-mail: roshafima@cheme.utm.my

R. Pogaku et al. (eds.), *Developments in Sustainable Chemical and Bioprocess Technology*, DOI: 10.1007/978-1-4614-6208-8_23,

Introduction

Bio-oil is produced by rapid quenching and condensation of vapors released from flash pyrolysis of biomass. It is a highly viscous liquid containing various oxygenated hydrocarbons such as acids, alcohols, aldehydes, esters, ketones, phenols, furans, alkenes, aromatics, and sugar. Bio-oil is also characterized by a relatively high water and oxygen content, low heating value, typically from 16 to 19 MJ/kg (wet basis), and high density and acidity. The density of the liquid bio-oil is about 1,200 kg/m^3, which is much higher than that of the original biomass. In addition, low sulfur and nitrogen content in the bio-oil compared to the feedstock results in a small impact on the environment when the oil is combusted (Bridgwater and Peacoke 2004; Acikgoz and Kockar 2007).

Since the 1990s, interest has developed in the use of bio-oil to substitute heavy fossil fuel oil in boilers to produce heat and electricity (Solantausta et al. 1993, 1994, 1995; Oasmaa et al. 1997, 2001; Gust 1997; Bridgwater and Peacoke 2004). Several researchers have performed combustion tests using different size of boilers (Solantausta et al. 1995), internal combustion engines (Solantausta et al. 1993, 1994, 1995), and gas turbines (Solantausta et al. 1995; Lopez and Monfort 2000). Experiments on bio-oil single-droplet combustion have also been conducted by thermogravimetric techniques (Garcia-Perez et al. 2006), by fiber-suspended droplets (D'Alessio et al. 1998; Calabria et al. 2007), and in entrained flow reactors (EFR) (Shaddix and Tennison 1998; Garcia-Perez et al. 2006).

Garcia-Perez et al. (2006) conducted thermogravimetric tests on two types of bio-oils heated at room temperature up to 750 °C at a heating rate of 10 °C/min in nitrogen and air flow rate of 150 mL/min. Three major stages were identified from the TG–DTG curves during the bio-oil combustion, which begins with the evaporation of light oil fractions (up to 300 °C), followed by the cracking of heavy oil fractions (above 300 °C) in the second stage. Then, in the final step, combustion of a solid residue (400–500 °C). They found that almost similar thermal behavior of all tested bio-oil samples was observed in the thermogravimetric analysis (TGA) experiments.

The objective of this chapter is to present the results of experimental study on combustion of pyrolysis oils derived from wheat straw and pine wood. In this study, the technique of TGA was applied to analyze the thermal treatment of the pyrolysis oils under well-controlled temperature in an oxidative (O_2) and non-oxidative (N_2) environment.

Methodology

Flash Pyrolysis Oil Production

In this study, two types of bio-oils generated from flash pyrolysis of pine wood and wheat straw were used. Pyrolysis experiments were carried out in a bench-scale pyrolysis centrifuge reactor (PCR) as shown in Fig. 1. The PCR facility consists of a screw-type feeder, reactor, condenser, coalescer, and char separator. The principle of the process is to apply tangentially feeding of solid biomass particles into the horizontally oriented Ø 82 × 200 mm tubular reactor. The centrifugal force created by the rotation of a three-blade rotor keeps the particles sliding on the heated wall while passing through the reactor. The solid particles pyrolyze while moving down the reactor pipe and left suspended in the gas through the tangential outlet. Larger char particles are removed by a change-in-flow separator, whereas fines are collected by a cyclone. Vapors are condensed in a direct water-cooled condenser (bubble chamber) filled with previously produced bio-oil. Aerosols that are not retained by the condenser are collected in a coalescer filled with rockwool (mineral fibers).

The biomass feedstock was fed to the reactor at a flow of 20–23 g/min and with an approximately gas reactor residence time of 0.3–0.5 s, and an initial solid heating rate of approximately 250–1,000 °K/s. All measurements are taken to determine the yield of char, gas, water, and oil for every single experiment.

The acidity of the bio-oils was measured using a MeterLab standard pH meter PHM210 instrument. A calibration using pH 4.005 and 7.000 standard buffer solutions was carried out prior to the tests. Only room temperature measurements were taken.

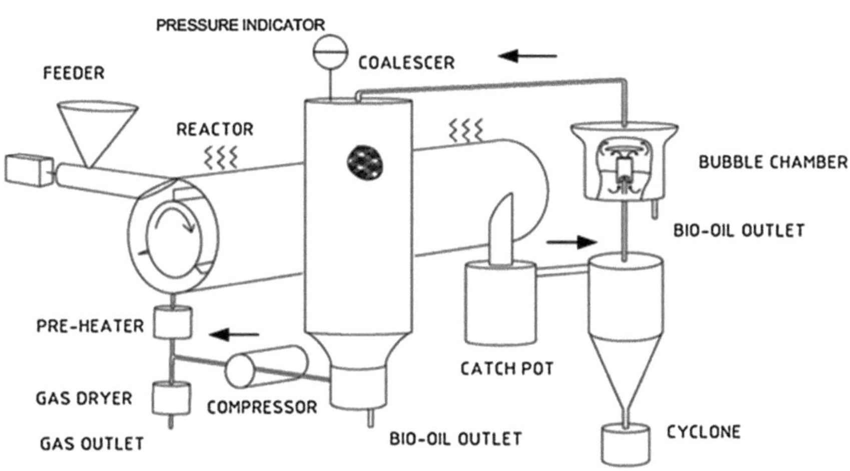

Fig. 1 Schematic diagram of pyrolysis centrifuge reactor (PCR)

Table 1 Flash pyrolysis oils physicochemical properties

Properties	Units	Pine wood oil	Wheat straw oil
Water content	wt.%	16.4	17.8
pH		3.04	3.20
Ash	wt.%	1.2	1.6
Density (20 °C)	g/cm^{-3}	1.23	1.20
Dynamic viscosity	cP		
At 20 °C		195.8	91.4
At 40 °C		47.8	27.4
Heating value (wet basis)	MJ/kg	16.7	16.4
Elemental analysis	% (db)		
C		50.7	50.2
H		6.4	6.2
O		42.9	43.6

The water content of the bio-oils was analyzed by a Karl Fischer titration (KFT) method using a Metrohm 701 KF Titrino analyzer.

The viscosity of the bio-oils was measured using an automated microviscometer AMV 200 instrument at 20 and 40 °C. The viscosity measurement temperatures of 20 and 40 °C were chosen for comparison with the petroleum fuel oil specifications.

The higher heating value (HHV) of the bio-oil sample was measured using a bomb calorimeter IKA C200 instrument with the DIN 51900-1 standard. The vessel was pressurized at 30 bar of oxygen. The uncertainty of the measurements is about 120 J/g. For calculated elementary analysis (C, H, O), the correlation method proposed by Bech et al. (2009) was employed.

The physical and chemical properties of the wheat straw bio-oil and pine wood bio-oil are summarized in Table 1.

Simultaneous Thermal Analysis

The thermal behaviors of pine wood oil and wheat straw oil were carried out using a NETZSCH simultaneous thermal analysis (STA) 449F1 thermogravimetric analyzer (TGA) coupled with differential scanning calorimetry (DSC). The TGA–DSC facilitated the acquisition of weight loss and heat flow data as a function of temperature or time. The experimental works were conducted at a heating rate of 10 °C min^{-1}, and the samples were heated at room temperature up to 700 °C in a 10 % of oxygen and nitrogen gas flow of 150 mL/min.

Results and Discussion

TGA is an analysis technique that is widely used to evaluate the thermal behavior of fuels in terms of evaporation and combustion characteristic. Figures 2 and 3 show the TG weight loss signal and DTG curves for the bio-oils from pine wood and wheat straw, respectively, heated at a heating rate of 10 °C/min under nitrogen (pyrolysis) and 10 % of oxygen (combustion) environment. TGA curves display weight losses as a function of temperature in controlled atmosphere, whereby DTG curves emphasize the zone of reaction where various reaction steps are taking place over the entire temperature range.

Figure 2 shows the TG curve of weight change of bio-oil samples as a function of temperature. Almost similar thermal behaviors were obtained for both bio-oil samples as seen in Figs. 2a and b. In general, three different regions can be distinguished from a TG curve and shown by the dotted vertical lines. The first phase (temperature below 200 °C) corresponds to the removal of water and light volatiles, where the weight loss of approximately 40 % was observed for both tested bio-oils. This behavior indicates that these bio-oils are characterized by a higher content of light fractions that is easily vaporized at temperature lower than 200 °C. In the second phase, occurring at temperature between 200 and 500 °C, the drop in weight of the samples is observed which is due to the release of light components by cracking processes of heavy hydrocarbons (carboxylic acids, furfural, hydroxyacetaldehyde, methoxy- and dimethoxyphenol compounds) (De Wild et al. 2008). At this stage, weights of the bio-oils are reduced to below 40 %. Finally, at the temperature of above 500 °C, the combustion of char residue occurred in the presence of oxygen (Phase III).

The DTG curves (Fig. 3) confirm these observations, showing two maximum distinct peaks. The first maximum temperature peak rate of bio-oil from straw and wood occurred at almost similar temperature of 162 °C, and the second maximum peak rate of straw bio-oil and wood bio-oil attained at the temperature of 495 and 470 °C, respectively. The first peak represents the evolution of water and volatilization of light hydrocarbons from the bio-oil, whereas the second peak

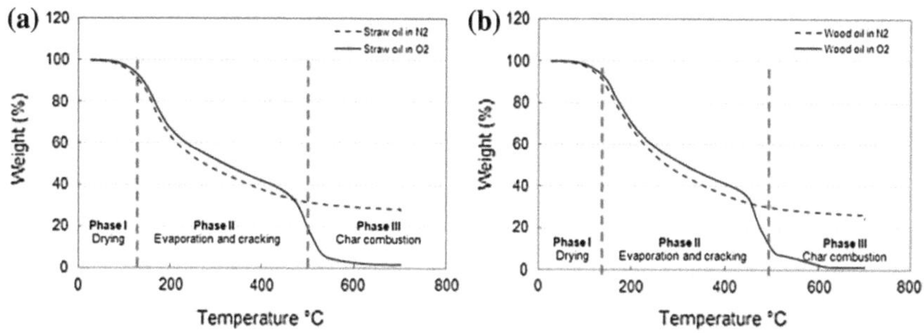

Fig. 2 TGA curves of the pyrolysis oils **a** straw oil and **b** wood oil, respectively, in N_2 and O_2 environment

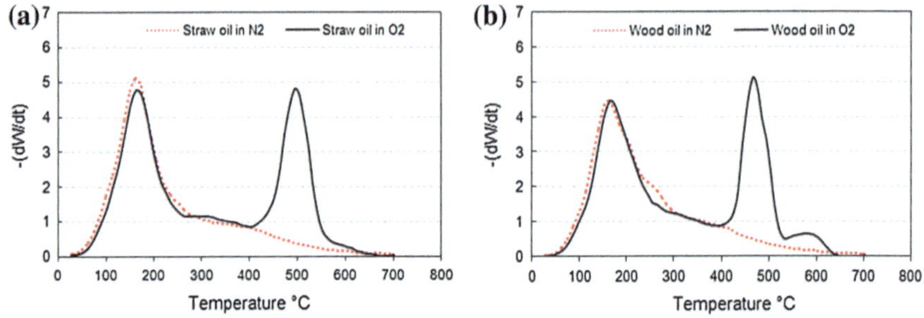

Fig. 3 DTG curves of the pyrolysis oils **a** straw oil and **b** wood oil, respectively, in N_2 and O_2 environment

corresponds to the combustion of the residue. The combustion profile (in oxygen) diverges initially from the evaporation profile (in nitrogen) at 400 °C.

At the final temperature of 700 °C, the remaining solid residues were about 1.7 and 1.3 wt.% for the straw oil and the wood oil, respectively, in the presence of oxygen. On the other hand, the solid residues remaining after 700 °C in nitrogen gas were higher (i.e., 27 wt.% for the straw oil and 25 wt.% for the wood oil). In addition, the remaining solid residue for straw oil is higher as compared to the wood oil, indicating that the straw oil contains a higher amount of ash.

Conclusions

In this study, the TGA was applied to analyze the thermal treatment of the pyrolysis oils under well-controlled temperature in an oxidative (O_2) and non-oxidative (N_2) environments. Drying, evaporation of light components, and cracking of heavy fractions and char combustion processes were identified through TGA–DTG curves for the tested bio-oil samples. An almost similar thermal behavior was observed for the both bio-oil samples in TGA experiments. It can be seen that the removal of water and evaporation of light hydrocarbons occurred below 200 °C, release of light components by cracking of heavy hydrocarbons in the bio-oil occurs at the temperature above 200 up to 500 °C, and, finally, the char combustion occurs at temperatures above 500 up to 700 °C. On the other hand, the solid residues remaining after 700 °C in non-oxidative environment were higher compared to those in oxidative environment. In addition, the remaining solid residue for straw oil is higher as compared to the wood oil, indicating that the straw oil contains a higher amount of ash.

Acknowledgments The author acknowledges the financial support of the Universiti Teknologi Malaysia (UTM) (Q.J130000.2642.05J10) and Ministry of Higher Education (MOHE) of Malaysia (R.J130000.7842.4F129).

References

Acikgoz, C., & Kockar, A. M. (2007). Flash pyrolysis of Linseed (Linum usitatissimum L.) for production of liquid fuels. *Journal of Analytical and Applied Pyrolysis, 78*, 406–412.

Bech, N., Jensen, P. A., & Dam-Johansen, K. (2009). Determining the elemental composition of fuels by bomb calorimetry and the inverse correlation of HHV with elemental composition. *Biomass and Bioenergy, 33*, 534–537.

Bridgwater, A. V., & Peacoke, G. V. C. (2004). Fast pyrolysis processes for biomass. *Renewable and Sustainable Energy Reviews, 4*, 1–73.

Calabria, R., Chiariello, F., & Massoli, P. (2007). Combustion fundamentals of pyrolysis oil based fuels. *Experimental Thermal and Fluid Science, 31*, 413–420.

de Wild, P. J., den Uil, H., Reith, H., Kiel, J. H. A., & Heeres, H. J. (2008). Biomass valorisation by staged degasification. A new pyrolysis-based conversion option to produce value-added chemicals from lignocellulosic biomass. *Journal of Analytical and Applied Pyrolysis, 85*, 124–133.

D'alessio, J., Lazzaro, M., Massoli, P., & Moccia, V. (1998). Thermo-optical investigation of burning biomass pyrolysis oil droplets. Twenty-seventh Symposium (International) on Combustion/The Combustion Institute (pp. 1915–1922).

Garcia-Perez, M., Lappas, P., Hughes, P., Dell, L., Chaala, A., Kretschmer, D., et al. (2006). Evaporation and combustion characteristics of biomass vacuum pyrolysis oils. *IFRF Combustiion Journal, 200601*, 1–27.

Gust, S. (1997). Combustion experiences of flash pyrolysis fuel in intermediate size boilers. In A. V. Bridgwater & D. G. B. Boocock (Eds.), *Developments in thermochemical biomass conversion* (pp. 481–488). London: Blackie Academic & Professional.

Lopez, J. G., & Monfort, S. J. J. (2000). Preliminary test on combustion of wood derived fast pyrolysis oils in a gas turbine combustor. *Biomass and Bioenergy, 19*, 119–128.

Oasmaa, A., Leppamaki, E., Koponen, P., Levander, J., & Tapola, E. (1997). *Physical characterization of biomass-based pyrolysis liquids*. Application of standard fuel oil analyses; VTT Publication 306. Espoo, Finland: VTT.

Oasmaa, A., Kytö, M., & Sipilä, K. (2001). Pyrolysis oil combustion tests in an industrial boiler. In Bridgwater, A. V. (Eds.), *Progress in thermochemical biomass conversion*. Oxford, UK: Blackwell Science.

Shaddix, C. R., & Tennison, P. J. (1998). Effects of char content and simple additives on biomass pyrolysis oil droplet combustion. Twenty-seventh Symposium (International) on Combustion/The Combustion Institute (pp. 1907–1914).

Solantausta, Y., Nylund, N. O., Westerholm, M., Koljonen, T., & Oasmaa, A. (1993). Wood pyrolysis liquid as fuel in a diesel power plant. *Bioresource Technology, 46*, 177–188.

Solantausta, Y., Nylund, N. O., & Gust, S. (1994). Use of pyrolysis oil in a test diesel engine to study the feasibility of a diesel power plant concept. *Biomass and Bioenergy, 7*, 297–306.

Solantausta, Y., Bridgwater, A. V., & Beckman, D. (1995). Feasibility of power production with pyrolysis and gasification systems. *Biomass and Bioenergy, 9*, 257–269.

Identification of Tetrathionate Hydrolase from *Thiobacillus Ferrooxidans*: An Enzyme Responsible for Enzymatic Devulcanization of Waste Rubber Products

M. D. Azratul and Y. Faridah

Abstract Malaysia is one of the largest producers of rubber and as a result produces million tonnes of vulcanized rubber waste from the rubber-based industries each year. Currently, less than 13 % of the waste tires generated annually are recycled by any means since the processes are costly and very inefficient. In the tire sector, many recycling methods have been adopted such as converting them into other useful rubber-based products, burnt as fuel or retreaded. Retreading is carried out via devulcanization process. The methods of devulcanizations adopted are chemical, mechanical, ultrasonic, microwave, and microbial or more specifically, enzymatic. However, none of the techniques previously developed have been proven to be commercially successful. Enzymatic devulcanization is one of the newest techniques to prepare waste rubber products for recycling. The process is deemed to be more efficient and less expensive compared to other devulcanization methods. Microbial devulcanization using *Thiobacillus ferrooxidans* provides some avenue of solving the problems associated with the other conventional processes. This microbe secretes an enzyme to utilize the elemental or organic sulfur together with iron available in the environment for its growth. The proposed enzyme, tetrathionate hydrolase, is responsible for degrading the sulfur cross-links in the devulcanized rubber rendering it to be more acceptable for recycling. The objective of the study is to identify secreted tetrathionate hydrolase which is responsible for devulcanization of rubber waste products. The condition needs to promote the growth of bacteria is by incubating the bacteria at 25 °C, initial pH buffer of 4, and at an agitation speed of 125 rpm,

M. D. Azratul (✉)
Biomolecular and Bioprocess Engineering Research Unit(BPMERU), Department of Biotechnology Engineering, Faculty of Engineering, International Islamic University Malaysia, P. O. Box 10, 50728 Kuala Lumpur, Malaysia
e-mail: zatulshima_2884@yahoo.com.my

Y. Faridah
Department of Biotechnology Engineering, Faculty of Engineering, International Islamic University Malaysia, P. O. Box 10, 50728 Kuala Lumpur, Malaysia
e-mail: yfaridah@iium.edu.my

R. Pogaku et al. (eds.), *Developments in Sustainable Chemical and Bioprocess Technology*, DOI: 10.1007/978-1-4614-6208-8_24,

with salt concentration as follows: in (g/l) of KH_2PO_4 = 4 g, $(NH_4)_2SO_4$ = 4 g, and $MgSO_4.7H_2O$ = 0.3 g. After preparing the initial sample, it was found that the specific activity of extracellular and intracellular tetrathionate hydrolase occurred on Day 1 and 2 is 0.070 μmol $min^{-1}mg^{-1}$(U/mg) and 0.274 μmol $min^{-1}mg^{-1}$ (U/mg), respectively, at pH 4, 25 °C. Future work includes reaction of purified tetrathionate hydrolase with used rubber to observe the devulcanization process. This work is important mainly because it is an environmental friendly and safe method of devulcanization and with thorough study, it is hoped that the objective to recycle higher percentage of waste rubber can be achieved.

Keywords Enzymatic devulcanization · *Thiobacillus ferrooxidans* · Tetrathionate hydrolase

Introduction

After Thailand and Indonesia, Malaysia is the largest producer of rubber and has greatly contributed to the expansion of rubber technology in the present world. The technology of rubber creates a variety range of products such as rubber gloves, storage tanks, and insulation of electrical devices; however, the most important one is its usage in tire industries. The advancement of tire industries started in 1839 when Charles Goodyear discovered that by boiling crude rubber along with sulfur in solution produced an improved rubber material (Lewin 2004) which results in the production of vulcanized rubber tires. The advantages are good tensile strength, resistance to temperature changes and organic solvents as well as extreme elasticity. However, after these tires are no longer in use, they will be disposed to the environment. The biggest challenges are when the discarded vulcanized rubber tires are not biodegradable due to their sulfur cross-linking and have become one of the major sources of pollution. Disposal of waste tires and its recycling give a big task to mankind because the accumulation of discarded waste tires leads to environmental pollution (Gulzad 2011).

To date, there is no environmentally sound and economically viable method to recycle or dispose-off the vulcanized scrap tires. The only environmentally sound way to dispose vulcanized scrap tires in large quantity is to burn it in the cement kiln (E-Waste Management 2008). Vulcanized scrap tires virtually indestructible if left in the dump site as it cannot biodegrade even in 50 years. Environmental laws had forbidden the burning of scrap tires because burning tires emits a lot of black smoke and harmful chemicals. Dump site for scrap tire heaps is space consuming, fire hazard and health hazards due to mosquito breeding (U.S. Environmental Protection Agency 1993).

Therefore, there is an urgent need to search for a safe recycling method to reduce the scrap tire threats. One of the methods is retreading the tires which it is carried out via devulcanization process which comprises of chemical, mechanical, ultrasonic, microwave, and microbial or enzymatic methods. Many researchers have attempted to discover the most effective way of recycling scrap tires in which they focus on the usage of bacteria to degrade rubber. This study is also known as microbial or enzymatic devulcanization. According to Romine et al. (1997), enzymatic devulcanization happened when a surface of a vulcanized rubber particle is exposed to at least one enzyme, maintaining the exposure for a time sufficient to convert sulfur to sulfoxide or sulfone, and halting conversion or preventing further degradation of the sulfoxide or sulfone. These two products are reactive, thus it is important to stop the reaction once sulfoxide and sulfone have been produced.

The manipulation of bacteria strains in devulcanization process will offer an environmentally and economically effective method of waste tires recycling. The main agents, namely the bacteria from the genus Acidithiobacillus species, have unique properties to catalyze aerobic oxidation of sulfide. These bacteria have been used in many decades in biomining industry for recovery of metals from sulfidic low grade ores and concentrates. *Thiobacillus ferrooxidans* are obligately aerobic, gram-negative, chemoautotrophic organism that generates its energy and reducing power for CO_2 fixation from the oxidation of inorganic iron and reduced sulfur compounds (Ingledew 1982). This microbe has the ability to secrete an enzyme known as tetrathionate hydrolase which is responsible to oxidize organic sulfur with iron that is available in the environment (Rawlings 2005). In addition, this enzyme is able to break the sulfur cross-link in vulcanized rubber which results in the tires being recyclable.

The objective of the study is to identify the secreted tetrathionate hydrolase by *T. ferrooxidans* which is responsible for devulcanization of rubber. This work is based on the previous work where the salt concentration as well as several process conditions such as temperature, initial pH of buffer, and agitation speed for the bacteria had been optimized by Yusof and Ahmad (2010) to promote higher growth of bacteria.

Methodology

Materials and Methods

Propagation of Bacteria

T. ferrooxidans (ATCC® 19859™) was secured from American type culture collection (ATCC), USA. These bacteria were first propagated to increase its amount using the method provided by ATCC, ATCC® MEDIA. The growths of the bacteria were observed within 1–2 weeks, when yellow–orange iron deposited.

Cultivation of Bacteria

T. ferrooxidans was cultivated in a sulfur-based media containing 4.0 g KH_2PO_4, 0.3 g $MgSO_4.7H_2O$, 4.0 g $(NH_4)_2SO_4$, 0.25 g $CaCl_2$, and 5.0 g $Na_2S_2O_3.5H_2O$ in 1.0 L distilled water. The medium was prepared without thiosulphate. The pH was adjusted to 4 and autoclaved at 121 °C for 15 min. Thiosulphate was filter sterilized and aseptically added after autoclaving. After that 100 ml of sulfur medium was transferred into several shake flasks. 1.0 ml of propagated bacteria was aseptically withdrawn and added to the 100 ml of sulfur medium.

Finally, all the shake flasks were placed in incubator at 25 °C and shaken at a speed of 125 rpm. All the shake flasks were examined every day to monitor the qualitative tetrathionate hydrolase activity for 14 days. 1 ml of each sample were collected, everyday, during the 14 days and stored at −20 °C for further analysis.

Analytical Analysis

Optical Density

200 μl of each cultivation sample was transferred into microtiter plate for optical density (OD) by using microplate reader. All samples were analyzed at 440 nm every day from Day 1 to Day 14.

Protein Assay and Protein Content Determination

The protein assay was conducted based on Bradford method (1976). Bovine serum albumin (BSA) was used in estimating the protein concentration using the standard curve.

The analysis of protein content was done on extracellular and intracellular protein. A 1 ml of sample was transferred from shake flask into 2.0 ml centrifuge tube from Day 1 until Day 14. The sample was centrifuged at room temperature at 5,000 g for 30 min producing pellet and supernatant. Then, the supernatant was transferred into another 2.0 ml centrifuge tube. The first supernatant was analyzed as extracellular protein. 100 μl of protein sample was withdrawn and mixed with 1 ml of Bradford solution. OD value of the supernatant was measured at 595 nm (Bradford 1976). For control, sample will be replaced by media without inoculum along with 1 ml of Bradford solution.

For intracellular protein, the pellet was added with 1 ml culturing media. Next, this sample was sonicated at 50 % amplitude and 30 s break in 2 min. After that, 2–3 drops of ethanol 95 % was pipetted into the centrifuge tube to stop the reaction. Again, the sample was centrifuged at room temperature at 5,000 g for

30 min. The pellet was discarded and 100 μl of protein supernatant was withdrawn and mixed with 1 ml Bradford solution. OD value of the supernatant was measured at 595 nm (Bradford 1976). For control, sample was replaced by media without inoculum along with 1 ml of Bradford solution. Lastly, protein concentrations were determined from the extrapolation of the standard curve of BSA.

Tetrathionate Hydrolase Assay

Tetrathionate analysis was done based on de Jong et al. (1997) method. This analysis was conducted on extracellular enzyme and intracellular enzyme. 1 ml of sample from shake flask was transferred into 2.0 ml centrifuge tube. The sample was centrifuged at room temperature at 5,000 g for 30 min. The first supernatant was transferred into another 2.0 ml centrifuge tube and analyzed as extracellular enzyme. For intracellular enzyme, the pellet was collected and added with 1 ml culturing media. The sample was sonicated at 50 % amplitude and 30 s break in 2 min. Next, about 2–3 drops of ethanol 95 % was added to stop the reaction at room temperature and finally, the sample was centrifuged at room temperature at 5,000 g for 30 min.

Enzyme activity was measured by addition of 40 μl 50 mM potassium tetrathionate ($K_2S_4O_6$) into 1,300 μl 1.5 M ammonium sulfate. 500 μl of enzyme sample was added with 160 μl distilled water. The sample was mixed thoroughly and incubated for 10 min at 30 °C. Then, 5 μl 1 M NaOH was added to stop the reaction. The absorbance was measured using UV-spectrometer set at 290 nm. Control sample was prepared by replacing the enzyme sample with media without inoculation. The remaining samples were stored at −20 °C.

Results and Discussions

Absorbance at 440 nm for the Growth of T. ferrooxidans

The growth rate of *T. ferrooxidans* is influenced by several factors such as media composition, temperature, pH, oxygen supply, and the mixing times. The highest growth rate of *T. ferrooxidans* was on Day 1. The bacteria were well supplemented with the elemental sulfur in the medium that serves as the energy source. Ammonium sulfate serves as the nitrogen source while calcium, ferric chloride, and magnesium sulfate supply inorganic ions. Potassium dihydrogen phosphate buffers the medium against pH change. Even though *T. ferrooxidans* has a pH optimum of about 1.5, it also grew well at pH 4 (de Jong et al. 1997).

Protein Content Determination

Extracellular and intracellular protein secretion and accumulation were determined based on the extrapolation of BSA standard curve. All protein contents were calculated from Day 1 to Day 14. The highest extracellular protein content occurred on Day 11 was 9.814 μg/ml. For the intracellular protein content, Day 6 gave the highest value at 10.593 μg/ml.

Tetrathionate Hydrolase Activity

Tetrathionate hydrolase is a periplasmatic enzyme with pH optimum below 2.5. Based on the optimized condition by Yusof and Ahmad (2010), the pH was done at 4 and it was found in this work the highest production of extracellular and intracellular of Tetrathionate hydrolase was at Day 6 and 5, respectively. The enzyme activity had been measured by calculating the changes in absorbance reading divided by minute and total enzyme in ml at 25 °C.

For example,

$$\text{Activity of enzyme} = \frac{\Delta \text{A}}{\text{min. } 0.5\,\text{ml}}$$

The reading of time was taken at 1 min. The maximum activity of extracellular and intracellular enzyme was found to be 0.0920 μmol min^{-1}/ml (U/ml) and 0.268 μmol min^{-1}/ml (U/ml), respectively. It had also been proven by de Jong et al. (1997) in their work where maximum tetrathionate hydrolyzing activity by *T. ferrooxidans* occurred at pH 4.

Sugio et al. (2009) mentioned in their work that tetrathionate is a reduced sulfur compound soluble and stable in the water acidified with sulfuric acid that is easily used by *T. ferrooxidans* cells as energy source. In *T. ferrooxidans*, thiosulfate is oxidized through the S4-intermediary pathway (Pronk et al. 1990; Kelly et al. 1997; Ghosh and Dam 2009). During this reaction, two molecules of thiosulfate undergo condensative oxidation to produce four sulfur atoms intermediate tetrathionate catalyzed by thiosulfate dehydrogenase. In the second step of this pathway, tetrathionate was hydrolyzed by tetrathionate hydrolase to produce sulfate and disulfane monosulfonic acid. According to de Jong et al. (1997), tetrathionate hydrolase will generate sulfate, thiosulfate, and elemental sulfur at the end of the reaction due to the high activity of the hydrolysis products.

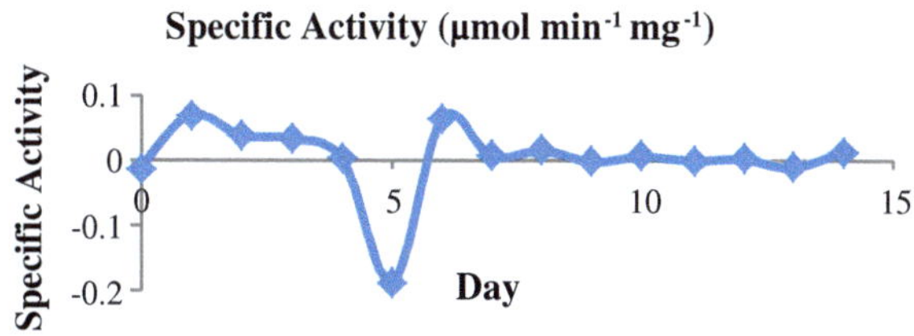

Fig. 1 Specific activity of extracellular tetrathionate hydrolase was measured during 14 days of incubation

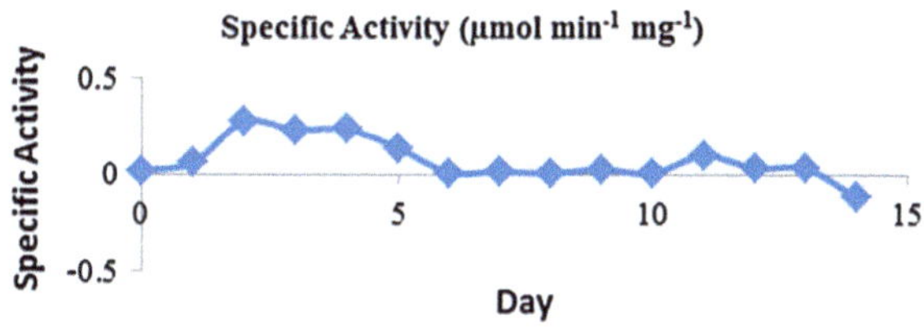

Fig. 2 Specific activity of intracellular tetrathionate hydrolase was measured during 14 days of incubation

Specific Activity of Tetrathionate Hydrolase

The specific activity of enzyme had also been studied. This measurement refers to the activity of enzyme per milligram of total protein (μmol $min^{-1}mg^{-1}$) at specific pH and temperature. The maximum specific activity of extracellular and intracellular enzyme occurred on Day 1 and 2 was 0.070 μmol $min^{-1}mg^{-1}$ (U/mg) and 0.274 μmol $min^{-1}mg^{-1}$ (U/mg), respectively, at pH 4, 25 °C (Figs. 1 and 2). This work is in accordance to the work done by Sugio et al. (2009), where they mentioned that the tetrathionate-grown *T. ferrooxidans* cells have high oxidizing activity at the early log phase where tetrathionate hydrolase produces Fe^{2+} from tetrathionate during growth on tetrathionate.

Conclusions

From this study, several findings could be concluded. Firstly, *T. ferrooxidans* was well grown by incubating the bacteria at 25 °C, initial pH buffer of 4 and at an agitation speed of 125 rpm, with salt concentration as follows: in (g/l) of KH_2PO_4 = 4 g, $(NH_4)_2SO_4$ = 4 g, and $MgSO_4.7H_2O$ = 0.3 g. With these conditions, the activities of extracellular and intracellular enzyme are 0.0920 μmol min^{-1} and 0.268 μmol min^{-1}, respectively. The specific activity of extracellular and intracellular tetrathionate hydrolase occurred on Day 1 and 2 is 0.070 μmol $min^{-1}mg^{-1}$(U/mg) and 0.274 μmol $min^{-1}mg^{-1}$ (U/mg), respectively, at pH 4, 25 °C.

Acknowledgements Authors would like to thank the Ministry of Science, Technology, and Innovation for providing fund to conduct this project.

References

Bradford, M. M. (1976). A rapid and sensitive method for the quantitation of microgram quantities of protein utilising the principle of protein-dye binding. *Analytical Biochemistry, 72*, 248–254.

De Jong, G. A. H., Hazeu, W., Bos, P., & Kuenen, J. G. (1997). Polythionate degradation by tetrathionate hydrolase of *Thiobacillus ferrooxidans. Microbiology, 143*, 499–504.

E-Waste Management. (2008). Special process recycle rubber (SPR).

Ghosh, W., & Dam, B. (2009). Biochemistry and molecular biology of lithotrophic sulfur oxidation by taxonomically and ecologically diverse bacteria and archaea. *FEMS Microbiology Reviews, 33*, 999–1043.

Gulzad, A. (2011). Recycling and pyrolysis of scrap tire. Report training visit on the development of human resource capacity of Kabul Polytechnic University.

Ingledew, W. J. (1982). *Thiobacillus ferrooxidans:* the bioenergetics of an adicophilic chemolithotroph. *Biochimica et Biophysica Acta, 683*, 89–117.

Kelly, D. P., Shergill, J. K., Lu, W. P., & Wood, A. P. (1997). Oxidative metabolism of inorganic sulfur compounds by bacteria. *Antonie van Leeuwenhoek, 71*, 95–107.

Lewin, S. Z. (2004). *Rubber. Microsoft Encarta Online Encyclopedia* [http://encarta.msn.com]. [Accessed 10 September 2012].

Pronk, J. T., Meulenberg, R., Hazeu, Z., Bos, P., & Kuenen, J. G. (1990). Oxidation of inorganic sulphur compounds by *Acidophilic thiobacilli. FEMS Microbiology Reviews, 75*, 293–306.

Rawlings, D. E. (2005). Characteristics and adaptability of iron- and sulphur-oxidizing microorganisms used for the recovery of metals from minerals and their concentrates. *Microbial Cell Factories*, 4–13.

Romine, R., & Snowden-Swan, L. (1997). Method for the addition of vulcanized waste rubber to virgin rubber products. United States Patent No.: 5,597,851.

Sugio, T., Taha, T. M., & Takeuchi, F. (2009). Ferrous ion production mediated by tetrathionate hydrolase in tetrathionate, sulfur and iron-grown *Acidithiobacillus ferrooxidans* ATTC 27230 cells. *Bioscience, Biotechnology, and Biochemistry, 73*(6), 1381–1386.

U.S.Environmental Protection Agency. (1993). Scrap tire technology and markets Noyes Data Corporation, NJ. USA.

Yusof, F., & Ahmad, A. A. (2010). Enzymatic devulcanization of waste rubber products in preparation for recycling. In: Extended Abstract for National Biotechnology Seminar, 24–26 May, PWTC, Kuala Lumpur, Malaysia.

Cross-linked Lipase in Hybrid Matrix for Biodiesel Production from Crude Jatropha Curcas Oil

R. Abdulla and P. Ravindra

Abstract Immobilization of lipases is gaining much attention these days due to the wide variety of reactions catalyzed by them. Moreover, the lipases play an important role in biodiesel production. In this study, lipase from *Burkholderia cepacia* was first cross-linked with glutaraldehyde followed by entrapment into hybrid matrix of equal proportions of alginate and κ-carrageenan natural polymers. The immobilized lipase gave promising results with stability parameters like pH, temperature, solvent, storage, enzyme leakage, and hydrolytic activity. A significant reduction of 65.76 % of enzyme leakage was obtained with this immobilized lipase. Moreover, a 100 % yield of biodiesel was produced from crude *Jatropha curcas* oil using this immobilized lipase.

Introduction

In the hunt of alternative, renewable energy resources; biodiesel has gained immense potential as a clean, environmental friendly reliable fuel. Nonedible oils represent a promising feedstock for biodiesel production over the edible vegetable oil counterpart. Among nonedible oils, *Jatropha curcas* (*J. curcas*) oil occupies the top position due to several reasons. *J. curcas* is a small perennial tree which belongs to Euphorbiaceae family. The plant is very easy to propagate as it can grow in almost all soil conditions. Due to the presence of some anti-nutritional factors, oil from Jatropha is rendered unsafe for cooking purposes. This property is exploited as biodiesel feedstock (Rahmath et al. 2011).

In general, various immobilization methods for enzymes have been classified into four: Adsorption, covalent binding, cross-linking, and entrapment. Cross-

R. Abdulla · P. Ravindra (✉)
Chemical Engineering Programme, School of Engineering and Information Technology, University Malaysia Sabah, Jalan UMS 88400 Kota Kinabalu, Sabah, Malaysia
e-mail: dr_ravindra@hotmail.com

R. Pogaku et al. (eds.), *Developments in Sustainable Chemical and Bioprocess Technology*, DOI: 10.1007/978-1-4614-6208-8_25,

linking with bifunctional agents like glutaraldehyde is one of the most frequently used method of lipase immobilization. But the smaller size of the aggregates is considered as a drawback in biodiesel production. Entrapment involves easier procedure, mild conditions, and minimum denaturation of enzyme (Jegannathan et al. 2008). The significance of using natural polymers like κ–carrageenan has been highlighted in the field of lipase immobilization (Jegannathan et al. 2008). These natural polymers are easy to handle and environmental friendly also. Interpenetrating polymer networks (IPNs) are defined as a combination of two cross-linked polymers. The potential of κ-carrageenan related IPNs as immobilization matrix for enzymes was shown by a research team (Sahin et al. 2005). The objective of our research was to design a simple environmental friendly hybrid matrix for lipase immobilization with decreased enzyme leakage and increased catalytic activity for biodiesel production.

Methodology

Immobilization of Lipase

Burkholderia cepacia lipase was first cross-linked with 25 % glutaraldehyde for 3 h. At the same time, 1.5 % (w/v) solution of κ-carrageenan in water was prepared by dissolving it at 80 °C for an hour and later the temperature was brought down to 35 °C. To this equal weight of sodium alginate was added and stirred to obtain uniform distribution. After this, the cross-linked lipase was mixed into this and again stirred for 15 min for uniform distribution. The mixture was dropped into 0.1 M solution of calcium chloride ($CaCl_2$) by using a syringe pump (Thermo Model TE-331, Japan) of 50 ml/h flow rate to form the desired immobilized lipase. These immobilized lipases in the collection flask were stirred for 10 min and later was stored in 0.1 M $CaCl_2$ solution overnight at 4 °C for hardening. Further, the immobilized lipase was washed with distilled water, air dried, and used for characterization studies followed by biodiesel production.

The lipase activity of free and immobilized lipase were calculated using the following formula:

$$\text{Lipase activity} = (A_{410\,\text{nm}}/\Delta\varepsilon) \times (V/v) \times (1/T) \times 10^6\ \mu\text{mol/min}$$

where,

$A_{410\ nm}$—Absorbance at 410 nm
$\Delta\varepsilon$—Molar extinction coefficient of nitrophenol (15000 M^{-1} cm^{-1})
V—Total volume of reaction
v—Volume of enzyme
T—Time of reaction.

One unit of lipase activity (U) is defined as the amount of enzyme required to hydrolyze 1 μmol/min of *p*-NPP under the similar assay conditions.

Leakage of Entrapped Lipase

To determine the extent of leakage of immobilized enzyme, two types of immobilized lipase were produced. One immobilized in κ-carrageenan/alginate matrix with glutaraldehyde cross-linking and other one immobilized in the same matrix but without cross-linking. Soon after formation, both the above said immobilized lipase was incubated at 25 °C, 200 rpm in 0.01 M phosphate buffer of pH 7. The enzyme leakage was studied by withdrawing the phosphate buffer every 1 h and recording the absorbance at 280 nm using a UV-Vis spectrophotometer.

Reusability

In order to evaluate the reusability, immobilized lipase was repeatedly used for hydrolysis of *p*-Nitro Phenyl Palmitate (*p*-NPP). Followed by hydrolysis, the immobilized lipase was washed with buffer solution and suspended again in fresh aliquot of the substrate (*p*-NPP) to measure the enzymatic activity.

Transesterification Process Parameters

The transesterification of crude *Jatropha curcas* oil (CJO) with ethanol was carried out in a baffled conical flask. Optimized process parameters for processing 10 g Jatropha oil were: 35 °C, 1:10 molar ratio of oil to ethanol, 1 g water, 5.25 g immobilized lipase, 300 rpm, and 24 h reaction time. At the end of the reaction, 5 ml of reaction mixture was centrifuged at 10,000 rpm for 5 min and the upper layer containing biodiesel was analyzed with GC–MS for the determination of ethyl ester content. Biodiesel yield is calculated as the total weight of FAEE to the total weight of oil used for transesterification.

Results and Discussion

Lipase Immobilization

Cross-linking and entrapment are the simple easy methods used for lipase immobilization. In cross-linking, intermolecular linkages are formed between enzyme molecules by means of bifunctional or multifunctional reagents such as glutaraldehyde. The cross-linked lipase is more stable due to strong interactions between the lipase and cross-linking agent. It has been proven that application of cross-linked lipase for biodiesel production gave considerable yield in much shorter time, but the smaller size of these aggregates creates problem for reuse (Cao et al. 2003). On the other hand, entrapment of lipase involves capture of lipase within a polymer matrix. Benefits of entrapment include mild operating conditions with cheap, fast, and easy procedure. Application of entrapped lipase has shown average results in biodiesel production. The reason can be attributed to the mass transfer limitations of entrapped lipase and erosion of enzyme from the surface of the support of the matrix (Xavier 1990). Moreover, the reaction time with entrapped lipase is much longer.

Thus, in an attempt to design a better biocatalyst for biodiesel production, the following approaches were done:

- The *B. cepacia* lipase powder was cross-linked with glutaraldehyde prior to entrapment. This will give rise to an aggregate which can be retained more inside the matrix.
- Equal proportions of sodium alginate and κ-carrageenan were used as hybrid matrix for entrapment in order to reduce the leakage of lipase. Hybrid matrices have been shown to be more effective in reducing enzyme leakage after entrapment.

Leakage of Entrapped Lipase

Enzyme leakage is a common problem in immobilized lipase systems. But here, it was observed that there was 65.76 % reduction in enzyme leakage from the hybrid matrix after cross-linking as compared to that from noncross-linked lipase (Fig. 1). Thus, it is understood that cross-linking with glutaraldehyde followed by entrapment in hybrid matrix can significantly reduce enzyme leakage. The reasons for the reduced enzyme leakage can be attributed to the increased molecular weight of the cross-linked enzyme and efficient interaction of enzyme moieties with the polymers of alginate and κ-carrageenan.

Reusability

The immobilized lipase was repeatedly used for hydrolysis of *p*-NPP in order to evaluate its reusability property. From Fig. 2, it can be seen that the immobilized lipase retained 75.54 % of its initial activity upon 10th hydrolysis of *p*-NPP.

Fig. 1 Enzyme leakage as a function of time

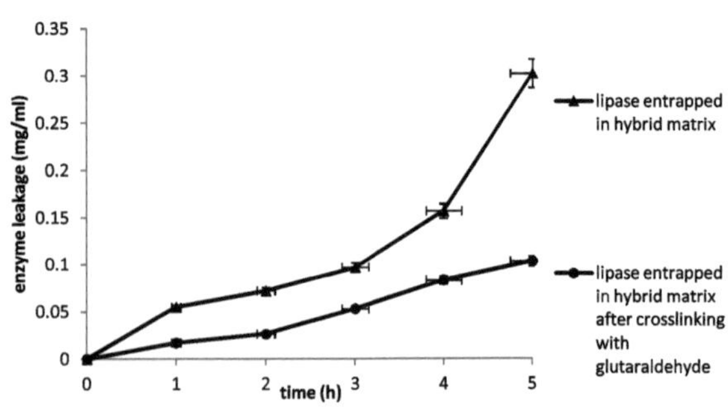

Fig. 2 Reusability of immobilized lipase

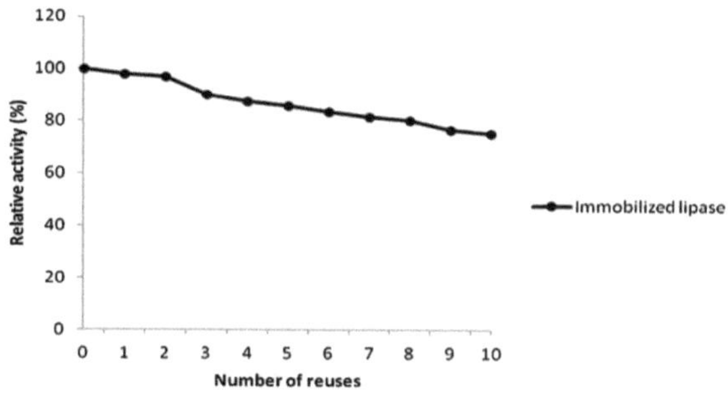

Fig. 3 Production of biodiesel (ethyl ester) from crude *Jatropha curcas* oil with respect to time

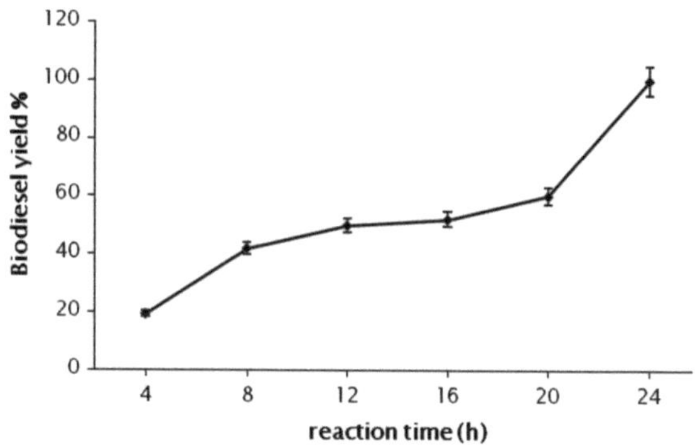

Immobilized Lipase in Biodiesel Production

The catalytic activity of immobilized *B. cepacia* lipase was determined in terms of transesterification activity. The immobilized lipase gave a biodiesel yield of 100 % with crude *Jatropha curcas* oil and the optimized process parameters are as follows: 10 g Jatropha oil: 35 °C, 1:10 molar ratio of oil to ethanol, 1 g water, 5.25 g immobilized lipase, 300 rpm, and 24 h reaction time. The result (Fig. 3) shows that the process parameters involved did not alter the catalytic activity of lipase.

Conclusions

Here, an attempt was done toward the design of a better biocatalyst for biodiesel research which can retain high catalytic activity and can be reused several times. Thus, in this study, lipase from *Burkholderia cepacia* was immobilized first by cross-linking and later entrapped in hybrid matrix of alginate and κ-carrageenan. The immobilized lipase showed better transesterification activity with 100 % biodiesel yield from crude *Jatropha curcas* oil.

The lipase immobilization method mentioned in this study is a way closer in designing of an efficient, environmental friendly, biocatalyst for biodiesel production. The main factors for immobilization such as better activity, lesser enzyme leakage, good stability of immobilized lipase, and low cost of carriers used are satisfied here. Moreover, this work also highlights the importance of using natural polymers such as sodium alginate and κ-carrageenan in biodiesel production since it makes the complete process green and environmental friendly.

Acknowledgments Authors would like to thank the University Malaysia Sabah and Ministry of Science, Technology and Innovation (MOSTI) for funding this project, FGRG0244-TK-1/2010.

References

Cao, L. Q., Langen, V. L., & Sheldon, R. A. (2003). Immobilized enzymes: carrier-bound or carrier-free? *Current Opinions in Biotechnology, 14*, 387–394.
Jegannathan, K. R., Sariah, A., Denis, P., Chan, E. S., & Ravindra, P. (2008). Production of biodiesel using immobilized lipase—a critical review. *Critical Reviews in Biotechnology, 28*, 253–264.
Rahmath, A., Chan, E. S., & Ravindra, P. (2011). Biodiesel production from *Jatropha curcas*: a critical review. *Critical Reviews in Biotechnology, 31*, 53–64.
Sahin, F., Demirel, G., & Tumturk, H. (2005). A novel matrix for the immobilization of acetylcholinesteras. *International Journal of Biological Macromolecules, 37*, 148–153.
Xavier, M. F., Hector, R. R., Hugo, S. G., Charles, G. H., & Clyde, H. A. (1990). Immobilized lipase reactors for modification of fats and oils—a review. *Journal of American Oil Chemists Society, 67*, 890–910.

Culture Conditions for Optimal Growth of Actinomycetes from Marine Sponges

M. A. A. Bukhari, A. N. Thomas and N. K. Wong

Abstract Actinomycetes are filamentous gram-positive bacteria that can be found abundantly in both terrestrial and marine environment. These bacteria are known as producers of many bioactive compounds through the production of secondary metabolites for their survival and adaptation in nature and have been widely used today as therapeutic agents. Marine actinomycetes have been the focus of research over the past decade for new drugs discovery due to its unique adaptation in the harsh sea environment. It is believed that marine actinomycetes could produce compounds that are rare and unique compared to the terrestrial actinomycetes. Despite its potential, marine actinomycetes are critically difficult to culture in laboratory because these actinomycetes live in extreme environment in the sea with high salt concentration, high pressure, low temperature, and constant pH changes of seawater in its natural environment. Hence, in this study, optimum condition to culture marine actinomycetes was achieved by culturing the marine actinomycetes from marine sponges on different culture condition such as different types of isolation media, pH, seasalt concentration, temperature, and incubation time. Starch casein agar (SCA) is shown to be the best isolation media compared to actinomycetes isolation agar (AIA) and Kuster agar (KUA). The growth of marine actinomycetes is optimum at pH 7, 40 % of seasalt concentration, 20–30 °C and 7–10 days of incubation time.

Introduction

Actinomycetes are gram-positive bacteria that known for its ability to produce novel secondary metabolites, which have been used widely today as antibiotics and anticancer drug (Olano et al. 2009). This unique ability has made these

M. A. A. Bukhari · A. N. Thomas · N. K. Wong (✉)
Biotechnology Programme, School of Science and Technology, Universiti Malaysia Sabah, 88400 Sabah, Kota Kinabalu, Malaysia
e-mail: nkwong@ums.edu.my

R. Pogaku et al. (eds.), *Developments in Sustainable Chemical and Bioprocess Technology*, DOI: 10.1007/978-1-4614-6208-8_26,

microbes as one of the major resources for drug-screening research (Lam 2006). Despite of the enormous attention on actinomycetes, marine actinomycetes are relatively underexplored as it is a lot easier to isolate and culture the terrestrial actinomycetes compared to its marine counterpart. On the other hand, actinomycetes strains from the marine environment live in an extremely different environment, they have their own unique requirement for growth and show a different physiological trait that have not been previously reported on terrestrial actinomycetes (Fenical et al. 2008). The extreme living conditions to which marine actinomycetes have to adapt to are reflected in the genetic and metabolic diversity of marine actinomycetes, which enable them to produce a complete different set of bioactive compounds as compared to terrestrial actinomycetes.

Despite of the previous assumption that marine actinomycetes might be originated from terrestrial origin, many researches proved that indigenous marine actinomycetes indeed exist, these include members from of various genera such as *Dietzia, Rhodococcus, Streptomyces, Salinispora, Marinophilus, Solwaraspora, Salinibacterium, Aeromicrobium marinum, Williamsia maris, and Verrucosispora* (Lam 2006). Among these, the first obligates new actinomycetes genus discovered was *Salinispora* (formerly known as *Salinospora*). More than 2,500 new genus of actinomycetes from marine sources have been found so far and astonishingly, the majority of it showed to have therapeutic and anticancer activity (Fleming et al. 2006). Salinosporamide A, for example, was isolated from marine actinomycetes genus *Salinospora,* and it shows to be a potential anticancer agent (Fenical et al. 2008). *Abyssomicins*, *Aureoverticillatam*, *Caprolactones*, *Chinicomycins, IB-00208*, *Trioxacarcins* etc. are among the other therapeutic novel metabolites found from marine actinomycetes. Indeed, marine environment is now regarded as a new source of rare actinomycetes for novel bioactive compounds. The exploration of the ocean, in search of new strains of actinomycetes is challenging yet rewarding as ocean constitutes approximately three quarters of the earth surface and diversity of the microorganisms in the sea is predicted far greater than the diversity of life forms in the rainforests (Sogin et al. 2006).

Semporna is located on the east coast of Sabah, Malaysia, and it is well known for its vast and beautiful sea area with most of it still in its natural condition. This undisturbed environment will serve as a perfect source of rare actinomycetes isolation as most part of it still retain in its natural condition which might be a preferable habitat of some indigenous marine actinomycetes. These marine actinomycetes can be found in sea water, sediments and also associated with other marine organisms such as fish, seaweeds, starfishes, sponges, and mollusks (Sahu et al. 2005). In this study, sponges have been used as a source of marine actinomycetes isolation due to the fact that sponges live symbiotically with various microorganisms in nature (Li 2009).

Although marine actinomycetes have a lot of potential compared to terrestrial actinomycetes, the major challenge in exploiting their potential is the difficulty in isolating these microbes from its environment (Murphy et al. 2009). The harsh conditions in the ocean such as high salt concentration, high pressure, anaerobic condition, and low temperature have made it difficult to manipulate and maintain in the laboratory. Hence, in this chapter, the study is focused on the effect of several growth parameters such as temperature, pH, and salt concentration on marine actinomycetes to determine its optimum isolation condition.

Methodology

Parameter Study

Six marine sponges were collected around Semporna coastal area (4°06′52.92″N and 118°38′04″E) at a depth between 10 and 15 m. The samples are labeled with a code from MS1 to MS6. The sampling area is shown in the red circle of Fig. 1.

The mesophyll part of the sponge was excised and homogenized for the parameter study to determine the optimum culture condition for marine actinomycetes from sponges. Parameters studied include the isolation media [starch casein agar (SCA), Kuster's agar (KUA), and actinomycetes isolation agar (AIA)], pH (5, 6, 7, 8, and 10), seasalt concentration (0, 20, 40, 60, 80, and 100 %), incubation time (3, 5, 7, 10, and 14 days) and incubation temperature (15, 20, 25, 30, 35, and 40 °C). The resulting colonies were further purified on (Oatmeal) OA media. Marine actinomycetes were then identified by using non-staining KOH method as well as observing its sporulation appearance on OA media.

Results and Discussion

The Influence of Isolation Media on the Growth of Marine Actinomycetes

In this study, six different marine sponges (MS1, MS2, MS3, MS4, MS5, and MS6) from different genus have been used as the isolation sources for marine

Fig. 1 Sabah sea area (*Red circle* shows the sampling area)

actinomycetes. The isolation media used were SCA, KUA, and AIA, of which SCA demonstrated to have the highest number of isolates. Out of six marine sponge samples tested, four samples (MS1, MS3, MS4, and MS5) showed to have the highest number of isolates on SCA. The growth of marine actinomycetes is relatively good on AIA, and two samples (MS2 and MS6) showed to have the highest number of isolates on AIA. All sponge samples yielded the lowest number of marine actinomycetes isolates on KUA. This study showed that SCA is the most favorable culture media, and this might due to the composition of nutrient especially in the form of carbon and nitrogen sources is most suitable for the growth of marine actinomycetes strains. Concentration of nutrients and organic matter can influence the growth of marine bacteria and also in its natural environment, the abundance and growth rates of bacteria are very much dependent on the subtle changes of the environmental variables such as temperature, nutrient, and energy supplies (Church et al. 2000). It is interesting to note that KUA was shown to be the best isolation media for marine actinomycetes when sediments, seaweeds mollusks, and fish were used as the sampling source (Sahu et al. 2005). This happened due to the specific interaction between microbes and host in which certain types of microbes often colonizing the same species of host regardless the location where the samples are collected (Taylor et al. 2004).

The Influence of pH on the Growth of Marine Actinomycetes

Figure 2b showed that pH range of 6–8 is the best pH condition to yield the highest number of isolates, while pH 7 is the optimum pH for the growth of marine actinomycetes on SCA media for all samples (MS1 to MS6). This pH range is much similar to the pH of natural seawater which ranges from 6.5 to 8.4 (Ross 1995). A suitable pH range is crucially important for bacterial growth since bacteria are really sensitive to the hydrogen ion concentration in their environment. The unsuitable pH condition may affect the large bioproteins (i.e., enzymes) within the metabolism of actinomycetes, and this will halt the metabolism in the cells or production of bioactive compounds. Despite all of that, a small number of strains can still grow at pH outside of this range such as pH 5 (acidic) and pH 10 (alkaline). This might be due to the adaptation of certain strains to live in an extreme large pH range in nature to ensure their survival.

The Influence of Incubation Temperature on the Growth of Marine Actinomycetes

All of the six sponge samples, except MS4, yielded the highest number of marine actinomycetes isolates at 30 °C as seen in Fig. 2c. MS4 produced the highest number of marine actinomycetes isolates at 25 °C. This indicated that 30 °C is

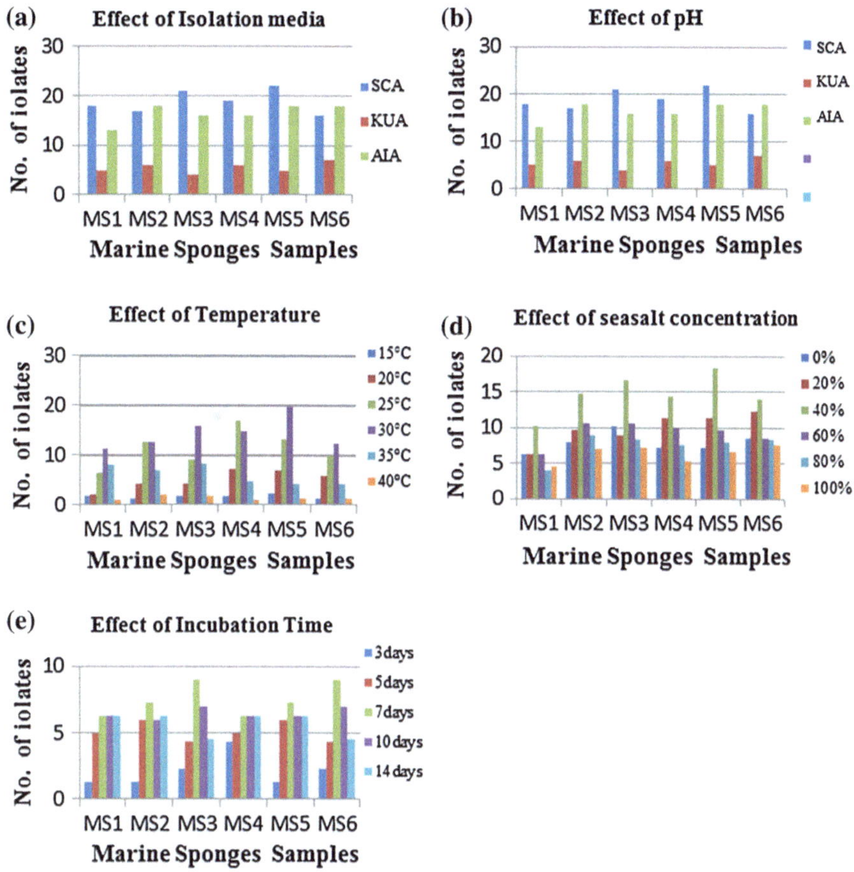

Fig. 2 Parameter study for the isolation of marine actinomycetes from marine sponges. **a** The effect of isolation media. **b** The effect of pH. **c** The effect of temperature. **d** The effect of seasalt concentration. **e** The effect of incubation time

optimum for the growth of most actinomycetes strains in this study. The data also showed that most of the isolates are able to grow well in temperature ranging from 20 to 30 °C of all samples. This temperature range is considered to be the most suitable condition for the marine actinomycetes to grow as it nearly resembles the seawater temperature in its natural habitat. The temperature of tropical oceans typically exceeds 20 °C and this temperature remains constant throughout the years (Ross 1995). The data also showed that a small number of marine actinomycetes strains can grow at temperature as low as 15 °C to the highest of 40 °C in all sponge samples. The ability to grow in a certain temperature range is mainly due to the adaptation of marine actinomycetes to the regular changes of seawater temperature throughout the year in its natural habitat. For the matter of fact, microorganisms can be found abundantly everywhere in a marine environment regardless of its temperature, they live abundantly from a high-temperature environment such as hydrothermal vents in the deep sea to the icy seawater in the Antarctica. Temperature can be a limiting factor for the bacterial growth because it can affect the chemical and biochemical processes in the cell. Every bacterium phylotype has its own minimal, maximal, and optimal temperatures that allow its growth and each bacteria can grow in a certain range of temperature. However, the most favorable growth only can be achieved at the optimal temperature (Pomeroy and Wiebe 2001).

The Influence of Seasalt Concentration on the Growth of Marine Actinomycetes

The data in this study showed that the growth of marine actinomycetes increases in the presence of artificial seawater. As shown in Fig. 2d, seasalt concentration of 40 % is the optimal seasalt concentration for all six sponge samples to yield the highest number of isolates. The numbers of isolates yielded rapidly increased as the seasalt concentration increased from 0 to 40 %, and the numbers of isolates decreased when the seasalt concentration is more than 40 %. The need of seawater for the growth of isolates is critically important especially during the isolation of marine bacteria from its natural sources. The isolation media should resemble its natural environment as closely as possible in order to the marine bacteria to grow on the isolation plates. The usage of isolation media prepared by using seawater is one of the ways to provide a suitable growth condition for marine bacteria. There is a difference between marine and terrestrial bacteria in term of salt requirement because the growth of marine bacteria is more preferable in media with seawater (McLeod 1965). The data also clearly showed that the marine actinomycetes can grow in all seasalt concentration ranging from 20 to 100 % although the number of isolates is different in all seasalt concentration. The effect to which salt concentration causes changes in bacterial growth depends on the osmotic balance required for such growth. Some bacteria require an astonishingly high level of salt to begin growth, whereas other bacteria would be immediately killed in high levels of salt. This is because marine actinomycetes in particular are different in salt tolerance (Kuster and Neumeier 1981).

The Influence of Incubation Time on the Growth of Marine Actinomycetes

The number of marine actinomycetes isolates from sponges with incubation time of 3, 5, 7, 10, and 14 days is shown in Fig. 2e. The data showed that the actinomycetes colony started to appear on SCA media only after 3 days of incubation time at 28 °C on SCA media. For all of the samples (MS1 to MS6), most isolates were observed after 5–10 days of incubation times, in which 7 day incubation time seems to yield the highest isolates of all sponge samples. Out of six sponge samples, four samples (MS1, MS2, MS4, and MS5) yielded a relatively high number of isolates even after 14 days of incubation time. This incubation time was slightly longer compared to the terrestrial actinomycetes which usually take only 3–5 days to grow on SCA media. Longer incubation time is needed for marine actinomycetes because marine actinomycetes are really slow growing in nature (Lam 2006) and that may contribute to their metabolites diversity.

Conclusions

Optimization of culture condition for isolating marine actinomycetes is crucial due to its potential as the producers of novel bioactive secondary metabolites over terrestrial actinomycetes. Marine actinomycetes were successfully isolated from marine sponges collected from Semporna coastal area, close distance from Sipadan Island of Sabah, Malaysia. The optimum yield of marine actinomycetes is obtained on culture condition of SCA media supplemented with 40 % seasalt concentration at pH 7 and incubated at 30 °C for 7–10 days. It is worth noting that the numbers of marine actinomycetes obtained also varied among all sponge samples from MS1 to MS6 on all of the isolation media, SCA, KUA and AIA, pH condition, temperature, seasalt concentration, and incubation time. The optimum growth conditions possibly varied among genus resulting in different number of strains. Nonetheless, this study provides a solid optimum cultural condition for the growth of actinomycetes from marine sponges.

References

Church, M. J., Hutchins, D. A., & Ducklow, H. W. (2000). Limitation of bacterial growth by dissolved organic matter and iron in the Southern Ocean. *Applied and Environment Microbiology, 66*, 455–466.

Fenical, W., Jensen, P. R., Palladino, M. A., Lam, K. S., Lloyd, G. K., & Potts, B. C. (2008). Discovery and development of the anticancer agent salinosporamide A(NPI-0052). *Bioorganic and Medicinal Chemistry, 17*, 2175–2180.

Fleming, L. E., Broad, K., Clement, A., Dewailly, E., Elmir, S., Knap, A., et al. (2006). Oceans and human health: Emerging public health risks in the marine environment. *Marine Pollution Bulletin, 53*, 545–560.

Kuster, E. & Neumeier, W. (1981). Halotolerance in some streptomycetes producing tetracyclines. *Zentralbl Bakteriol Suppi 11*, 315–319.

Lam, K. S. (2006). Discovery of novel metabolites from marine actinomycetes. *Current Opinion in Microbiology, 9*, 245–251.

Li, Z. (2009). Advances in marine microbial symbionts in the China sea and related pharmaceutical metabolites. *Marine Drugs Journal, 15*, 219–224.

McLeod, R. A. (1965). The question of the existence of specific marine bacteria. *Bacteriological Reviews, 29*, 9–23.

Murphy, T., Maloney, H. N. & Fenical, W. (2009). *Natural products from marine microorganisms, in Phytochemistry and Pharmacognosy, in Encyclopedia of Life Support Systems (EOLSS), Developed under the Auspices of the UNESCO*. Oxford,UK: Eolss Publishers.

Olano, C., Méndez, C., & Salas, J. A. (2009). Antitumor compounds from marine actinomycetes. *Marine Drugs, 7*(2), 210–248.

Pomeroy, L., & Wiebe, W. (2001). Temperature and substrates as interactive limiting factors for marine heterotrophic bacteria. *Aquatic Microbial Ecology, 23*, 187–204.

Ross, D. A. (1995). *Introduction to oceanography*. New York: Harper Collins College Publishers.

Sahu, M. K., Sivakumar, K., & Kannan, L. (2005). Isolation of actinomycetes from various samples of the Vellar estuary, southeast coast of India. *Pollution Research, 24*, 45–48.

Sogin, M. L., Morrison, H. G., Huber, J. A., Welch, D. M., Huse, S. M., Neal, P. R., et al. (2006). Microbial diversity in the deep sea and the underexplored "rare biosphere". *Proceedings of the National Academy of Sciences, 103*, 12115–12120.

Taylor, M. W., Schupp, P. J., Dahllof, I., Kjelleberg, S., & Steinberg, P. D. (2004). Host specificity in marine sponge-associated bacteria, and potential implications for marine microbial diversity. *Environmental Microbiology, 6*, 121–130.

Genetic-Algorithm-Based Optimisation for Exothermic Batch Process

M. K. Tan, H. J. Tham and K. T. K. Teo

Abstract The aim of this chapter is to optimise the productivity of an exothermic batch process, by maximising the production of the desired product while minimising the undesired by-product. During the process, heat is liberated when the reactants are mixed together. The exothermic behaviour causes the reaction to become unstable and consequently poses safety issues. In the industries, a dual-mode controller is used to control the process temperature according to a predetermined optimal reference temperature profile. However, the predetermined optimal reference profile is not able to limit the production of the undesired by-product. Hence, this work proposed a genetic-algorithm-based controller to optimise the batch productivity without referring to any optimal reference profile. From the simulation results, the proposed algorithm is able to improve the production of the desired product and reduce the production of the undesired by-product by 15.3 and 34.4 %, respectively. As a conclusion, the genetic-algorithm-based optimisation performs better in raw materials utilisation as compared to the predetermined optimal temperature profile method.

Introduction

Batch process has been applied in the manufacturing industry due to its flexibility to handle various productions of high value-added product, such as specialty chemicals, agrochemicals, and pharmaceuticals. Although batch process is used

M. K. Tan (✉) · K. T. K. Teo
Modelling, Simulation and Computing Laboratory, School of Engineering and Information Technology, Universiti Malaysia Sabah, Jalan UMS 88400 Kota Kinabalu, Sabah, Malaysia
e-mail: msclab@ums.edu.my

H. J. Tham
Chemical Engineering Programme, School of Engineering and Information Technology, Universiti Malaysia Sabah, Jalan UMS 88400 Kota Kinabalu, Sabah, Malaysia
e-mail: hjtham@ums.edu.my

R. Pogaku et al. (eds.), *Developments in Sustainable Chemical and Bioprocess Technology*, DOI: 10.1007/978-1-4614-6208-8_27,

for various types of production, quite often the common aim is to optimise the production of the desired product while at the same time minimising the waste. Since there is no inflow and outflow during the process, the raw materials utility is fully relied on the reactor temperature especially for the exothermic reaction (Nisenfeld 1996). In exothermic reactions, heat is liberated during the process hence consequently increases the reactor temperature. If the liberated heat is higher than the plant cooling capacity, the reaction will become unstable and, hence, poses a safety issue to the plant personnel (Hazard Investigation 2002).

Previously, batch productivity optimisation for exothermic process was obtained by controlling the reactor temperature according to a predetermined optimal temperature profile (Mujtaba et al. 2006; Sujatha and Pappa 2012; Tan et al. 2011). Although they can control the reactor temperature to follow the desired trajectory effectively, the aim of minimising the waste may not be achieved. Besides, the global price competition and escalating raw materials costs have also urged the batch industries to consider an effective way of utilising the raw materials (Fernandez et al. 2012).

For these reasons, genetic algorithm (GA) is introduced to optimise the exothermic batch productivity. This work reports the performance of a GA in exothermic batch process optimisation without referring to any reference values. The performance of the proposed GA is then compared with the conventional dual-mode controller (DM) which uses the predetermined optimal temperature profile.

Methodology

Batch Process Modelling

A benchmark batch process model, developed by Cott and Macchietto (1989), is used in this study. It is assumed that a two-parallel, well-mixed, and irreversible liquid-phase exothermic reactions occur in the process, as shown in below.

$$\text{Reaction 1}: \mathrm{A} + \mathrm{B} \rightarrow \mathrm{C} \qquad \text{Reaction 2}: \mathrm{A} + \mathrm{C} \rightarrow \mathrm{D}$$

where A and B are the raw materials, C is the desired product, and D is the undesired by-product. Initially, all the raw materials are charged into the reactor and left to react for 120 min. The jacket surrounding the reactor is used to control the reactor temperature. Figure 1 illustrates the schematic of the batch process system.

The dynamic modelling of a batch process can be divided into two parts: component balance and energy balance. The law of conservation of mass is applied to model the component balance of the reactor contents. The production/ consumption rate of all substances and each component balance are described in (1) and (2), respectively.

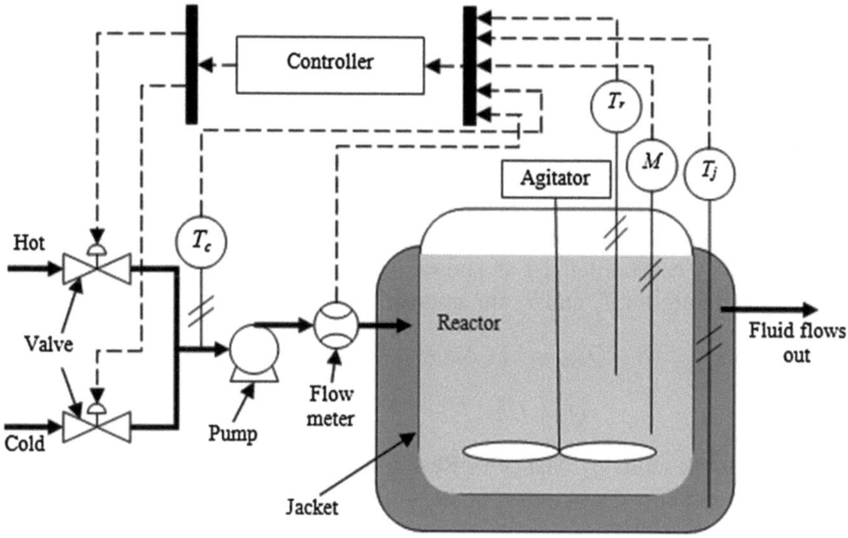

Fig. 1 Schematic of batch process system

$$\begin{bmatrix} R_1 \\ R_2 \end{bmatrix} = \begin{bmatrix} k_1 & M_A & M_B \\ k_2 & M_A & M_C \end{bmatrix} \quad (1)$$

$$\begin{bmatrix} \dot{M}_A \\ \dot{M}_B \\ \dot{M}_C \\ \dot{M}_D \end{bmatrix} = \begin{bmatrix} -1 & -1 \\ -1 & 0 \\ 1 & -1 \\ 0 & 1 \end{bmatrix} \cdot \begin{bmatrix} R_1 \\ R_2 \end{bmatrix} \quad (2)$$

where R_i is the reaction rate of Reaction i, k_i is the reaction rate constant of Reaction i, M_j is the molar concentration of substance n, and $\dot{M}_n$is the changing rate of substance n.

Since both the reactions are temperature dependence, the reaction rate constant for both reactions can be modelled using Arrhenius equation, as shown in (3).

$$k_i = k_{oi} \exp\left(\frac{-E_{ai}}{R \cdot T_r}\right), \text{ where } i = 1, 2 \quad (3)$$

where k_{oi} is the frequency factor of Reaction i, E_{ai} is the activation energy of Reaction i, R is the gas constant, T_r is the reactor temperature in unit Kelvin.

On the other hand, the energy balance of the reactor and jacket are shown in Eqs. (4) and (5), respectively.

$$\frac{\mathrm{d}T_r}{\mathrm{d}t} = \frac{(Q_{\text{exo}} + Q_j)}{M_r C_{\rho r}} \quad (4)$$

$$\frac{dT_j}{dt} = \frac{F_j \rho_j C_{\rho j}(T_c - T_j) - Q_j}{V_j \rho_j C_{\rho j}} \tag{5}$$

where Q_{exo} and Q_j are the exothermic heat released and heat transferred from jacket to reactor, respectively, M_r is total molar concentration of reactor contents, $C_{\rho r}$ is the heat capacity of reactor contents, T_j and T_c are the jacket and fluid temperature, respectively, F_j is the fluid flow rate, ρ_j is the fluid density, $C_{\rho j}$ is the heat capacity of the fluid, and V_j is the volume of the jacket.

Q_{exo} and Q_j can be defined as shown in (6) and (7), respectively. Here, the initial temperature for T_r and T_j are assumed to be 25 °C.

$$Q_{exo} = \Sigma(-\Delta H_i R_i), \text{where } i = 1, 2 \tag{6}$$

$$Q_j = UA_r(T_j - T_r) \tag{7}$$

where ΔH_i is the enthalpy change of Reaction *s*, U is the heat transfer coefficient between the jacket and reactor, A_r is the surface area of reactor conducts with the jacket.

Other important physical variables of the process are described in (8) and (9), respectively, where n = A, B, C, D. The plant parameters can be taken from Tan et al. (2011).

$$M_r = \Sigma M_n \tag{8}$$

$$C_{pr} = \frac{\Sigma(C_{\rho n} \cdot M_n)}{M_r} \tag{9}$$

where $C_{\rho n}$ is the heat capacity of substance n.

Genetic Algorithm Modelling

The GA proposed for the optimisation is based on the biological evolution theory and Darwin's natural selection concept: "survival of the fittest". In this case, it is used to manipulate the jacket inlet fluid temperature, $T_{c,}$ with a sampling time of 60 s. First, the real number chromosome representation technique is applied to represent the potential solutions (fluid temperature). After a few trials, results showed that a population size of 50 is enough in this work. The manipulated variables are bounded in the range of 0–120 °C.

Each chromosome is then evaluated by a fitness function in order to distinguish their suitability to the process optimisation. In this work, the fitter chromosome is able to maximise the production of the desired product while minimise the undesired product. The optimisation function, J, as defined in (10) is applied to evaluate the fitness of each chromosome. The chromosome with high fitness value will receive preferential treatment in procreation process later. Ranking selection technique is implemented to select the chromosomes into a mating pool so that the mating pool will not be dominated by those high fitness value chromosomes.

$$\max_{T_c} J = \int \Delta M_c dt - 6.25 \int \Delta M_D dt \quad (10)$$

During the crossover operation, two chromosomes (parents) are randomly selected from the mating pool. Then, with a probability of 90 %, both selected parents will exchange some of their information with each other and create two new chromosomes (offspring), whereby they have 10 % chance to duplicate into new generation. The blending (arithmetic) technique is employed in this crossover operation, as expressed in (11) and (12). The first offspring generated using this technique is merely the compliment of the second offspring.

$$x_1 = \beta \cdot P_1 + (1 - \beta) \cdot P_2 \quad (11)$$

$$x_2 = (1 - \beta) \cdot P_1 + \beta \cdot P_2 \quad (12)$$

where x_i is the offspring i, P_i is the parent i, and β is the random number in [0, 1].

The newly created offspring will have a 1 % chance to be mutated in order to avoid the potential solutions being trapped in the local maxima. During the mutation, a new chromosome is randomly selected from the entire solution space. The evolutionary process is stopped when 10th generation is reached, and it will return the optimal fluid temperature to the plant.

Results and Discussions

In this study, the consumption rate of the limiting reactant (substance A) is limited to 7.855 kmol so that the optimisation performance of the proposed GA and the DM can be compared equitably. The configuration of the DM is taken from Cott and Machietto (1989), which is well tuned for this case. The optimisation performance of GA and DM are shown in Fig. 2a and b, respectively.

In general, one cycle of batch process can be divided into three stages. At the beginning of the process (first stage), the reactor contents should be heated up to a certain temperature to enable the chemical reactions to take place. This happens during 0–10 min when full heating is given to the reactor and before the production starts. It can be observed from Fig. 2a-ii that the fluid temperature determined by GA is not a smooth straight line during this period compared to the DM, as shown in Fig. 2b-ii. This is because GA is a stochastic search method that searches the optimal solution through the entire solution space. In some cases, such as a batch process, there is not only one optimal solution in the solution space, and hence, it affects the GA output response to have a small fluctuation.

The second stage can be categorised as the stage where the reaction rate increases rapidly. This is the critical stage due to the huge amount of exothermic heat being released and causes the reactor temperature to increase rapidly. Hence,

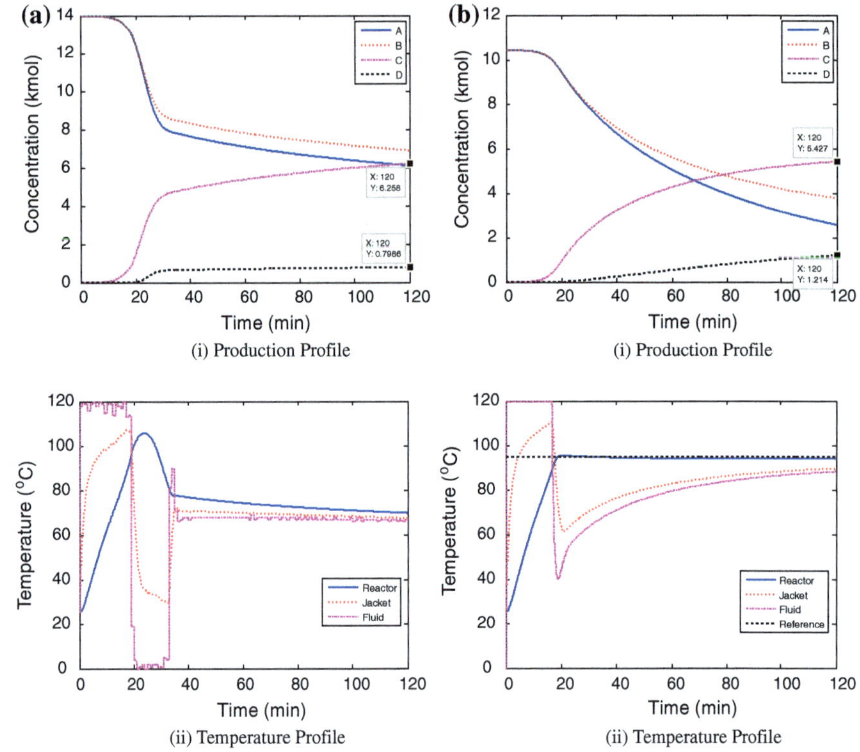

Fig. 2 Performance of genetic algorithm and dual-mode controller

the cooling system plays an important role in order to avoid the thermal runaway. The results show that the reaction starts when the reactor temperature is around 60 °C, as shown in Fig. 2a-ii and b-ii at 10 min. During the first part in this stage, both controllers continue to give full heating to the reactor whereas the cooling is given during the second part of this stage. For GA, the full heating is given during 10–20 min in order to speed up the production rate using high temperature, as shown in Fig. 2a, whereas the DM only gives full heating during 10–18 min with the purpose of increasing the reactor temperature up to the reference point in the shortest time, as shown in Fig. 2b. It can also be observed from the results that the waste production started at 20 min, when the temperature is high. In order to limit the waste production, GA proposes full cooling during 20–30 min. Conversely, DM tried to maintain the reactor temperature at the desired trajectory, rather than limit the waste production. Therefore, the waste is produced linearly from 20 min until the batch ends, as shown in Fig. 2b-i.

In the last stage, the GA seems to be slowly reducing the reactor temperature in order to limit the waste production, and at the same time, ensuring the production of desired product is increased, as shown in Fig. 2a.

Table 1 Performance of the proposed genetic algorithm and the conventional dual-mode controller in optimising the batch productivity

	Genetic algorithm	Dual-mode controller
Product C (kmol)	6.258	5.427
Undesired product D (kmol)	0.799	1.214

The overall performance show that the GA is able to harvest 6.258 kmol of desired product C and 0.796 kmol of undesired by-product D, whereas the DM only harvests 5.427 kmol of desired product C and 1.214 kmol of undesired by-product D. The results show that the desired product harvested by the GA is 15.3 % more than the DM, whereas the waste produced by the GA is 34.4 % less than the DM. Table 1 summaries the performance of the proposed GA and the conventional DM controller to optimise the batch productivity.

Conclusions

In this chapter, GA is proposed to optimise the raw materials utility for an exothermic batch process without referring to any optimal temperature profile. The performance of the developed GA is examined using a benchmark exothermic batch process model. The results show that the proposed method can perform better than the conventional DM which follows an optimal temperature profile. In future, the work will be focusing on optimising the GA development in handling various uncertainties and to improve the robustness of GA.

Acknowledgments The authors would like acknowledge the financial support of Universiti Malaysia Sabah (UMS) under Postgraduate Scholarship Scheme.

References

Cott, B. J., & Macchietto, S. (1989). Temperature control of exothermic batch reactors using generic model control. *Industrial and Engineering Chemistry Research, 28*(8), 1177–1184.

Fernandez, I., Renedo, C. J., Perez, S. F., Ortiz, A., & Manana, M. (2012). A review: Energy recovery in batch processes. *Renewable and Sustainable Energy Reviews, 16*(4), 2260–2277.

Hazard Investigation. (2002). Hazard Investigation: Improving reactive hazard management. U.S. Chemical Safety and Hazard Investigation Board. Washington, U.S.

Mujtaba, I. M., Aziz, N. & Hussain, M. A. (2006). Neural network based modelling and control in batch reactor. *Chemical Engineering Research and Design 84*(A8), 635–644.

Nisenfeld, E. (1996). *Batch control: Practical guides for measurement and control.* North Carolina: American Technical Publishers.

Sujatha, S., & Pappa, N. (2012). Performance of gain scheduled generic model controller based on BF-PSO for a batch reactor. *Asian Journal of Scientific Research, 5*(2), 31–44.

Tan, M. K., Tham, H. J. & Teo, K. T. K. (2011). PID-based temperature control for exothermic chemical reactor using hybrid QL-GA. In: 18th Regional Symposium on Chemical Engineering, 27–28 Oct, Ho Chi Minh City, Vietnam.

Q-Learning-Based Controller for Fed-Batch Yeast Fermentation

H. S. E. Chuo, M. K. Tan, H. J. Tham and K. T. K. Teo

Abstract Industrial fed-batch yeast fermentation process is a typical nonlinear dynamic process that requires good control technique and monitoring to optimize the yeast production. This chapter explores the applicability of Q-learning in determining the feed flow rate in a fed-batch yeast fermentation process to achieve multiobjectives optimization. However, to develop such control system, the complex nature of the yeast metabolism that will affect the system stability has to be considered. Q-learning is well known for its interactive properties with the process environment and is suitable for the learning of system dynamic. Therefore, the utilization and performance of Q-learning to seek for the optimal gain for the controller is studied in this chapter. Meanwhile, the performance of Q-learning under the process disturbance is also tested.

Introduction

The fed-batch application in industrial bioprocess is mainly for the substrate feeding control. A good control on substrate inflow to the bioreactor system is able to avoid serious overflow metabolism and increase the cell productivity

H. S. E. Chuo (✉) · M. K. Tan · H. J. Tham · K. T. K. Teo
School of Engineering and Information Technology, University Malaysia Sabah, Jalan UMS, 88400 Kota Kinabalu, Sabah, Malaysia
e-mail: hsechuo@gmail.com

M. K. Tan
e-mail: mktan@ieee.org

H. J. Tham
e-mail: hjtham@ums.edu.my

K. T. K. Teo
e-mail: kenteo@ums.edu.my

R. Pogaku et al. (eds.), *Developments in Sustainable Chemical and Bioprocess Technology*, DOI: 10.1007/978-1-4614-6208-8_28,

(Sonnleitner and Henes 2007). The overflow metabolite, i.e., alcohol in yeast fermentation, if highly concentrated, would be toxic and therefore suppress the cell growth. For a fed-batch fermentation process in which product concentration precision is highly required, a normal three-mode controller is bound by the sensors problem to maintain the control action (Smets et al. 2002).

In recent years, the research focus on artificial intelligence in fermentation optimization has brought about breakthroughs in classic control methods. For example, differential evolution technique (Yüzgeç 2010; Kapadi and Gudi 2004) has been applied for the stochastic search of product optimization. This approach requires the knowledge of effective optimal range in order to minimize the computation time and avoid trap in local optimization. Neural-network-based model training for model predictive control has faced challenges in training and mapping the large number of data until small-error model of satisfactory can be obtained (Ławryńczuk 2011).

The Q-learning algorithm, well known for its explorative and interactive properties with the process environment, is a worth-trying alternative to handle the optimization of the dynamic fermentation. In this chapter, the applicability and the development of Q-learning to adapt the controller gain in a fed-batch yeast fermentation is of the major interest for the multiobjectives optimization. The robustness of Q-learning-based controller under the influence of disturbance will also be discussed in this chapter, in comparison with nominal exponential feeding (Chuo et al. 2011; Teo et al. 2010) and scheduled-gain proportional control under the same process basis.

Dynamic Metabolism of Baker's Yeast Fermentation

The dynamic in the yeast fermentation process is represented by the summarized and re-arranged material balances using the parameters in Yüzgeç (2010):

$$\begin{aligned} \mathrm{d}C_s = {} & \frac{F}{V}(S_0 - C_S) \\ & - C_X\left(1.08547Q_S - 0.08547Q_{S,\mathrm{OX}} + Q_M + 1.228457Q_{E,\mathrm{OX}} + Q_M\right) \end{aligned} \tag{1}$$

$$\mathrm{d}C_X = -\frac{F}{V}C_X + C_X\left(0.05Q_S + 0.535Q_{S,\mathrm{OX}} + 0.7187Q_{E,\mathrm{OX}}\right) \tag{2}$$

$$\mathrm{d}C_E = -\frac{F}{V}C_E + C_X\left[0.4859\left(Q_S - Q_{S,\mathrm{OX}}\right) + Q_{E,\mathrm{OX}}\right] \tag{3}$$

$$\mathrm{d}C_O = -\frac{F}{V}C_O + C_X\left(0.3857Q_{S,\mathrm{OX}} + 0.8896Q_{E,\mathrm{OX}}\right) + k_La\left(C_O^* - C_O\right) \tag{4}$$

$$\mathrm{d}V = F \tag{5}$$

where dC_S, dC_X, dC_E, dC_O, and dV denote the change of concentration of substrate, yeast, ethanol, oxygen, and bioreactor volume, respectively. Assume Monod growth kinetics, the substrate consumption rate, Q_S, can go through either oxidative, $Q_{S,\mathrm{OX}}$ or fermentative, $Q_{S,\mathrm{RED}}$ consumption. The rate of oxidation of substrate in the cell, $Q_{S,\mathrm{OX}}$, is determined by the smallest rates at which glucose and oxygen are taken up by the cells. On the other hand, the rate of oxidation of ethanol, $Q_{E,\mathrm{OX}}$, in cell is determined by the smallest of the rates at which ethanol and limited oxygen are taken up by the cells. Specific growth rate, μ, and respiratory quotient, RQ, expressed in terms of Q_S, $Q_{S,\mathrm{OX}}$, and $Q_{E,\mathrm{OX}}$ are shown as below.

$$Q_S = 2.943\frac{C_S}{0.612 + C_S}\left(1 - e^{-\frac{t}{2}}\right) \tag{6}$$

$$Q_{O,\mathrm{lim}} = 0.255\frac{C_O}{9.6\times10^{-5} + C_O}\frac{3.5}{3.5 + C_e} \tag{7}$$

$$Q_{E,\mathrm{UP}} = 0.238\frac{C_e}{0.1 + C_e}\frac{3.5}{3.5 + C_S} \tag{8}$$

$$Q_{S,\mathrm{OX}} = \min\begin{pmatrix} C_S \\ 0.359 \\ Q_{O,\mathrm{lim}}/0.3857 \end{pmatrix} \tag{9}$$

$$Q_{E,\mathrm{OX}} = \min\begin{pmatrix} Q_{e,\mathrm{up}} \\ 1.1236Q_{O,\mathrm{lim}} - 0.4334Q_{S,\mathrm{OX}} \end{pmatrix} \tag{10}$$

$$\mu = 0.05Q_S + 0.535Q_{S,\mathrm{OX}} + 0.7187Q_{e,\mathrm{OX}} \tag{11}$$

$$RQ = \frac{0.1124Q_{S,\mathrm{OX}} + 0.462Q_S + 0.645Q_{e,\mathrm{OX}}}{0.3857Q_{S,\mathrm{OX}} + 0.8896Q_{e,\mathrm{OX}}} \tag{12}$$

Methodology

In a Q-learning algorithm, a learning agent actively interacts with the dynamic environment. The agent decides the best action that causes transition of state to the environment and the latter responds in a new state to the agent. A desirable action is taken based on the reward function which specifies the overall objective of the learning. A reward function assigns rewards or penalties depending on the incorporated preference indices that tell the agent the best way to achieve the goal (Syafiie et al. 2008). The difficulty of designing the reward function by putting it in the shoe of different dynamic situation lies in the determination of these preference indices and the effective range.

Multistep Action Q-Learning

Multistep action (MSA) Q-learning (Schoknecht and Riedmiller 2003) considers the inborn delay nature of the fermentation system to react to the substrate feeding. In MSA, the algorithm will look into the effect (resulting rewards) after executing a sequence of *m* number of actions.

In the proposed algorithm, constant action is executed throughout *m* number of steps before the agent decides on the next action for the next round of multistep. The execution of MSA, a^m, causes transitions to the environment, resulting in the state, s^m. The maximum Q-value comes from the execution of the best action, $(a^m)'$, determined by the reward function, r^m, resulting in the maximum state, $(S^m)'$. The Q-value is updated using the Q-learning function:

$$Q_{t+1}(s^m, a^m) \leftarrow (1-\alpha)Q_t(s^m, a^m) + \alpha\left[r^m + \gamma^m \max_{a_n \in A} Q\big((s^m)', (a^m)'\big)\right] \tag{13}$$

where $Q_t\,(s^m, a^m)$ is the Q-value for state-action at time *t*, α is the learning rate, and γ is the discount factor. The learning rate determines the importance of past experience, and the discount factor weighs the importance of near term rewards (Chuo et al. 2011).

Q-Learning-Based Controller

In this chapter, Q-learning (QL)-based controller is used to react to the gain, K_P, that multiplies the feedback error, E_t, to the substrate feed flow rate, *F*, as shown in Fig. 1.

The various states at time *t* and the calculated error, E_t, are related to the process gain, K_P, by the agent using reward function, r^m, and the Q-learning function. The argument maximum Q-value determines the optimized K_P used to tune the magnitude of the error to determine the next feed flow rate, *F*, for the *m* steps time.

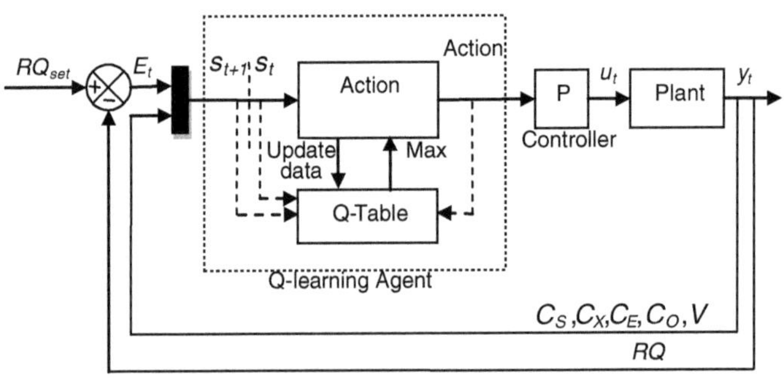

Fig. 1 Q-learning-based control process flow diagram

$$u_t = F_0 K_p E_t \tag{14}$$

$$E_t = RQ_{\text{set}} - RQ \tag{15}$$

The fermentation process was run for 10 h with a total number of 500 actions, each for m steps time. The process step size is 0.001 h and m is 0.02 h or 72 s. Within m steps, the amount of accumulated yeast (q_x) and ethanol (q_e) was stored. The accumulated values instead of the final concentrations at m steps were taken because the dynamicity of the system within a period should not be represented by the final concentration alone. The difference of average specific growth rate at m steps with the critical specific growth rate at 0.21 h^{-1} (q_μ) was also calculated for reward calculation using the reward function:

$$r^m = \beta_1 q_x - \beta_2 q_e - \beta_3 q_\mu \tag{16}$$

where β_1, β_2, and β_3 indicates the weight of the preference index qx, qe, and $q\mu$, respectively. In this case, $\beta_1 = 1$, $\beta_2 = 100$, and $\beta_3 = 10$. More yeast production will result in higher r^m, and more ethanol production will result in lower r^m. Therefore, only actions with higher rewards will be chosen, which has represented the multiobjectives optimization. The K_P with the highest Q-value will be chosen as the gain for the QL-based proportional controller at each m step.

Results and Discussion

For the nominal case, the initial concentration of glucose (C_s), yeast (C_x), ethanol (C_e), and oxygen (C_o) are 3.5, 9.0, 0, and 0.008 g/l, respectively. The substrate feeding stream has a concentration of 325 g/l, initial volume of 50,000 l, and initial feed flow rate 100 l/h. The following results showed the concentration profiles of glucose, yeast (to be maximized), and ethanol (to be minimized) for (1) nominal exponential feeding, $F = 100e^{0.3t}$ (Fig. 2a), (2) scheduled-proportional control, with linear K_P increment of 3 for each m steps, starting at initial $K_P = 0$ (Fig. 2b), and (3) QL-based control (Fig. 2c and d).

Under optimum scheduled-proportional control, the final yeast and ethanol concentrations are 32.91 g/l and negligible, respectively, more robust compare to nominal exponential feeding of 28.21 and 0.0025 g/l, respectively. Under the same initial conditions and using the range of gain, K_P, suggested by trial-and-error tuning scheduled-proportional controller, i.e., $0 \leq K_P \leq 1{,}500$, Q-learning-based controller can adjust and seek for the optimal gain. Maximum final yeast concentration and minimum final ethanol concentration, i.e., 40.42 g/l and negligible, respectively, are obtained for Q-learning-based control. On the other hand, disturbance in substrate feeding concentration can happen in fermentation and overfeeding of substrate tends to trigger the overflow metabolism. In this case, it is introduced as substrate concentration change of 525 g/l at $1 < t \leq 1.5$ h. Scheduled-proportional controller is unable to control the increasing ethanol

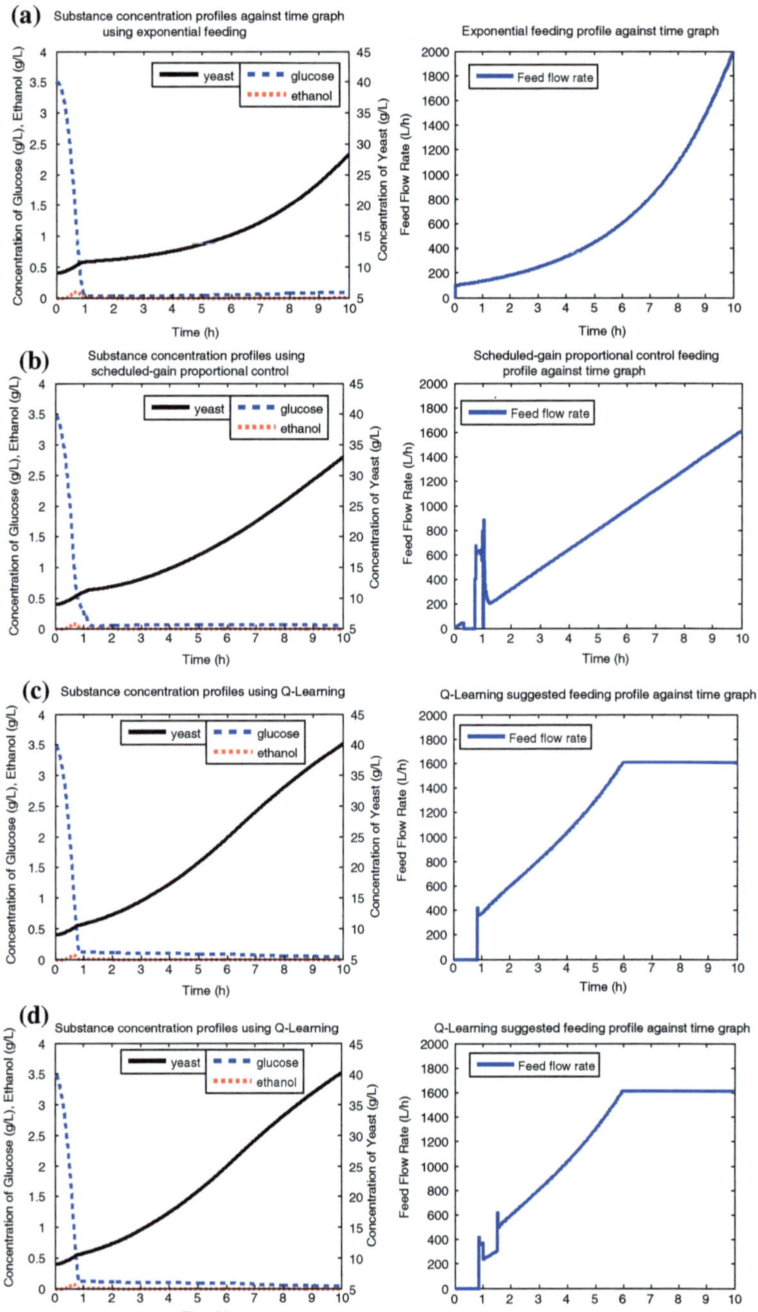

Fig. 2 Concentration profiles of glucose, yeast, and ethanol (*left*) and the substrate feeding profile (*right*) for **a** nominal exponential feeding; **b** scheduled-proportional control; **c** Q-learning-based control and **d** Q-learning-based control under disturbance of $S_0 = 525$ g/l at $1 < t \leq 1.5$

concentration in the case of disturbance. The final concentrations of yeast and ethanol are 33.13 and 19.02 g/l, respectively, for scheduled-proportional controller; 40.49 g/l, negligible, respectively, for Q-learning-based control, as shown in Fig. 2d. Q-learning is able to adapt the process gain when necessary and achieve its goal of maximizing yeast and minimizing ethanol even under the disturbance, as shown in the feeding profile in Fig. 2d.

Conclusions

In this study, Q-learning has shown result of satisfactory in tuning the controller gain to achieve multiobjectives optimization, compare to nominal exponential feeding and scheduled-proportional controller. Meanwhile, Q-learning is able to seek for optimal process gain that can reduce the disturbance effect in the substrate feeding stream rather than increasing-gain scheduled-proportional controller.

Acknowledgments Gratitude is expressed to the Universiti Malaysia Sabah and the Ministry of Higher Education, Malaysia, for their financial support on this research.

References

Chuo, H. S. E., Tan, M. K., Tham, H. J., & Teo, K. T. K. (2011). Optimization of fed-batch baker's yeast fermentation process using learning algorithm. In *Proceedings of 18th Regional Symposium on Chemical Engineering*, 27–28 Oct, Ho Chi Minh City, Vietnam.

Kapadi, M. D., & Gudi, R. D. (2004). Optimal control of fed-batch fermentation involving multiple feeds using differential evolution. *Process Biochemistry, 39*, 1709–1721.

Ławryńczuk, M. (2011). Online set-point optimization cooperating with predictive control of a yeast fermentation process: A neural network approach. *Process Engineering Applications of Artificial Intelligence, 24*, 968–982.

Schoknecht, R., & Riedmiller, M. (2003). Learning to control at multiple time scales. In *Proceedings of ICANN 2003*, 26–29 June, Istanbul, Turkey.

Smets, I. Y., Bastin, G. P., & van Impe, J. F. (2002). Feedback stabilization of fed-batch bioreactors: non-monotonic growth kinetics. *Biotechnology Progress, 18*, 1116–1125.

Sonnleitner, B., & Henes, B. (2007). Controlled fed-batch by tracking the maximal culture capacity. *Journal of Biotechnology, 132*, 118–126.

Syafiie, S., Tadeo, F., & Martinez, E. (2008). Model-free learning control of chemical processes. In C. Weber, M. Elshaw & N. M. Mayer (Eds.), *Reinforcement learning: theory and applications.* Vienna: I-Tech Education and Publishing.

Teo, K. T. K., Tham, H. J., & Tan, M. K. (2010). Optimization of substrate feed flow rate for fed-batch yeast fermentation process. In *Second International Conference on Computational Intelligence, Modelling and Simulation*, 28–30 Sept, Bali, Indonesia.

Yüzgeç, U. (2010). Performance comparison of differential evolution techniques on optimization of feeding profile for an industrial scale baker's yeast fermentation process. *ISA Transactions, 49*, 167–176.

In Vitro Antimicrobial Activity of *Cynodon dactylon* (L.) Pers. (bermuda) Against Selected Pathogens

S. Abdullah, J. Gobilik and K. P. Chong

Abstract *Cynodon dactylon* (L.) Pers. is a type of perennial grass that possesses great medicinal values. In this study, the antimicrobial activity of the plant crude extract from seven different solvents (acetone, chloroform, diethyl ether, ethanol, ethyl acetate, methanol, and n-pentane) was investigated against some pathogens (*Bacillus cereus, Bacillus subtilis, Escherichia coli, Klebsiella* spp., *Pseudomonas aeruginosa, Staphylococcus aureus, Streptococcus pyogenes,* and *Streptococcus pneumonia*) using disc diffusion method and thin-layer chromatographic (TLC) bioassay for plant-SPE extracts against *Aspergillus niger*. Crude extraction showed that ethanolic extraction produced highest yield (7.065 %) followed by methanolic (5.420 %) and chloroform (3.550 %) extraction. The lowest yield was obtained from n-pentane extraction (0.500 %). The antimicrobial study revealed broad spectrum of antimicrobial activity from ethanol (7.0–10.0 ± 0.0–1.0 mm) and ethyl acetate (7.0–12.0 ± 0.0–1.0 mm) extracts against all of the bacterial pathogens. Both methanol and acetone extracts showed activity to *B. cereus* (8.0 ± 0.0 mm) and *B. subtilis* (7.0 ± 0.0 mm), while chloroform extract showed activity to *B. subtilis* (7.0 ± 0.0 mm) and *S.pyogenes* (8.3 ± 0.6 mm), respectively. Diethyl ether extraction showed activity only to *S. pyogenes* (7.3 ± 0.6 mm), while no activity was observed from n-pentane extraction. Great antimicrobial activity were observed for both ethyl acetate and ethanol SPE-based extracts (SBE) with size of inhibition ranging from 8.0 ± 0.0 mm to 15.7 ± 0.6 mm for ethyl acetate SBE and 8.0 ± 0.0 mm to 13.0 ± 0.0 mm for ethanol SBE. No significant antimicrobial activity was observed from thin-layer chromatographic bioassay against *A. niger*.

S. Abdullah · K. P. Chong (✉)
Sustainable Palm Oil Research Unit (SPOR), School of Science and Technology, Universiti Malaysia Sabah, 88400 Kota Kinabalu, Sabah, Malaysia
e-mail: chongkp@ums.edu

J. Gobilik
School of Sustainable Agriculture, Universiti Malaysia Sabah, Sandakan Campus, Mile 10, Sg. Batang 90000 Sandakan, Sabah, Malaysia
e-mail: jgobilik@ums.edu.my

R. Pogaku et al. (eds.), *Developments in Sustainable Chemical and Bioprocess Technology*, DOI: 10.1007/978-1-4614-6208-8_29,

Keywords *Cynodon dactylon* • Crude extract • SPE-based extract • TLC bioassay • Antimicrobial

Introduction

Medicinal plants play a very important role in pharmaceuticals industry in developing alternative drugs to overcome the pitfalls possessed by the synthetic drugs (Yadav et al. 2010). The development of drug-resistant pathogens that are mostly involved in nosocomial infection has raised concern among medicinal practitioners (Adalet et al. 2011). It was believed that the intense used of a number of synthetic antimicrobial drugs which contributed to the development. Besides the tougher jobs to search for more effective drugs against the pathogens, it also created the problem in controlling the growth of infectious diseases caused by the pathogens. Meanwhile, most of the synthetic drugs possess side effect to the consumers (Chaudhari et al. 2011). Given the alarming incidence of antibiotic resistance in pathogens raises concern among the medical practitioners, there is a constant need for new and effective therapeutic agents. Hence, there is a need to develop alternative antimicrobial drugs for the treatment of infectious diseases from medicinal plants. Plants produce large number of organic compounds such as alkaloids, flavonoids, glycosides, tannins, terpenoids, and phenolics as secondary metabolites which are used as defensive mechanism against respective pathogens (Francisco and Pinotti 2000; Zwenger and Basu 2008; Chong et al. 2011). These compounds possess medicinal values and become an attractive subject for researchers to develop new antibiotics (Joy et al. 2001). Previous studies show that many plant secondary metabolites act as bioactive compounds, chemotherapeutic, bactericidal, and bacteriostatic agents (Singh and Gupta 2008; Pongsak and Parichat 2010; Syahriel et al. 2012). As a result, considerable amount of attention of antimicrobial substances derived from higher plants is immerging in recent years. However, the assessments on plant-derived antibiotics are still under investigation. *Cynodon dactylon* is a perennial, pan-tropical species of grass which belongs to the family Poaceae. It is found almost everywhere in tropical, sub-tropical, and even in semi-arid climates (Watson and Dallwitz 1992). It is not only widespread, but also widely used by human. Scientifically, it is tested to have antidiabetic effect, diuretic activity, antioxidant, anticancer potentials, antiulcer activity, right heart failure protective activity, and allergic effect (Singh et al. 2007; Albert-Baskar and Ignacimuthu 2010). Traditionally, *C. dactylon* is used as a rejuvenator and for wound healing (Mangathayaru et al. 2009). Besides used in remedy, the plant is also used as forage for animals and as turfgrass (Wu et al. 2007). This study will report the antimicrobial activity of *C. dactylon* against some common pathogens such as *Escherichia coli, Staphylococcus aureus,* and *Pseudomonas aeruginosa,* which are highly associated with nosocomial infection.

Methodology

Plant Collection (Cynodon Dactylon)

Wild ecotype of *C. dactylon* was collected near the Universiti Malaysia Sabah's (UMS) area in Kota Kinabalu, Malaysia, sowed, and maintained in a mini-nursery in campus. Voucher (jgobilik 1090/2011) was kept at School of Sustainable Agriculture, UMS, and a duplicate was submitted to BORH Herbarium, Institute of Tropical Biology and Conservation, UMS. Samples were collected in the evening during the daylight time. Prior to extraction, the plant samples were cleaned thoroughly with distilled water to remove soil and dirt.

Plant Extraction

The whole part of *C. dactylon* which was cleaned by distilled water was shade-dried for 24 h in a drying chamber at 40–50 °C and powdered using a mechanical blender (Waring® Commercial Blender). Approximately, 100 g of plant powder was later soaked into 200 mL of different solvents (acetone, chloroform, diethyl ether, ethanol, ethyl acetate, n-pentane, and methanol–Merck) and shaken on a platform shaker (LabCompanion™) at 150 rpm with temperature of 25 °C to obtain various plant extracts. The soaking process was repeated three times for each extraction to obtain a complete extraction. The extracts obtained were then evaporated and concentrated under reduced pressure (768–7 mmhg) using Rota-Vapor™ (BUCHI) to obtain 1 ml of extract per 10 g of plant sample. Aliquots were then kept in −20 °C temperature for further use.

Solid-Phase Extraction

Strata™-X 33um Polymeric Sorbent reverse-phase (200 mg/6 mL) (Phenomenex) cartridges with 12-cartridge manifold system was used. Methanol absolute (1 ml) was used to activate the sorbent and further equilibrated with sterile deionised distilled water (1 ml). Samples were then loaded into the cartridges and washed with 1 % methanol (1 ml) to remove any impurities from the samples and finally eluted with 2 ml of methanol:acetonitrile (1:1; v/v). The aliquots were taken to dryness using purified nitrogen gas. Dried aliquots were stored in −20 °C for further bioassays. In this study, only ethyl acetate and ethanol extracts were used for SPE-based extract bioassay.

Test Microorganisms

Pure cultures of eight different pathogens (*Bacillus cereus, Bacillus subtilis, E. coli, Klebsiella* spp., *P. aeruginosa, S. aureus, Streptococcus pyogenes,* and *Streptococcus pneumonia*) were obtained from Queen Elizabeth Hospital, Kota Kinabalu, Malaysia. The microbial cultures were preserved in 30 % glycerol stock solution at −85 °C temperature. Prior to antimicrobial activity study, the test microorganisms were subcultured on nutrient agar (NA), incubated at 37 °C for 24 h before grown in nutrient broth (NB) and adjusted according to McFarland standard to achieve approximately 1×10^8 CFU/ml before introducing to the test media. For thin-layer chromatographic (TLC) bioassay, *Aspergillus niger* was obtained from stock culture of UMS. The *A. niger* was cultured on potato dextrose agar (PDA) for approximately 7–10 days, before it is suspended in potato dextrose broth (PDB) prior to TLC bioassay.

Antimicrobial Activity Study

Thin-layer chromatographic bioassay: Thin-layer chromatography (TLC) (Silica Gel 60 F_{254}—Merck) bioassay coupled with *A. niger* culture was used as preliminary screening for antimicrobial activity. Approximately, 2 mg of *C. Dactylon* ethanol and ethyl acetate extracts were applied on TLC plate which developed in ethyl acetate:hexane (1:9) solvent system and then examined under UV light (265 and 325 nm) to spot the presence of compounds of interest. The plates later were sprayed with 7–10 day-old *A.niger* which was suspended in PDB. TLC plates were incubated in a moist chamber for 48 h at 25–27 °C to spot any antimicrobial property of the separated compounds from the plant.

Disc diffusion bioassay: Antimicrobial activity of the plant extracts was evaluated using disc diffusion method according to Kirby-Bauer method as described by Lalitha (2011). Sterile discs of Whatman® No.3 paper (6 mm diameter) were prepared for the disc diffusion bioassay. All extract concentrations were standardised to 100 mg/mL, and 60 μl (6 mg) of the extracts were loaded to each disc. For the negative control discs, 60 μL of the respective solvents was loaded to separate discs. Chloramphenicol antibiotic discs (1 mg/ml, 30 μg/disc) were used as positive control to compare the antimicrobial activity. Mueller-Hinton Agar (MHA) medium was prepared and sterilised in an autoclave at 121 °C for 15 min at 15 psi before transferring it to sterilised petri dish. Approximately, 0.1 ml culture of bacterial pathogens adjusted according to McFarland standard was placed on the MHA media and spread throughout the plate using spread plate technique. The discs loaded with test extracts, their corresponding solvents, and the antibiotic discs were placed on the media with the help of a sterile forceps carefully with adequate spacing between each other. The plates were kept at room temperature for 30 min, which helps to diffuse the extract into the medium.

The plates were then incubated at 37 °C for 24 h in an incubator to determine the antibacterial activity of the respective solvent extraction of *C. dactylon*. After incubation, zone of inhibition in diameter was measured and recorded.

Results

Extraction Yield

Different solvents have different resolving strength towards the plant constituents which resulted in different yield as shown in Table 1. From all of the extractions, ethanolic extraction produced the highest yield (7.065 %) followed by methanolic extraction (5.420 %) and chloroform extraction (3.550 %). Most of the extractions using non-polar solvents produced lower yield. The lowest yield was obtained from n-pentane extraction with only 0.500 % of yield. The other extraction yield results are as shown in the Table 1.

Antimicrobial Activity

Thin-layer chromatographic bioassay: Separation of bioactive compounds from *C. dacylon* for ethanol and ethyl acetate extracts was the best in ethyl acetate:hexane (1:9) solvent system which produces higher resolution and greater separation, as shown in Fig. 1. However, the antimicrobial properties from both ethanol and ethyl acetate extracts were not detected (Fig. 2). The extracts were unable to inhibit the growth of *A. niger* after incubated for 48 h.

Disc diffusion bioassay (crude extract): Ethanol and ethyl acetate extracts of the plant exhibited broad spectrum of antimicrobial activity. From Table 2, both ethanol and ethyl acetate extracts showed significant effect to almost all the tested pathogens with the size of inhibition between 7.0 ± 0.0 mm and 10.0 ± 1.0 mm for ethanol extract and 7.0 ± 0.0 and 12.0 ± 1.0 mm for ethyl acetate extract. The greatest activity observed was against *S. pyogenes* (10.0 ± 1.0 mm) for ethanol extract while *B. cereus* (12.0 ± 1.0 mm) for ethyl acetate extract. The least activity observed was against *B.subtilis* (7.0 ± 0.0 mm) for ethanol extract while *S. pneumonia* (7.0 ± 0.0 mm) for ethyl acetate extract. Methanol extract showed very weak activity against *Streptococcus pneumoniae* (7.0 ± 0.0 mm), while both methanol and acetone showed weak activity against *B. cereus* (8.0 ± 0.0 mm) and

Table 1 Extraction yield of *C. dactylon* with different solvents

Solvent	Acetone	Chloroform	Diethyl ether	Ethanol	Ethyl acetate	Methanol	n-pentane
Yield (%)	2.9605	3.550	0.923	7.065	1.218	5.420	0.500

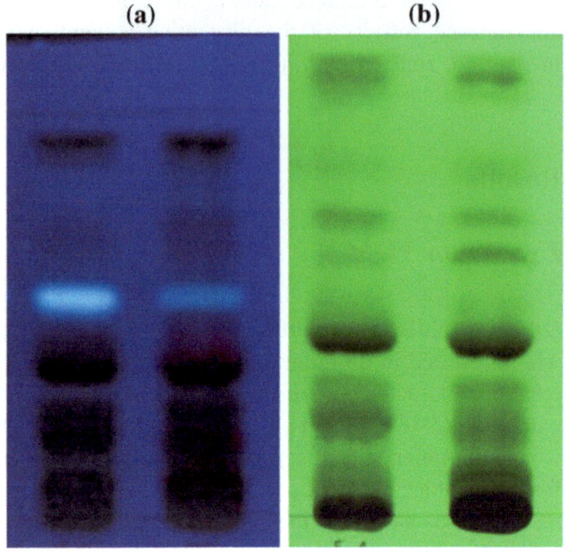

Fig. 1 Separation of bioactive compounds from *C.dactylon* by TLC, E.A:Hexane (2:18) solvent system, observed under UV light (i = 254 nm; ii = 365 nm). **a** Ethyl acetate extract; **b**: Ethanol extract

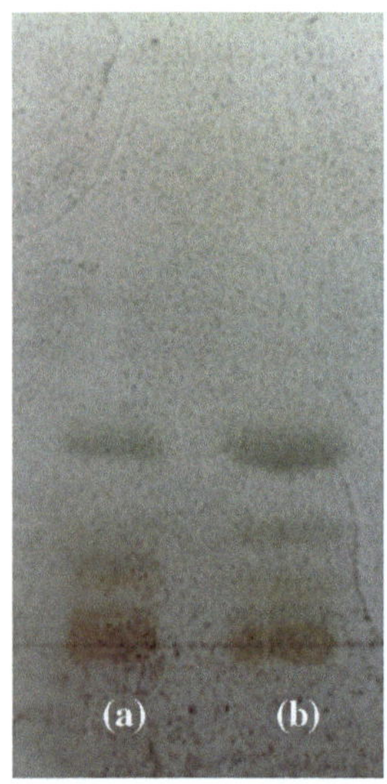

Fig. 2 Thin-layer chromatographic bioassay of ethyl acetate (E.A) and ethanol (EtOH) extracts of *C. dactylon*. *A. niger* was able to grow on the plate after incubated for 48 h. **a** E.A extract; **b** EtOH extract

B. subtilis (7.0 ± 0.0 mm). Chloroform extract showed inhibition to two pathogens tested, *B. subtilis* (7.0 ± 0.0 mm) and *S. pyogenes* (8.3 ± 0.6 mm). Diethyl ether extract exhibited the least activity only to *S. pyogenes* (7.3 ± 0.6 mm). Meanwhile, n-pentane extract did not inhibit any of the tested pathogens. In the present study, *B. subtilis* (except for the diethyl ether and n-pentane extracts) and *S. pyogenes* (except for acetone and n-pentane extracts) were more sensitive towards most of the extracts. Meanwhile, *E. coli*, *P. aeruginosa*, *S. aureus,*

Table 2 Antimicrobial activity of *C.dactylon*crude extracts with the respective solvents against pathogens

Tested microbial pathogens	Size of inhibition zone (mm) of the plant extract with the respective solvents*							
	Acetone	Chloroform	Diethyl ether	Ethanol	Ethyl acetate	Methanol	n-pentane	CHL
Bacillus cereus	8.0 ± 0.0	n.d	n.d	9.0 ± 0.0	12.0 ± 1.0	8.0 ± 0.0	n.d	24.0 ± 1.0
Bacillus subtilis	7.0 ± 0.0	7.0 ± 0.0	n.d	7.0 ± 0.0	8.0 ± 0.0	7.0 ± 0.0	n.d	22.3 ± 0.6
Escherichia coli	n.d	n.d	n.d	8.3 ± 0.6	8.0 ± 0.0	n.d	n.d	23.3 ± 0.6
Klebsiella spp.	n.d	n.d	n.d	8.3 ± 0.6	8.0 ± 0.0	n.d	n.d	20.0 ± 1.0
Pseudomonas aeruginosa	n.d	n.d	n.d	8.0 ± 0.0	9.3 ± 0.6	n.d	n.d	24.0 ± 1.0
Staphylococcus aureus	n.d	n.d	n.d	9.0 ± 1.0	10.0 ± 0.0	n.d	n.d	22.3 ± 0.6
Streptococcus pyogenes	n.d	8.3 ± 0.6	7.3 ± 0.6	10.0 ± 1.0	8.0 ± 0.0	7.0 ± 0.0	n.d	24.0 ± 0.0
Streptococcus pneumoniae	n.d	n.d	n.d	7.3 ± 0.6	7.0 ± 0.0	n.d	n.d	21.0 ± 1.0

*Values presented are means of three replicates, ± stand. dev.; each disc loaded with approx. 60 ul or 12 mg/disc of plant extract
n.d = not detected
C CHL = Chloramphenicol (1 mg/ml, 30 μg/disc)

Table 3 Antimicrobial activity of *C. dactylon* EtOH- and EA SPE-based extraction against pathogens after 17-h incubation

Tested microbial pathogens	Size of inhibition zone (mm)*				
	A	B	C	D	CHL
Bacillus cereus	8.0 ± 0.0	8.0 ± 0.0	15.7 ± 0.6	13.0 ± 0.0	23.7 ± 0.6
Bacillus subtilis	7.0 ± 0.0	7.0 ± 0.0	11.0 ± 0.0	8.0 ± 0.0	20.0 ± 0.0
Escherichia coli	n.d	n.d	9.0 ± 0.0	8.0 ± 0.0	24.3 ± 0.6
Klebsiella spp.	7.0 ± 0.0	8.0 ± 0.0	8.0 ± 0.0	7.0 ± 0.0	25.7 ± 0.6
Pseudomonas aeruginosa	7.0 ± 0.0	7.0 ± 0.0	10.0 ± 0.0	7.0 ± 0.0	19.0 ± 1.0
Streptococcus pyogenes	7.0 ± 0.0	n.d	8.0 ± 0.0	11.0 ± 1.0	19.0 ± 0.0
Streptococcus pneumonia	n.d	8.0 ± 0.0	10.0 ± 0.0	8.0 ± 0.0	18.0 ± 0.0
Staphylococcus Aureus	n.d	7.0 ± 0.0	10.0 ± 0.0	9.0 ± 0.0	18.0 ± 1.0

*Values presented are means of three replicates, ± stand. dev.; each disc loaded with approx. 60 ul or 12 mg/disc of plant extract; *A* EA SPE-based extract (elute portion), *B* EtOH SPE-based extract (elute portion), *C* EA SPE-based extract (flush portion), *D* EtOH SPE-based extract (flush portion); n.d = not detected; EA = Ethyl acetate, EtOH = Ethanol, SPE = Solid-Phase Extraction
CHL = Chloramphenicol (1 mg/ml, 30 μg/disc)

Klebsiella spp, and *S. pneumoniae* were only sensitive to ethanol and ethyl acetate extracts. No growth or activity was observed on negative discs. Strong inhibition to all the pathogens' growth observed on standard chloramphenicol disc proved none of the microorganisms tested were chloramphenicol resistant.

Disc diffusion bioassay (for solid-phase extraction-based extract, SBE): Flush portion from both ethyl acetate and ethanol SPE-based extracts exhibited greater antimicrobial properties in contrast to the elute portion of the extracts. As shown in Table 3, both ethyl acetate and ethanol SBEs showed significant activity to almost all the tested pathogens with size of inhibition ranging from 8.0 ± 0.0 to 15.7 ± 0.6 mm for ethyl acetate SBE and from 13.0 ± 0.0 mm to 8.0 ± 0.0 mm for ethanol SBE. In accordance to the bioassay of the ethyl acetate crude extract, the greatest activity for ethyl acetate SBE was observed against *B. cereus* with 15.7 ± 0.0 mm of inhibition diameter, higher than the crude extract (12.0 ± 1.0 mm) followed by *B. subtilis* (11.0 ± 0.0 mm), *P. aeruginosa, S. pneumonia,* and *S. aureus* with 10.0 ± 0.0 mm of inhibition diameter, respectively. The weakest activity was observed against *Klebsiella* spp. and *S. pyogenes* with 8.0 ± 0.0 mm of inhibition diameter, respectively. The greatest activity for ethanol SBE was against *B. cereus* with 13.0 ± 0.0 mm of inhibition diameter, in contrast to ethanol crude extract in which the greatest activity was observed against *S. pyogenes* (10.0 ± 1.0 mm). Ethanol SBE also showed activity against *S. pyogenes*, with 11.0 ± 1.0 mm of inhibition diameter, and some activity against *S. aureus* (9.0 ± 0.0 mm). The weakest activity for ethanol SBE was observed against *Klebsiella* spp. and *P. aeruginosa* with 7.0 ± 0.0 mm of inhibition diameter, respectively.

Discussion

Higher plants consist of wide range of bioactive compounds, such as alkaloids, terpenoids, flavonoids, saponins, tannins, etc. which were utilised by the plants themselves as a defensive mechanism and to maintain the plant biological activities (Zwenger and Basu 2008). From the extraction yield in Table 1, polar solvents were able to produce higher yield. This implies that most of the plant constituents are polar compounds such as saponins and the phenolics. Previously, the plant has been studied to contain many bioactive constituents which contributed to its antimicrobial activity (Syahriel et al. 2012). Biological actions are primarily due to these components in a very complicated concert of synergistic or antagonistic activities. Mixtures of such chemicals show a broad spectrum of biological effects and pharmacological properties. To a large extent, the age of the plant, percentage humidity of the harvested material, situation and time of harvest, and the method of extraction are possible sources of variation for the chemical composition, toxicity, and bioactivity of the extracts (Mandal et al. 2007). From the previous study, the abundance of glycoside and its derivatives were identified from the plant (Syahriel et al. 2012). Cardiac glycosides from *C. dactylon* were previously studied to possess antiarrhythmic activity against ischaemia or reperfusion-induced arrhythmias and cardioprotective properties in tested rat (Albert-Baskar and Ignacimuthu 2010; Najafi et al. 2008). A large number of constitutive plant compounds have been reported to have antimicrobial activity. Due to the great extent of pharmacological effects exert by the plant glycoside, the antimicrobial activity exhibited by the plant in the present study could be induced by the derivatives from this carbohydrate constituent. Carbohydrate and fatty acid derivatives from natural source have been proven to possess broad-spectrum antimicrobial activity (Nobmann et al. 2009). The antibacterial activity of ethanol extract was believed to be due to the presence of active principle in the extracts such as saponins, phenolics, and terpenoids which might be responsible for the broad spectrum of antibacterial activity compared to the other extracts (Kafaru 1994; Singh and Gupta 2008). Higher resolving strength of ethanol in regards to its yield percentage consequently enables it to resolve comparatively more bioactive compounds which might explain the considerable antimicrobial activity compared to the other solvents. Meanwhile, other solvents such as ethyl acetate were able to resolve other trace bioactive constituents which are not being able to be resolved by ethanol in greater amount, explaining the significant antimicrobial activity which levelled to the ethanolic extraction. Generally, gram-negative bacteria were more resistant to antibiotics than gram-positive bacteria. The resistance is due to the differences in their cell wall composition. In gram-negative bacteria, the outer membrane acts as a great barrier to many environmental substances including antibiotics. Presence of thick murine layer in the cell wall prevents the entry of the inhibitors (Madigan et al. 2009). The present study revealed that there is no significance difference between gram-negative and gram-positive bacteria in terms of susceptibility to the crude extracts although for the extracts other than ethanol and ethyl acetate extracts, the activity was mostly observed against gram-positive bacteria. However, no significant difference in the antimicrobial activity between gram-positive and gram-negative bacteria was observed for both ethanol and ethyl acetate extracts. It may be due to the presence of broad-spectrum antibiotic compounds in *C. dactylon* which are able to penetrate or deteriorate the defensive or growth mechanism of the microorganisms. The potential of antimicrobial compound sources from the plant extract is worth to be further explored by identifying the bioactive constituents responsible for the action.

Acknowledgments The authors acknowledge their profound gratitude to Ministry of Higher Education (MOHE), Malaysia for funding the research (Grant code FRG0288) and the School of Science and Technology, Universiti Malaysia Sabah for providing the facilities for research work. We are indebted to Dr. Katrina and the team, Department of Pathology, Queen Elizabeth Hospital for their support in supplying the biological specimens.

References

Adalet, A., Cenk, A., Azap, A., Onder, E., & Ismail, B. (2011). The impact of a nationwide antibiotic restriction program on antibiotic usage and resistance against nosocomial pathogens in Turkey. *International Journal of Medical Science, 4*, 339–344.

Albert-Baskar, A., & Ignacimuthu, A. (2010). Chemopreventive effect of *Cynodon dactylon* (L.) Pers. extract against DMH-induced colon carcinogenesis in experimental animals. *Experimental and Toxicology Pathology, 62*, 423–431.

Chaudhari, Y., Mody, H. R., & Acharya, V. B. (2011). Antibacterial activity of *Cynodon dactylon* on different bacterial pathogens isolated from clinical samples. *International Journal of Pharmaceutical Studies and Research, 1*, 16–20.

Chong, K. P., Rossall, S., & Atong, M. (2011). HPLC Fingerprints and in vitro antimicrobial activity of syringic acid, caffeic acid and 4-hydroxybenzoic acid against *Ganoderma boninense*. *Journal of Applied Sciences, 13*, 2284–2291.

Francisco, I. A., & Pinotti, M. H. P. (2000). Cyanogenic glycosides in plants. *Brazilian Archives of Biology & Technology, 43*, 487–492.

Joy, P. P., Thomas, J., Mathew, J., & Skaria, B. P. (2001). Medicinal Plants. *In*: T. K. Bose, J. Kabir, P. Das, & P. P. Joy (eds.) *Tropical Horticulture, Vol. 2*. Calcutta: Naya Prokash.

Kafaru, E. (1994). *Immense help from Nature's Workshop*. Lagos: Elikat Health Services.

Lalitha, M. K. (2011). *Methods of antimicrobial susceptibility testing in Manual on antimicrobial susceptibility testing*. Vellore: Department of Microbiology, Christian Medical College.

Madigan, M. T., Martinko, J. M., Dunlap, P. V., & Clark, D. P. (2009). *The principles of bacterial cell wall In Brock Biology of Microorganisms* (12th ed.). San Francisco: Pearson Benjamin Cummings.

Mandal, V., Mohan, Y., & Hemalatha, S. (2007). Microwave assisted extraction—an innovative and promising extraction tool for medicinal plant research. *PHCOG REV: Review article. Pharmacognosy Reviews, 1*, 7–17.

Mangathayaru, K., Umadevi, M., & Reddy, C. U. (2009). Evaluation of the immunomodulatory and DNA protective activities of the shoots of *Cynodon dactylon*. *Journal of Ethnopharmacology, 123*, 181–184.

Najafi, M., Nazemiyeh, H., Ghavimi, H., Gharakhani, A., & Garjani, A. (2008). Effects of the hydroalcoholic extract of *Cynodon dactylon* (L.) pers. on ischemia/reperfusion-induced arrhythmias. *DARU: Journal of Pharmaceuticals Science, 16*, 233–238.

Nobmann, P., Bourke, P., Dunne, J., & Henehan, G. (2009). In vitro antimicrobial activity and mechanism of action of novel carbohydrate fatty acid derivatives against *Staphylococcus aureus* and MRSA. *Journal of Applied Microbiology, 108*, 2152–2161.

Pongsak, R., & Parichat, P. (2010). Contents and antibacterial activity of flavonoids extracted from leaves of *Psidium guajava*. *Journal of Medicinal Plants Research, 5*, 393–396.

Singh, R., & Gupta, A. (2008). Antimicrobial and antitumor activity of the fractionated extracts of Kuli musli (*Curculg oorchioides*). *International Journal of Green Pharmacy, 1*, 34–36.

Singh, S. K., Kesari, A. N., Gupta, R. K., Jaiswal, D., & Watal, G. (2007). Assessment of antidiabetic potential of *Cynodon dactylon* extract in streptozotocin diabetic rats. *Journal of Ethnopharmacology, 114*, 174–179.

Syahriel, A., Gobilik, G., & Chong, K. P. (2012). Preliminary phytochemical study and antimicrobial activity from various extract of *Cynodon dactylon* (L.) Pers. (Bermuda) against selected pathogens. *International Journal of Pharmacy Pharmaceutical Science, 4*, 227–230.

Watson, L., & Dallwitz, M. J. (1992). *The grass genera of the world.* Wallingford: CAB International.

Wu, Y. Q., Taliaferro, C. M., Martin, D. L., Anderson, J. A., & Anderson, M. P. (2007). Genetic variability and relationships for adaptive, morphological, and biomass traits in Chinese Bermuda grass accessions. *Crop Science, 47*, 1985–1994.

Yadav, Y. C., Jain, A., & Deb, L. (2010). Neuropharmacological screening techniques for pharmaceuticals—a review. *International Journal of Pharmacy Pharmaceutical Science, 2*, 10–11.

Zwenger, S., & Basu, C. (2008). Plant terpenoids: applications and future potentials. *Biotechnology and Molecular Biology Review, 3*, 1–7.

Part III
Separation Processes

Extraction of Glucose from Kenaf Core Using Mild Acid Treatment

S. M. Nurhafizah, M. Maizirwan, H. Anuar, R. Othman and M. N. Nur Aimi

Abstract Recently, many industries are aiming to reduce the usage of petroleum-based products and synthetic fibres due to increased environmental concerns regarding such materials. This has led to extensive research on natural fibres with the goal of producing eco-friendly products that will replace the existing petroleum-based ones. Natural fibres can be derived from many sources such as plants, animals, or minerals. However, plant fibres, such as kenaf, are more desirable to most researchers. Kenaf, scientifically known as *Hibiscus cannabinus. L*, has a complex structure because it consists of lignin, cellulose, and hemicelluloses. Due to this reason, kenaf needs to undergo a treatment process in order to remove lignin and hemicellulose, reduce the crystallinity of its constituent cellulose, and increase its porosity. In this study, acid treatment method was employed and the influence of several process parameters, such as temperature and time, for achieving high yield of glucose conversion was studied. The highest substantial glucose yield was 3.4 g/L produced at 200 °C after 60 min.

S. M. Nurhafizah (✉) · H. Anuar · M. N. Nur Aimi
Department of Manufacturing and Materials Engineering, Kulliyyah of Engineering, International Islamic University Malaysia, 53100 Kuala Lumpur, Malaysia
e-mail: nurhafizah.seeni@gmail.com

M. Maizirwan
Department of Biotechnology Engineering, Kulliyyah of Engineering, International Islamic University Malaysia, 53100 Kuala Lumpur, Malaysia

R. Othman
Herbarium Laboratory, Kulliyyah of Architecture and Environmental Design, International Islamic University Malaysia, 53100 Kuala Lumpur, Malaysia

R. Pogaku et al. (eds.), *Developments in Sustainable Chemical and Bioprocess Technology*, DOI: 10.1007/978-1-4614-6208-8_30,

Introduction

Lately, biopolymer industries are attempting to decrease their dependence on petroleum-based fuels and products due to increased environmental consciousness. This phenomenon leads to the need for replacing synthetic fibres with environment-friendly and sustainable materials. The tremendous increase in production and use of plastics in every sector of modern life led to huge plastic wastes. Disposal problems, as well as strong regulations and criteria for cleaner and safer environment, have directed scientific research towards eco-composite materials. Each individual type of fibre, among the different types of eco-composites which contain natural fibres and natural polymers, has a key role. Currently, the most viable way towards fabricating eco-friendly composites is through the incorporation of natural fibres as reinforcement.

Naturally, plant fibres, such as from kenaf, have three major components which consist of cellulose, lignin, and hemicellulose (Maya and Sabu 2008). These materials give plant fibres a complex structure that cannot be directly converted into end products like ethanol and lactic acid in biopolymer industry (Sun and Cheng 2002; Ren et al. 2009). Due to this reason, plant fibres require treatment in order to remove lignin and hemicelluloses, to reduce cellulose crystallinity as well as to increase the porosity of the materials (McMillan 1994; Sun and Cheng 2002; Klinke et al. 2004). There are many types of treatment processes such as physical, chemical (acid and alkaline), and biological treatments; however, acid treatment is commonly used by many researchers.

Sulphuric acid (H_2SO_4) or hydrochloric acid (HCl) is commonly utilized as the chemical for lignocellulosic material or plant fibre hydrolysis such as kenaf biomass. According to Xiang et al. (2003) and Hamelinck et al. (2005), there are two types of acid treatments: mild acid treatment and concentrated acid treatment. In dilute acid treatment, with concentration of 0.7–3.0 %, it requires high operating temperature around 160–250 °C. In contrast, concentrated acid treatment requires high amounts of acid about 34 %, and this can be considered as uneconomical since it leads to high operating costs as well as various environmental problems such as corrosion and pollution (Banerjee et al. 2010).

Kenaf is a biennial herbaceous plant which can grow up to 20 ft (Scott et al. 2007). It contains long, soft bast fibres, contributing 30–40 % of the dry weight of the stem. The central core of the stem contains weakly disbursed pith cells surrounded by a thick cylinder of short woody fibres.

The main objectives of this study are to establish analytical method of glucose content extraction from kenaf biomass as well as to reduce the crystallinity of cellulose in order to be used for producing various end products such as lactic acid and ethanol.

Methodology

Design of Experiment

The experiment was developed and designed by central composite design (CCD) using Design Expert v6.0.8 Software to determine the optimum yield of glucose content from kenaf biomass.

Sample Preparation

Kenaf biomass was obtained from Kenaf Fibre Industry Sdn Bhd, Kelantan, Malaysia, and was stored under room-temperature conditions upon arrival. Later, it was crushed and sieved using 100-μm nylon mesh.

Hydrolysis Treatment Process

In this process, 2 % (v/v) of H_2SO_4 was prepared in 100 ml working volume. The optimization process was conducted using a hot plate with 100 rpm of agitation speed and 3 g of kenaf biomass in order to obtain optimum glucose content by varying two parameters with three levels, which are mixing time and temperature. The experiments were carried out for 11 runs with two replications based on the CCD method. For the analysis of the optimization results, regression analysis was done by developing the regression equation.

Total Glucose Content Analysis

Total glucose content was analysed using high-pressure liquid chromatography (HPLC) (Agilent model 1200) that comprised of a quaternary pump with auto-sampler injector, microdegassers, column compartment equipped with thermostat, and a diode-array detector. The column used was a ZORBAX Eclipse XDB-C_{18} end-capped 5-μm, 4.6 × 150 mm reverse-phase column (Agilent Technologies, USA) and UV detector at 210 nm. The conditions of HPLC were mobile-phase acetonitrile/phosphate (1:9 v/v), flow rate of 1.0 ml min^{-1} for 15 min, temperature 20 °C, and injection volume 1 μL. Each injection was performed in duplicate.

Results and Discussion

This study investigates the efficacy of mild acid (chemical) treatment in extracting glucose from kenaf biomass which was used as the raw material. Two factors have been chosen for the optimization of the chemical treatment process, mixing time and temperature, in order to maximize the amount of glucose conversion. To effectively determine the maximum glucose conversion from kenaf biomass under optimum condition, the experiments were designed based on CCD by Design Expert v6.0.8 Software. The polynomial regression model relating the percentage of conversion (Y) with independent variables, mixing time (A), and temperature (B) is as follows:

$$Y = -13.9001 + 2.1586\mathrm{A} - 0.0526\mathrm{B} - 0.0282\mathrm{A}^2 + 7.8599 \times 10^{-4}\mathrm{B}^2 - 0.0126\mathrm{AB} + 1.6667 \times 10^{-4}\mathrm{A}^2\mathrm{B}$$

where Y = the predicted glucose content

The average of two glucose contents obtained was considered as the observed amount of glucose content. All the variables and observed and predicted results are recorded in Table 1.

The determination coefficient ($R^2 = 0.904$) in this study specifies high correlation between the observed and predicted values, and indicates the level of precision with which the glucose content can be predicted using the model equation and the values of the independent variables, mixing time, and temperature. This sample of variation implies that only 90.4 % of the glucose content observed can be attributed to the effect of the independent variables, whereas the remaining 9.6 % cannot be explained by the model. The analysis of variance (ANOVA) is presented in Table 2. From this table, it can be observed that the model is significant (p value is 0.0480), the lack of fit is also significant at 0.0394, and the R^2 for is 90.45 %.

Table 1 Observed and predicted values of the response

Run	Mixing time (min)	Temperature (°C)	Glucose (g/L)	
			Actual	Predicted
1	30.00	180.00	0.366	0.555176
2	60.00	180.00	0.754	0.944114
3	30.00	160.00	0.334	0.238706
4	45.00	200.00	0.457	0.64638
5	45.00	180.00	0.388	0.355657
6	60.00	200.00	3.397	3.30231
7	45.00	180.00	0.439	0.355657
8	30.00	200.00	0.374	0.27843
9	60.00	160.00	0.531	0.436694
10	45.00	180.00	0.620	0.355657
11	45.00	160.00	0.504	0.693713

Table 2 ANOVA for response surface quadratic model

Sources	Sum of squares	Degree of freedom	Mean square	*F* value	*p* value (Prob > *F*)
Model	7.15	6	1.19	6.31	0.0480*
A	2.17	1	2.17	11.49	0.0275*
B	1.105E-003	1	1.105E-003	5.848E-003	0.9427
A^2	0.39	1	0.39	2.08	0.2226
B^2	0.25	1	0.25	1.33	0.3137
AB	2.00	1	2.00	10.57	0.0313*
A^2B	0.75	1	0.75	3.97	0.1171
Residual	0.76	4	0.19		
Lack of fit	0.73	2	0.36	24.41	0.0394
Pure error	0.030	2	0.015		
Cor total	7.91	10			

R-Sq = 90.45 %
R-Sq(adj) = 76.11 %
* Significant

From the statistical analysis in Table 2, the variables that are significant are the linear effect of mixing time (A) and the interaction effect between mixing time and temperature (AB). Since, the linear and the interaction effect of mixing time are significant model parameters, mixing time can be considered as the limiting factor. A small variation in the mixing time value will have considerable effect on the value of the response and the amount of glucose production. Table 1 shows that the maximum glucose yield is in Run 6 which is 3.397 g/L (observed value) and also that the predicted value is very close at 3.30231 g/L. Run 6, therefore, shows the optimum condition: a mixing time of 60 min and a temperature of 200 °C. The successful chemical treatment using dilute H_2SO_4 indicates that the complex structure of kenaf biomass is composed of lignocellulosic material and that the acid effectively breaks down this complex structure into simple sugars, which can be consumed by microorganisms. By referring to the chemical treatment's results, it can be observed that the higher the heating rate (mixing time and temperature), the easier the separation of the carbohydrates from the lignin matrix of kenaf biomass. Thus, the optimum condition will lead to the lowest chemical destruction of fermented sugars which is a prime requirement for the production of any useful end product like ethanol or lactic acid.

In order to determine the optimum values of the variables, 3D surface and 2D contour plots are used as graphical representations of the regression equation (Tanyildizi et al. 2005) as shown in Fig. 1. The ultimate objective of the response surface is to investigate the optimum values of the variables such as the maximization of the response (Tanyildizi et al. 2005). In 2D plot, each contour curve signifies the infinite number of values of the two variables with the other three remaining at their respective zero level. The smallest ellipse from the confined surface in the contour plot is used to indicate the maximum predicted value. The perfect interaction between the independent variables will result in an elliptical

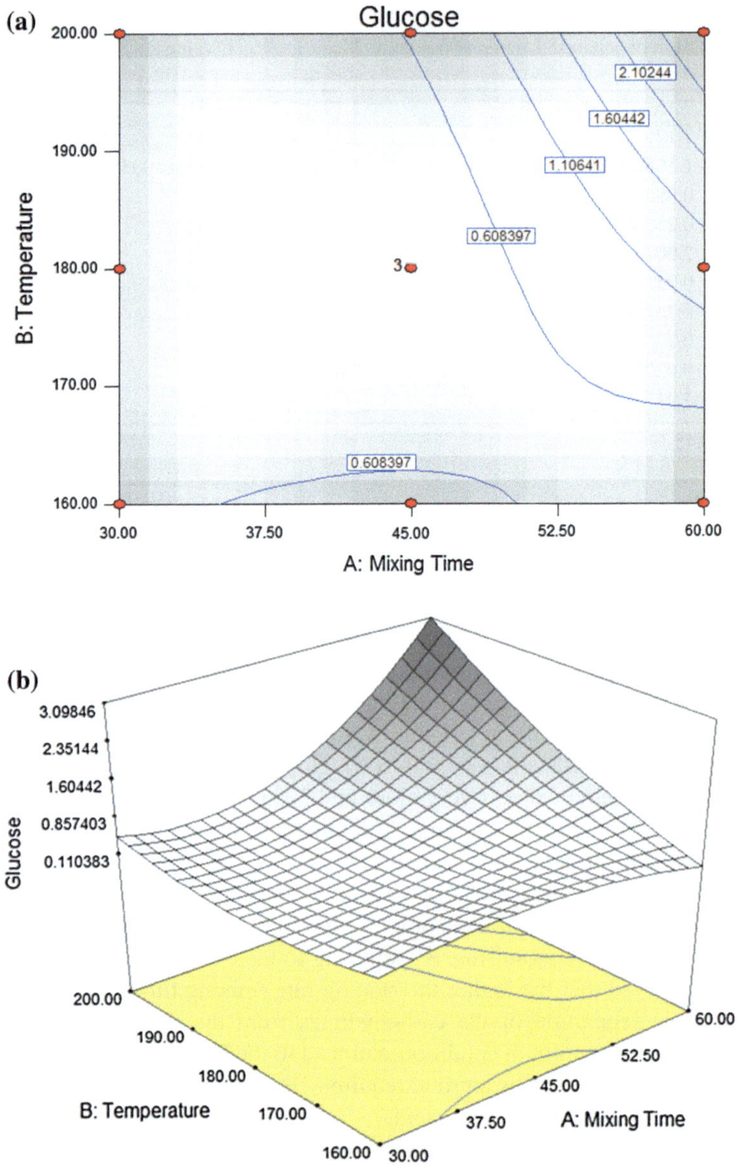

Fig. 1 **a** 2D contour plot and **b** 3D response surface showing the effect of time (min) and temperature (°C) on the glucose concentration (mass of fibre was 3 g and agitation speed was 100 rpm)

contour (Muralidhar et al. 2001). Figure 1 illustrates the response surface of the model equation in order to estimate glucose conversion against mixing time and temperature. The 3D surface plot was generated by the regression model in order to illustrate the effects of the independent variables and the combined effects of each independent variable upon the response variable (Duta et al. 2006). It can be observed that the highest percentage conversion to glucose was achieved at the combination of the longest time and highest temperature. According to Xiang et al. (2003), the hydrogen bonding in the hemicellulose and cellulose fraction are easily broken down at high temperature compared to low temperature, which results in the high yield of glucose.

Conclusions

This study established that the best chemical treatment conditions for optimum yield of glucose from kenaf biomass was 200 °C temperature and 60-min mixing time with a corresponding total glucose content of 3.4 g/L.

Acknowledgments The authors would like to express their gratitude to COMSTECH-TWAS for funding this research and International Islamic University Malaysia for support.

References

Banerjee, S., Mudliar, S., Giri, B., Satpute, D., Chakrabarti, T., & Pandey, R. A. (2010). Commercializing lignocellulosic bioethanol: Technology bottlenecks and possible remedies. *Biofuels, Bioproducts and Biorefining, 4*, 77–93.

Duta, F. P., de Franca, F. P., & Lopes, L. M. D. A. (2006). Optimization of culture conditions for exopolysaccharides production in Rhizobium sp. using the response surface method. *Journal of Biotechnology, 9*(4), 391–399.

Hamelinck, C. N., Hooijdonk, G., & Faaij, A. P. C. (2005). Ethanol from lignocellulosic biomass: Techno-economic performance in short-, middle- and longterm. *Biomass and Bioenergy, 28*, 384–410.

Klinke, H. B., Thomsen, A. B., & Ahring, B. K. (2004). Inhibition of ethanol—producing yeast and bacteria by degradation products produced during pre-treatment of biomass. *Applied Microbiology and Biotechnology, 66*, 10–26.

Maya, J. J., & Sabu, T. (2008). Review biofibres and biocomposite. *Carbohdyrate Polymer, 71*, 343–364.

McMillan, J. D. (1994). Pretreatment of lignocellulosic biomass. In M. E. Himmel, J. O. Baker, & R. P. Overend (Eds.), *Enzymatic conversion of biomass for fuels production* (pp. 292–324). Washington, DC: American Chemical Society.

Muralidhar, R. V., Chirumamila, R. R., Marchant, R., & Nigam, P. (2001). A response surface approach for the comparison of lipase production by Candida cylindracea using two different carbon sources. *Biochemical Engineering Journal, 9*, 17–23.

Ren, H., Huang, H.Z. & Zheng, J. (2009). Detoxifying and recylcing of washing solution used in pretreatment of lignocellulose-containing materials. http://www.faqs.org/patents/app/20090056889. Accessed on 27 Feb 2010.

Scott, K., Flores, D. M., Kobayashi, G., & Sonomoto, K. (2007). Direct L-lactic acid fermentation with sago starch by a novel amylolytic lactic acid bacterium, Enterococcus faecium. *Enzyme and Microbial Technology, 41*(1–2), 149–155.

Sun, Y., & Cheng, J. (2002). Hydrolysis of lignocellulosic materials for ethanol production: Are view. *Bioresource Technology, 83*, 1–11.

Tanyildizi, M. S., Ozer, D., & Elibol, M. (2005). Optimization of α-amylase production by Bacillus sp. using response surface methodology. *Process Biochemistry, 40*(7), 2291–2296.

Xiang, Q., Lee, Y. Y., Pettersson, P. O., & Torget, R. W. (2003). Heterogeneous aspects of acid hydrolysis of cellulose. *Applied Biochemistry and Biotechnology, 105–108*, 505–514.

Solid–Liquid Mass Transfer Coefficients in an Ultrasound-Irradiated Extraction of Iota-Carrageenan

D. Krishnaiah, D. M. R. Prasad, R. Sarbatly, A. Bono, S. M. Anisuzzaman and K. Krishnaiah

Abstract A 20-kHz intensity ultrasound was used in the extraction of iota-carrageenan from *Eucheuma denticulatum* seaweed by using water as solvent. In a simplified extraction procedure, the seaweed particle size of 0.7125E-03 m, 1.2E-03 m, and 1.7E-03 m diameter were studied at different ultrasonic amplitude levels, ranging from 2.08E-06 to 6.4E-06 m and the temperatures from 30 to 60 °C. The maximum iota-carrageenan yield obtained was 57.2 %. This study suggested that ultrasound intensity employed reduced the extraction time required and improved the yield of iota-carrageenan by 6 %. The particle diameter variation during the extraction was found to be a linear equation. The solid–liquid mass transfer coefficient was correlated for ultrasound extraction of iota-carrageenan in terms of operating conditions employed in this investigation as

$$k_L = 0.0027A^{0.866}D_p^{-0.533}(T)^{0.133}$$

Keywords Ultrasound intensity · Extraction · Particle swelling · Mass transfer coefficient

D. Krishnaiah (✉) · R. Sarbatly · A. Bono · S. M. Anisuzzaman
School of Engineering and Information Technology, Universiti Malaysia Sabah, Jalan UMS, 88400 Kota Kinabalu, Sabah, MALAYSIA
e-mail: krishna@ums.edu.my

K. Krishnaiah
Indian Institute of Technology, Madras, India

D. M. R. Prasad
Faculty of Chemical and Natural resources Engineering, Universiti Malaysia Pahang, 26300 Gambang, Pahang, Malaysia

R. Pogaku et al. (eds.), *Developments in Sustainable Chemical and Bioprocess Technology*, DOI: 10.1007/978-1-4614-6208-8_31,

Abbreviation

Nomenclature

A	Amplitude (m)
A	Interfacial area (m^2)
C_A	Concentration (kg/m^3)
C_A^*	Equilibrium concentration (kg/m^3)
D_p	Diameter of the particle (m)
F	Frequency (Hz)
I	Sound wave intensity (W/m^2)
k_L	Solid–liquid mass transfer coefficient (m/s)
p	Constant Eq. (8)
q	Constant Eq. (8)
T	Temperature (^oC)
t	Time (s)
V	Volume of the extractor (m^3)

Greek letters

ρ	Density of liquid (kg/m^3)
ω	Angular frequency (rad/s)
α, β, ψ, φ, δ, γ	Constants

Introduction

Iota-carrageenan is a biopolymer resource from *Eucheuma denticulatum* red seaweed. It is an important gel-forming polysaccharide useful in the food processing industries. Despite the similarity of carrageenan structures, their gels have different properties. Based on these properties, these are widely used as thinking, gelling agent or texture enhancers, and stabilizer (Michel et al. 1997). Carrageenans are conventionally separated by solvent extraction. It is a key processing step for the recovery and purification of active ingredients of plant material. The traditional techniques of solvent extraction of plant materials are mostly based on the appropriate choice of solvents and the use of heat and/or agitation to increase the solubility of components and the rate of mass transfer. Usually, these techniques require long extraction time and have low efficiency. Moreover, many natural products are thermally unstable and may degrade during thermal extraction (Wu et al. 2001). The objective of any extraction procedure is to achieve high extraction efficiency at a reduced processing time. In this respect, a procedure that could give the highest yield of the effective constituents in a shorter processing time would be the ultrasonic-assisted technology (Xia et al. 2006). Recently, numerous reports have documented on the application of intense ultrasound in the extraction of organic compounds from various plant materials, such as polysaccharides from *Salvia officinalis L.* (Hromádková et al. 1999), hemicellulose from buckwheat hulls (Hromádková and Ebringerová 2003), ginseng saponins from ginseng roots and cultured ginseng cells (Wu et al. 2001; Pal et al. 1997) oil from soybeans (Li et al. 2004), pyrethrins from pyrethrum flowers and oil from woad seeds (Romdhane and Gourdon 2002), and proteins and essential oils from various parts of plant and plant seeds. In all cases, acceleration of extraction kinetics and increase in yield were observed. Besides, Vinatoruet al. (1997) and Hromadkovaet al. (2002) documented its viability in achieving greater polysaccharide yield and reduced processing time. A comparative study conducted between the classical and ultrasonically assisted extraction concludes that the ultrasound improves the yield substantially in shorter time (Vinatoru et al. 1997). Ultrasonication is the application of high-intensity, high-frequency sound waves and their interaction with materials. The propagation and interaction of sound waves alter the physical and chemical properties of materials that are subjected to ultrasound (Mason 1992). The most probable mechanism for ultrasonic enhancement of extraction is the intensification of mass transfer and easy access of solvent to the cells. The collapse of cavitation bubbles near the cell wall is expected to produce cell disruption together with good penetration of the solvent into the cells, through the ultrasonic jet (Vinatoru et al. 2001). The high-intensity ultrasonication can accelerate heat and mass transport in a variety of food processing operations and has been successfully used to improve drying, mixing, homogenization, and extraction (Fairbanks 2001). The mechanical effect of acoustic cavitation is able to facilitate the release of intracellular compounds contained within the body of plants. This is mainly due to the disruption of cell walls which enhances mass transfer of cell contents. Moreover, other possible benefits of ultrasound in extraction are improved solvent penetration and capillary effects (Hromádková et al. 1999; Vinatoru et al. 2001).

In addition to the beneficial effect of ultrasound on the product yield, it allows change in the experimental conditions, for instance, decrease of temperature and pressure. Applying ultrasound at lower temperatures, thermal damage to extract and loss of volatile components can be avoided. Then, the quality of solute as in the case of i-carrageenan extraction from seaweed can be maintained. Thus, the factors that govern the action of ultrasound are as follows: pressure, temperature, sonication time, and ultrasound frequency (Romdhaneand Gourdon 2002).

The aim of this work was to evaluate the yield of iota-carrageenan from the seaweeds by using ultrasonic extraction, develop a new mathematical model for estimating the solid–liquid mass transfer coefficient at particle swelling, and to correlate the experimental data for solid–liquid mass transfer coefficient with different operating conditions such as amplitude, temperature, and particle size used in this investigation.

Methodology

Seaweed Particles Preparation

Raw seaweed *(E. denticulatum)* collected from Marine Borneo Research Institute, University Malaysia Sabah, was cleaned with double-distilled water by removing any foreign material such as sea grass and other plant leaves that could present after harvesting, drying, transportation, and storage and was dried at 70 °C for 24 h in vacuum oven. It was cut into small particles by using high-speed knife cutter (SDN, Malaysia) and sieved to provide the average particle size of 0.7125E-03, 1.2E-03, and 1.7E-03 m and was used for the extraction of i-carrageenan by ultrasound. These particles were packed in airtight plastic bags until processing.

Ultrasound-Assisted Extraction

Specific size of seaweed particles (3 g) was mixed with 500 ml double-distilled water in a 600-ml plastic beaker. The seaweed–water suspension was ultrasonicated, whereby, the sonicator probe horn was allowed into the sample with its tip dipped into the solution as shown in Fig. 1. The titanium probe (with a 13-mm-diameter tip) was connected to a 750-W high-intensity ultrasonic liquid processor (Sonics, Newtown, CT, USA) having a frequency of 20 kHz. It was operated at a constant pulse of 30:15. Suspensions were kept in a water bath at 30 °C during sonoextraction. The parameters chosen for the ultrasonic-assisted extraction are as

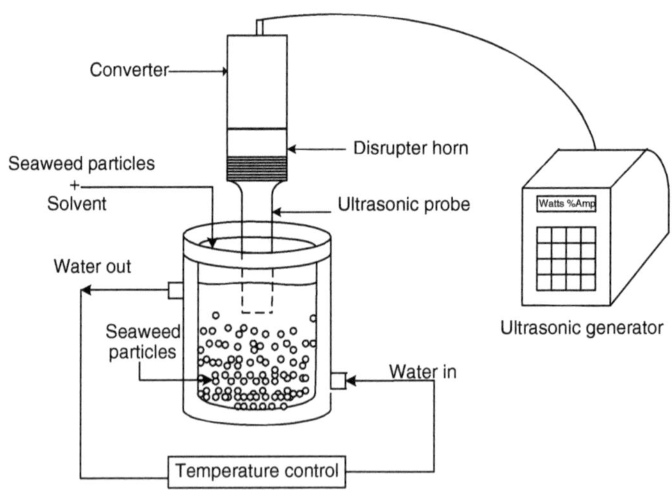

Fig. 1 Experimental apparatus used for ultrasound extraction

follows: average particle diameter of 0.7125E-03, 1.2E-03, and 1.7E-03 m and the amplitudes are 2.08E-06 m (20 %), 3.8E-06 m (40 %), 5.46E-06 m (60 %), and 6.4E-06 m (80 %). Samples were collected every 5 min and filtered (Whatman No. 5) to remove insoluble residue. The purified sample solutions were collected in the test tubes (10 ml volume) and preserved at low temperature (4 °C). In the similar way, the thermal extraction (Hilliou et al. 2006) has been conducted with different temperatures without ultrasonication.

Colorimetric Determination of Carrageenan

The extracted iota-carrageenan was estimated in the samples by colorimetric method (Soedjak 1994). The filtered carrageenan extract solution of 0.5 ml was pipetted out into a 100-ml volumetric flasks and 10 ml of 0.41 mM (0.18 mg/mL) methylene blue stock solution was added and made to 100 ml by adding double-distilled water. Then the solutions were mixed, and absorbance was measured at 559 nm. The absorbance of carrageenan solutions was determined with a UV Double-Beam Spectrophotometer (Perkin Elmer, Germany). The iota-carrageenan concentration was estimated from the standard curve established by plotting the absorbance at 559 nm versus the iota-carrageenan concentration.

Determination of Amplitude of Ultrasonic Waves

The power dissipation with horn tip was varied by changing the amplitude percentage in the ultrasound generator (Nilesh et al. 2001). It was calibrated from the watt meter. The intensity of the ultrasound (I) was calculated as

$$\mathrm{I} = \frac{\text{Power dissipation (W)}}{\text{Cross section area of horn}} \times 100 \tag{1}$$

Intensity of ultrasound (I) for the planar waves is characterized by the equation:

$$\mathrm{I} = \frac{1}{2}\rho c\mathrm{A}(\mathrm{A}2\omega 2) \tag{2}$$

The angular frequency (rad/s) is given as

$$\omega = 2\pi f \tag{3}$$

The estimated intensity (I) from the Eq. (1) was substituted in the Eq. (2) and calculated the value of amplitude (A).

Extraction Yield

The seaweed was dried in an oven until constant weight at 60 °C. The dried seaweed was used for extraction. After extraction, the extract was dried at 60 °C and weighed. The iota-carrageenan yield was calculated as

$$\text{Yield }\% = \frac{\text{Weight of carrageenan extracted}}{\text{Weight of the seaweed}} \times 100 \tag{4}$$

Results and Discussion

Effect of Ultrasonic Wave Amplitude on Iota-Carrageenan Yield

In the present study, the iota-carrageenan yield from dried *E. denticulatum was* 51.2 % (w/w) at 90 °C. It was found that the thermal extraction of seaweed with water in the temperature range of 115–120 °C resulted in the iota-carrageenan yield of 40–50 % (Aguilan and Boom 2003) which is similar to the present observation as shown in Fig. 2. However, with the ultrasonic-assisted extraction the i-carrageenan yield was in the range of 35–57.2 %. The overall yield increased by 6 % due to the breakdown of extracellular mass, thus the inner i-carrageenan diffused into the solvent. The cell walls are not completely opened in the thermal extraction. This could be the reason for low yield in the thermal extraction. The influence of different ultrasound amplitudes (2.08E-06, 3.8E-06, 5.46E-06 and 6.4E-06 m) on iota-carrageenan yield is shown in Fig. 3. Iota-carrageenan yield increased with increasing ultrasonic wave amplitude. The extraction rate curve is in the exponential phase for the first 20 min and later exhibits a gradual increase. The change in i-carrageenan yield could be due to the cavitational effects caused by the high-intensity ultrasound. The high-intensity ultrasound propagates through

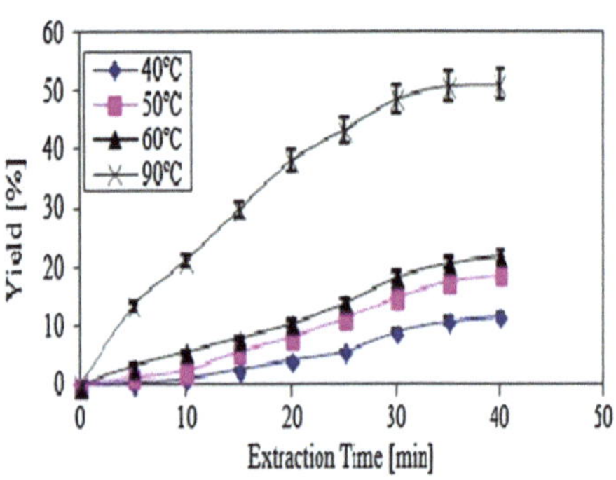

Fig. 2 i-carrageenan yield at various temperatures of 0.7125E-03-m-diameter particles from thermal extraction

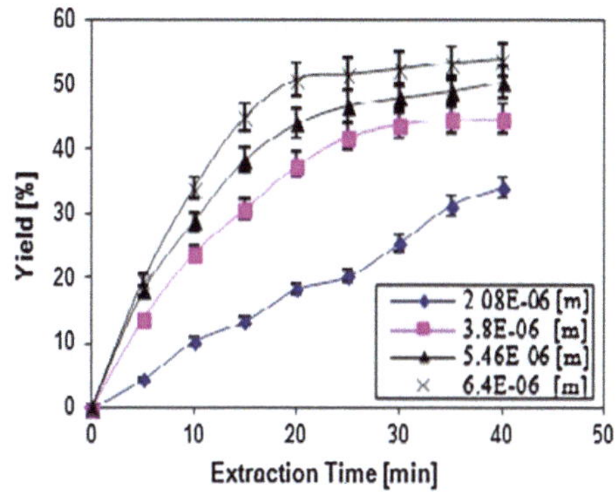

Fig. 3 i-carrageenan yield as a function of extraction time at different amplitudes of ultrasound at 30 °C and 0.7125E-03-m-diameter particles

the viscous liquid medium; it causes compression and shearing of the seaweed particles, which results in localized change in density and elastic modulus (Price et al. 1995). As a result, the sinusoidal compression and shear waves will be distorted into shock waves at a finite distance from the ultrasonic transducer. The abrupt decrease in pressure at the edge of the ultrasonic wave in the negative pressure cycle generates small bubbles. These bubbles will collapse in the positive pressure cycle and produce turbulent flow conditions, which is associated with high pressure and temperature (Hardcastle et al. 2000; Price 1990; Price 1993). The amplitude increases the cavitational bubble collapse and is more violent since the resonant bubble size is proportional to the amplitude of the ultrasonic wave (Suslick et al. 1987). Bubbles collapse near the plant membrane and may cause a strong shear force to be exerted on microfractures to be formed in the plant tissues.

Effect of Temperature on Iota-Carrageenan Yield

The ultrasonication was conducted in the hot water bath to supply additional temperature to the feed solution. It was varied from 30 to 60 °C by 10 °C increment. As the temperature rises from 30 to 60 °C, the yield varied from 54 to 57.2 %. The high temperature accelerated the rate of extraction because the net available energy increases the shear and tear of the seaweed particles, and also the additional temperature caused to soften the cell wall. This phenomenon typically experiences in few microseconds and later results gradual temperature increase in the medium since the heat transfer from the cavitational bubbles to the medium is small (Suslick and Price 1999). Therefore, the change in solvent (water) viscosity is moderate. The overall change in the yield was only 6 % (Fig. 4).

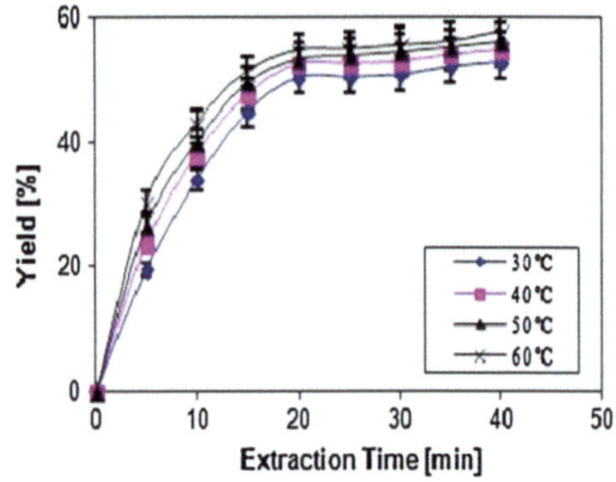

Fig. 4 i-carrageenan yield as a function of extraction time at different temperatures of 0.7125E-03-m-diameter particle and 6.4 E-06 m amplitude

Effect of Particle Diameter on Iota-Carrageenan Yield

The particle diameter affects the i-carrageenan yield with time. This is due to the overall surface area of the seaweed particles exposed to the high-intensity ultrasound. In the present work, different diameters of seaweed particles (0.7125E-03, 1.2E-03, and 1.7E-03 m) were used. The net change in the particle diameter by 0.987E-03 m effect's 5 percentage change in yield as shown in Fig. 5. In the extraction process, the dry seaweed particles were swelled and softened by absorbing water in the ultrasonicator. The swelling rate was measured with respect to time using the hand-held microscope with an accuracy of 0.02 mm (C and D microscope, UK). The swelling of seaweed particles during the ultrasound extraction increases with time. As shown in Fig. 6, the swelling diameter increases linearly, and the constants of Eq. 9 are shown in the Table 1 which is used in solid–liquid mass transfer coefficient estimation, and the equilibrium values are shown in Table 2.

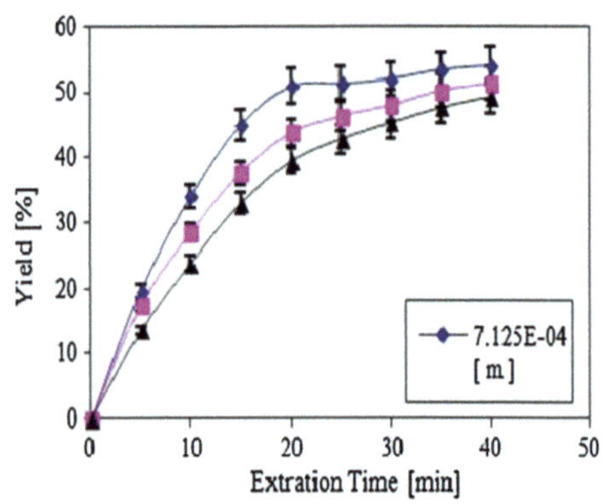

Fig. 5 i-carrageenan yield as a function of extraction time at different diameter of the seaweed particles at 30 °C

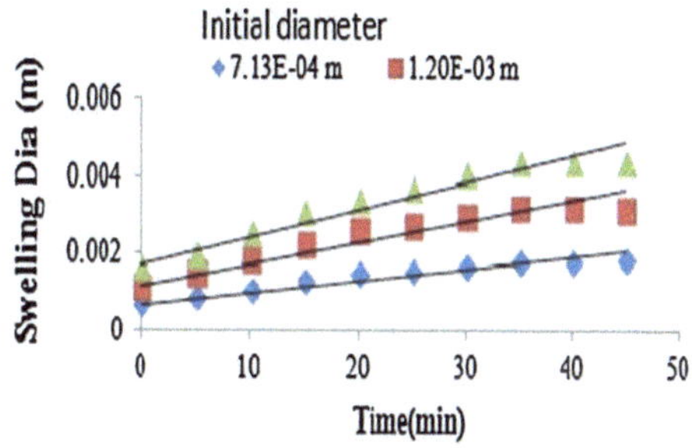

Fig. 6 Seaweed particles swelling

Table 1 p and q values of Eq. 9

Particle diameter (m)	p	q
7.10E-04	0.0007	0.00003
1.20E-03	0.0012	0.00005
1.70E-03	0.0017	0.00007

Table 2 Equilibrium yield

Temperature (°C)	Equilibrium yield (%)
30	52.2
40	54.5
50	56.3
60	58.1

Correlation of Solid–Liquid Mass Transfer Coefficient

Estimation of solid–liquid mass transfer coefficients based the dynamic behavior during solid–liquid extraction. The differential equation is based on Fick's second law of unidirectional diffusion of solute i-carrageenan into the solvent. In modeling the ultrasound extraction process, it is assumed that (1) the shape of the particle is spherical (2) mass transfer resistance exists only in the liquid film. Since the rate of extraction of solute depends upon the solid–liquid mass transfer coefficient and interfacial area, the mass balance equation can be written as

$$V\frac{dC_A}{dt} = k_L a(C^*_{A\_} C_A) \quad (5)$$

Rearranging the above equation gives

$$\frac{dC_A}{(C^*_A - C_A)} = k_L\left(\frac{a}{V}\right)dt \quad (6)$$

Since the particles are spherical, the interfacial surface area is

$$a = \pi D^2 \quad (7)$$

As the seaweed particles absorb water, swelling was observed. Then the interfacial area is a function of particle diameter and time t. Hence, to further compute the solid–liquid mass transfer coefficient, the ratio of interfacial area at $t = 0$, $a = a_o$, and $t = t$,

$$\mathrm{a} = \mathrm{a_t} \text{ is expressed as} \frac{\mathrm{a_t}}{\mathrm{a_o}} = \frac{\pi \mathrm{D_t^2}}{\pi \mathrm{D_0^2}} = (\frac{\mathrm{D_t}}{\mathrm{D_0}})^2 \tag{8}$$

The swelling diameter of the particles was measured experimentally for all the samples classified and plotted against time to obtain a linear equation as shown in the Fig. 6.

$$D_t = p + qt \tag{9}$$

Dividing with D_o gives

$$\frac{D_t}{D_0} = \frac{p}{D_0} + \left(\frac{q}{D_0}\right)t \tag{10}$$

Introducing $\alpha = \frac{p}{D_0}$ and $\beta = \frac{\mathrm{q}}{\mathrm{D_0}}$, the resulting equation becomes

$$\frac{\mathrm{D_t}}{\mathrm{D_0}} = (\alpha + \beta \mathrm{t}\ ) \tag{11}$$

Substituting Eq. (11) into (8), the interfacial area becomes

$$a_t = a_o(\alpha + \beta t)^2 \tag{12}$$

Since the interfacial area can be computed using the equation above, the Eq. 12 was substituted in Eq. 6

$$\frac{dC_A}{(C_A^* - C_A)} = k_L \frac{[a_o(\alpha + \beta t)^2]}{V} \mathrm{d}t \tag{13}$$

With the following boundary conditions: At $t = 0$, $C = 0$ and $t = t$, $C = C_A$ On integration of Eq. (12) gives

$$\int_o^{C_A} \frac{dC_A}{(C_A^* - C_A)} = \frac{k_L a_o}{V} \int_o^t (\alpha + \beta t)^2 \mathrm{d}t \tag{14}$$

Therefore,

$$-\ln\left(1 - \frac{C_A}{C_A^*}\right) = \frac{k_L a_0}{V}\left(\alpha^2 t + \alpha\beta t^2 + \left(\frac{\beta^2}{3}\right)t^3\right) \tag{15}$$

Substituting $X = \left(\alpha 2t + \alpha\beta t^2 + \left(\frac{\beta^2}{3}\right)t^3\right)$ into Eq. (14) results

$$-\ln\left(1-\frac{C_A}{C_A^*}\right)=\frac{k_L a_0}{V}X \tag{16}$$

Based on the Eq. 16, the solid–liquid mass transfer coefficient was determined for different particle sizes at various ultrasonic amplitudes and temperatures. In the present investigation, an attempt has been made to correlate mass transfer coefficient k_L with the operating variables using the numerical values of the mass transfer coefficients obtained from Eq. 16 with the following equations.

$$k_L \alpha A^{0.866} \tag{17}$$

$$k_L \alpha T^{0.133} \tag{18}$$

The solid–liquid mass transfer coefficient is proportional to amplitude and temperature. An increase in amplitude results in an increase in acoustic intensity and turbulence thus the rate of mass transfer increases. As the temperature is increased, its viscosity reduces. Therefore, it is easier to generate acoustic cavitation which increases the mass transfer rate.

$$k_L \alpha \frac{1}{D_p^{0.533}} \tag{19}$$

Equation 19 represents solid–liquid mass transfer coefficient variation with the particle diameter. The microstreaming event, which is the localized turbulence of the solid–liquid film (Thompson and Doraisamy 2000), and the collapse of the cavitational bubble near the cell wall were greater during the extraction of 7.125E-04 m particles. In addition, the collision of these particles had a higher frequency, and the ultrasonic waves propagated through the liquid medium. This behavior was apparent for all the particles and exhibited high extraction rates. Therefore, the results suggest that the ultrasound extraction of carrageenan with a smaller seaweed particle size was faster compared to the larger particle size. Although Eq. 19 satisfactorily represents the data, its application is within the 35 min since the swelling behavior of particles is found to be linear in this period as shown in Fig. 6. By combining the above Eqs. 17, 18, 19,

$$k_L = 0.0027A^{0.866}D_p^{-0.533}T^{0.133} \tag{20}$$

The predicted k_L was compared with the experimental k_L as shown in Fig. 7. It shows that almost 95 % of the data points are within 12 % error bands. The predicted values obtained by this model gives a very close approximate to the experimental data. As can be seen, the experimental data can be satisfactorily predicted by the theoretical model since the average coefficient of determination r^2 is 0.978. This indicates the validity of the mathematical model.

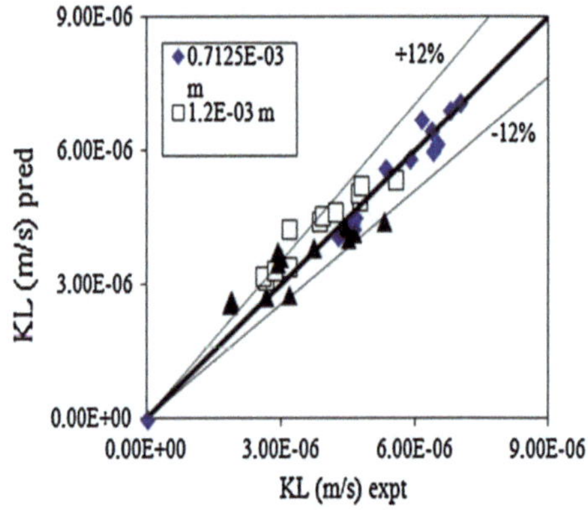

Fig. 7 Comparison of experimental and predicted mass transfer coefficients

Conclusions

The ultrasound extraction of i-carrageenan from seaweed has been studied by varying the parameters such as a particle diameter, different ultrasonic wave amplitudes, and temperature. The influence of ultrasound on the solid–liquid extraction process with regard to yield and mass transfer coefficient has also been studied. The net yield increased by 6.0 %. The diameter of particle varies linearly with time. A simple model for estimating the mass transfer coefficient was developed, and the mass transfer coefficient was correlated in terms of operating conditions viz., amplitude, diameter of particle, and temperature and compared with experimental data satisfactorily.

Acknowledgments The authors thank University Malaysia Sabah for supporting this work with a scholarship.

References

Aguilan, J. T., Boom, J. E. (2003). Structural analysis of carrageenan from farmed varieties of Philippine seaweed. *Botanica marina, 46*, 179–192.

Fairbanks, H. V. (2001). Drying powdered coal with the aid of ultrasound. *Powder Technology, 40*, 257–264.

Hardcastle, J. L., Ball, J. C., Hong, Q., Marken, F., Compton, R. G., Bull, S. D., et al. (2000). Sonoelectrochemical and sonochemical effects of cavitation: correlation with interfacial cavitation induced by 20 kHz ultrasound. *Ultrasonics Sonochemistry, 7*, 7–14.

Hromádková, Z., Ebringerová, A., & Valachovič, P. (1999). Comparison of classical and ultrasound-assisted extraction of polysaccharides from *Salvia Officinalis* L. *Ultrasonics Sonochemistry, 5*, 163–168.

Hilliou, L., Larotonda, F. D. S., Abreu, P., Ramos, A. M., Sereno, A. M., & Goncalves, M. P. (2006). Effect of extraction parameters on the chemical structure and gel properties of k/i-hybrid carrageenans obtained from Mastocarpus stellatus. *Biomolecular Engineering, 23*, 201–208.

Hromádková, Z., & Ebringerová, A. (2003). Ultrasonic extraction of plant materials—investigation of hemicellulose release from buckwheat hulls. *Ultrasonics Sonochemistry, 10*, 127–133.
Hromádková, Z., Ebringerová, A., & Valachovič, P. (2002). Ultrasound-assisted extraction of water soluble polysaccharides from roots of valerian *Valeriana officinalis L. Ultrasonics Sonochemistry, 9*, 37–44.
Li, H., Pordesimo, L., & Weiss, J. (2004). High intensity ultrasound-assisted extraction of oil from soybeans. *Food Research International, 37*, 731–738.
Mason, T. J., Paniwnyka, L. J., & Lorimera, J. P. (1996). The uses of ultrasound in food technology. *Ultrasonics Sonochemistry, 3*, S253–S260.
Mason, T. J. (1992). Industrial sonochemistry: potential and practicality. *Ultrasonics Sonochemistry, 30*, 192–196.
Michel, A. S., Mestdagh, M. M., & Axelos, M. A. V. (1997). Physico- chemical properties of carrageenan gels in presence of various cations. *International Journal of Biological Macromolecules, 21*, 195–200.
Nilesh, V. P., Parag, R. G., Vishwas, Y. D., & Pandit, A. B. (2001). Mixing time analysis of a sonochemical reactor. *Ultrasonics Sonochemistry, 8*, 23–33.
Pal, N., Korfiatis. P., & Patel. V. (1997). Sonochemical Extraction and biological treatment of pentachlorophenol contaminated wood. *Journal of hazardous materials, 53*,165–182.
Price, G. (1990). The use of ultrasound for the controlled degradation of polymer solutions. In T. J. Mason (Ed.), *Advances in sonochemistry*. London: JAI Press.
Price, G. J., White, A., & Clifton, A. A. (1995). The effect of high intensity ultrasound on solid polymers. *Polymer, 26*, 4919–4925.
Price, G. (1993). Applications of high intensity ultrasound in polymer chemistry. *Chemistry & Industry, 3*, 75–78.
Romdhane, M., & Gourdon, C. (2002). Investigation in solid-liquid extraction: influence of ultrasound. *Chemical Engineering Journal, 87*, 11–19.
Suslick, K. S., Casadonte, D., Green, M., & Thompson, M. (1987). Effects of high intensity ultrasound on inorganic solids. *Ultrasonics Sonochemistry, 25*, 56–59.
Suslick, K. S., & Price, G. J. (1999). Applications of ultrasound to materials chemistry. *Annual Review of Materials Science, 29*, 295–326.
Thompson, L. H., & Doraisamy, L. K. (2000). The rate enhancing effect of ultrasound by inducing super saturation in a solid-liquid system. *Chemical Engineering Science, 55*, 3085–3090.
Soedjak, H. S. (1994). Determination of carrageenan and other anionic hydrocolloids with methylene blue. *Analytical Chemistry, 66*, 4514–4518.
Vinatoru, M., Toma, M., Radu, O., Filip, P. I., Lazurca, D., & Tmason, T. J. (1997). The use of ultrasound for the extraction of bioactive principles from plant materials. *Ultrasonics Sonochemistry, 4*, 135–139.
Vinatoru, M., Toma, M., Paniwynk, L., & Mason, T. J. (2001). Investigation of the effects of ultrasound on vegetal tissues during solvent extraction. *Ultrasonics Sonochemistry, 8*, 137–142.
Wu, J., Lin, L., & Foo, T. C. (2001). Ultrasound-assisted extraction of ginseng saponins from ginseng roots and cultured ginseng cells. *Ultrasonics Sonochemistry, 8*, 347–352.
Xia, T., Shi, S., & Wan, X. (2006). Impact of ultrasonic-assisted extraction on the chemical and sensory quality of tea infusion. *Journal of Food Engineering, 74*, 557–560.

Supercritical Fluid Extraction of Black Pepper's Bioactive Compounds: Solubility and Mass Transfer

S. P. Then, F. Panau and Y. Samyudia

Abstract Mass transfer of black pepper's bioactive compound extracted using supercritical CO_2 was studied with the parameters such as range of temperature (45–55 °C), pressure (3,000–5,000 psi), CO_2 flow rate (5–10 ml/min), and pepper particle size (0.4–1 mm). The solubility-controlled period and diffusion-controlled period of the extraction curve were examined. Static extraction was performed to calculate solubility, and the value obtained varied from 4.8 to 10.3 kg oil/m^3 CO_2. Transition time analyzed through extraction curve varied from 18 to 43 min and increased with particle size. Mass transfer rates for solubility-controlled period and diffusion-controlled period were analyzed. The average values were—2.7×10^{-3} kg/m^{3}s and 1.04×10^{-4} kg/m^{3}s, respectively. Particle size was found to be the dominant factor during solubility-controlled period, while pressure was the key factor for the intra-particle diffusion. The fluid-phase mass transfer coefficients and diffusivity were calculated based on Sovova's model.

Introduction

The pharmaceutical value of black pepper's essential oil in enhancing drugs' bioavailability has increased the intention of the industry to develop black pepper extraction (Srinivasan 2009). Supercritical fluid extraction (SFE) appears to be the most promising alternative over the conventional processes as supercritical fluid preserves the extract from solvent contamination and thermal degradation (Salgin et al. 2006; Bruner 2005; Reverchon et al. 1999). In spite of great number of experimental data reported regarding SFE on black pepper (Li 2006; Bruner 2005; Perakis et al. 2005; Ferreira et al. 1999), the information and analysis that define

S. P. Then (✉) · F. Panau · Y. Samyudia
Chemical Engineering, School of Engineering, Curtin University of Technology, Sarawak Campus CDT 250 98000 Miri, Sarawak, Malaysia
e-mail: then.siewping@gmail.com

R. Pogaku et al. (eds.), *Developments in Sustainable Chemical and Bioprocess Technology*, DOI: 10.1007/978-1-4614-6208-8_32,

the process mass transfer constraints are still scarce. This study focuses on identifying the mass transfer rate and solubility of the SFE process of black pepper by using CO_2 as the solvent. The effects of pressure, temperature, solvent flow rate, and particle size were investigated. The transition time from solubility-controlled period (SC) to diffusion-controlled period (DC) was examined. The mass transfer coefficients were evaluated based on Sovova's model.

Mass Transfer Coefficients

Mass transfer in SFE is driven by solute concentration difference in fluid phase and solid phase. The transfer coefficients that describe the characteristics of fluid and solid phases for each extraction period can be calculated using the general mass balance equations developed by Sovova (2007):

$$k_s = \frac{(1-\varepsilon)\rho_f \Delta y}{G_s} \tag{1}$$

$$k_d = \frac{(1-\varepsilon)(1-r)\rho_s \Delta x}{G_d} \tag{2}$$

where k_s is fluid-phase mass transfer coefficient (s^{-1}); k_d is diffusivity (s^{-1}) of solute from particle core to surface; G_s and G_d are volumetric mass transfer rates during SC period and DC period, respectively (kg oil/m^3s); ε is particle bed porosity; ρ_f is CO_2 density (kg/m^3); ρ_s is solid density (kg/m^3); r is grinding efficiency which is defined by Perakis et al. (2005):

$$r = \frac{x_{\text{surface}}}{x_0}$$

In this equation, x_0 is total solute contained in the untreated particle bed (kg/kg particle bed) and x_{surface} is the amount of solute tabulated on the surface of the particles which are directly exposed to the solvent (kg/kg particle bed). x_0 was obtained experimentally by running the SFE process until all the extractable solute was depleted.

Δy is the solute's concentration difference in bulk CO_2 and particle surface (y^+) (kg/kg CO_2). The mean Δy is defined by the equation below (Ferreira et al. 1999):

$$\Delta y = \frac{y^+}{ln\frac{y^+}{y^+ - y_{\text{final}}}}$$

In this equation, y_{final} is the total amount of solute extracted during the SC period per solvent mass (kg/kg CO_2). As mass transfer resistance in broken cells is neglected, particle surface concentration is assumed to be constant as long as broken cells exist on the surface, and the magnitude is equal to its equilibrium solubility at solvent phase at that particular pressure and temperature, y_s (Sovova 2007):

$$y^{+} = y_s(T, P)$$

Δx is the solute's concentration difference between the particle surface and particle core (kg/kg solid):

$$\Delta x = x_{\text{core}} - x_{\text{surface}}$$

During the DC period of extraction, the fluid-phase concentration is much lower than the solute solubility in the solvent. If the assumption of equilibrium at a particle surface and constant solubility holds, the solute that arrives from particle core to its surface dissolves immediately in the solvent and the concentration in broken cells becomes practically zero (Sovova 2007). Consequently, Δx is equivalent to x_{core} and can be stated as:

$$\Delta x = x_0 - x_{\text{surface}}$$

Methodology

Materials

Sarawak black peppers were supplied by Malaysia Pepper Board. Liquid CO_2 (99.99 % purity) was purchased from Eastern Oxygen Industrial Sdn. Bhd.

SFE Equipment and Experimental Procedure

The experiments were carried out in a bench-scale apparatus of SFT-100 (Fig. 1). A fixed bed was formed in the extractor with 40 g of black pepper particles. Fiber glass wool (5 mm thick) was placed on the ends of the column to prevent blockage. The extractor was heated and pressurized with supercritical CO_2, and the extraction started. The extracted bioactive compounds were recovered by depressurizing the saturated CO_2. The solvent was continuously pumped to the extractor to maintain the pressure set point. Experiments were performed for various groups of parameters in the following ranges: pressure 3,000, 4,000, and 5,000 psi, temperature 45, 50, 55 °C, solvent flow rate 5, 7.5, 10 ml/min, and particle size 0.4, 0.8, 1 mm. Restrictor block was maintained at 70 °C to prevent extract agglomeration.

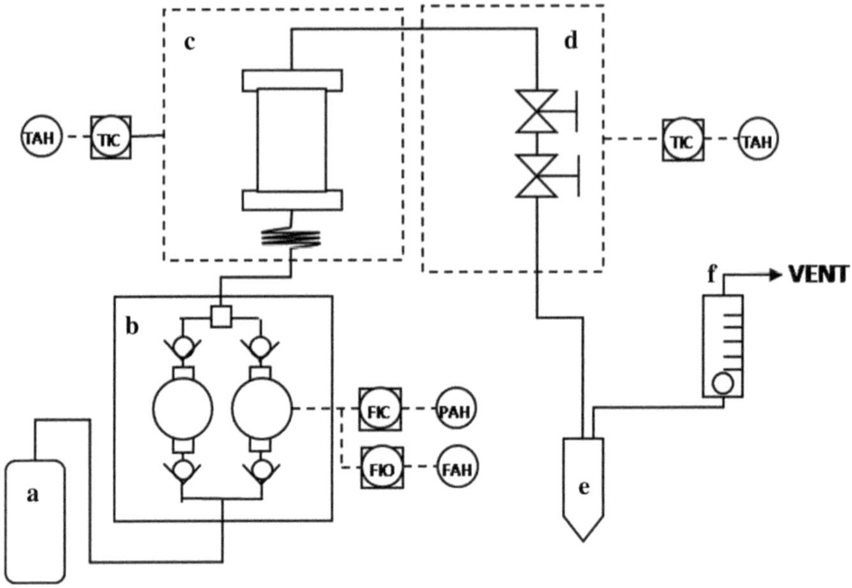

Fig. 1 SFE equipment setup. *A* Liquid CO_2; *B* cooler block; *C* extractor oven; *D* restrictor block; *E* collecting vial; *F* flow meter. *Source* SFT-100 user manual

Results and Analysis

Solubility Solubility of the black pepper's bioactive compound was obtained by performing static extraction for 15 min. This is a reliable method for measuring solubility as the easily accessible oil from broken cells is directly dissolved to solvent rapidly due to very low transfer resistance of solvent phase (Sovova 2007). Saturated solvent is achieved in 15 min, and the solubility is presented as mass of bioactive compounds extracted per volume of solvent used. Results are shown in Fig. 2.

Solubility is statistically correlated with pressure and temperature ($p = 0.0025$, $p = 0.112$). It is directly proportional to pressure while inversely proportional to temperature. The highest value obtained at 5,000 psi and 45 °C is 10.38 kg/m^3CO_2. Solubility is enhanced by 15 % for every pressure increase of 500 psi and 20 % for every temperature decrease of 5 °C (Fig. 2a) when a single factor is considered. This can be explained through the density of supercritical CO_2. As temperature decreases or pressure increases, density alters dramatically to the value approaching the optimum liquid density for the solubility of hydrocarbon in black pepper (Shi and Zhou 2007). Another significant factor for solubility is pressure–temperature interaction (Fig. 2b). The sensitivity of solubility toward changes of pressure is in fact dependent on solvent temperature. A linear relationship between pressure and solubility exists around 53.5 °C. The strength of the sensitivity increases as temperature decreases from 53.5 to 45 °C. The strongest

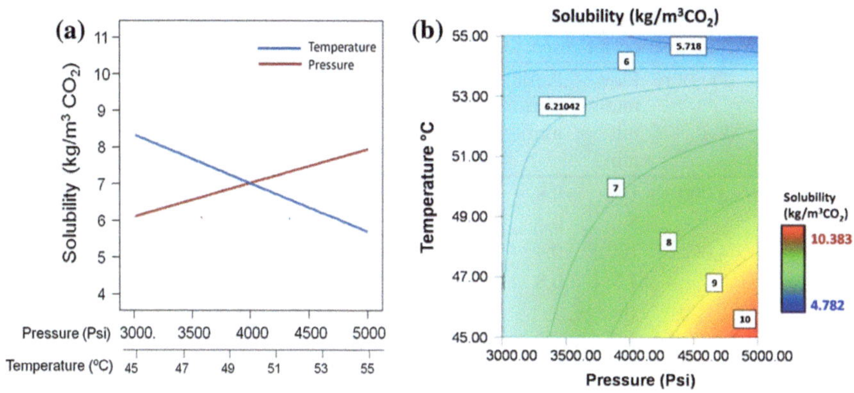

Fig. 2 **a** Effect of temperature at constant pressure of 4,000 psi and effect of pressure at constant temperature of 50 °C. **b** The interaction effect of pressure and temperature on solubility of black pepper's bioactive compounds in supercritical CO_2

interaction is found at 45 °C, where the solubility increases by 1 unit with every 500 psi step change in pressure.

The results are consistent with the findings of Kumoro et al. (2009) and Sovilj et al. (2010) who agreed that both temperature and pressure are the significant factors in solubility. The results contradict the finding of Ferreira et al. (1999) who stated that pressure is the only significant factor, and neither temperature nor the interaction between temperature and pressure affects solubility. This is due to the near-critical temperature range applied by Ferreira et al. (1999). They had found a decrease in solubility as temperature decreased from 40 to 30 °C. Yet, for the range beyond the critical point, when temperature decreased from 50 to 40 °C, the solubility increased. This is consistent with our results.

Mass transfer a quantitative analysis was performed to study the mass transfer of the SFE of black pepper. The transfer rate during SC period and DC period, transition time from SC period to DC period, and fluid mass transfer coefficients were investigated. The results are shown in Table 1.

The transition time was obtained from the experimental extraction curve, referring to the corresponding time where the curve of the line changes. Particle size is the only significant factor ($p < 0.0001$). The transition time varies from 18 to 40 min and increased with particle size. The shortest time was obtained when pepper particles of 0.4 mm diameter were used.

Mass transfer rate in SC period was taken as the total yield achieved during the first period of the extraction per unit volume of the particle bed over extraction time. The significant factors found are pressure, particle size, and the interaction of particle size and temperature ($p = 0.0262$, $p < 0.0001$, $p = 0.0309$). Higher extraction rate is achieved at higher pressure and smaller particle size. This is because as particle size decreases, the amount of easily accessible oil increases due to greater surface/volume ratios. The effect of temperature on mass transfer rate in

Table 1 Mass transfer rate and mass transfer coefficient for solubility-controlled period and diffusion-controlled period for respective experimental condition

Temperature °C	Pressure Psi	Flow rate ml/min	Particle size mm	Transition time s	G_s $\times 10^3$ kg/m^3 s	G_d $\times 10^4$ kg/m^3 s	k_s $\times 10^3$ per s	k_d $\times 10^4$ per s
45	3,000	10	0.8	27	2.2209	0.8964	0.9160	0.2176
		5	0.4	18	4.7077	0.5913	1.6353	0.1596
	4,000	10	0.4	24	5.7880	1.5269	1.0956	0.4459
		7.5	0.8	24	1.5931	0.8132	1.4493	0.2538
	5,000	10	1	28	1.9177	0.9802	1.6489	0.2857
		5	0.8	29	1.8604	1.3377	1.8215	0.3717
		7.5	0.4	20	7.5349	2.0073	1.2683	0.6756
50	3,000	10	1	32	0.9910	0.2988	1.2150	0.0849
		5	0.8	38	1.2334	0.7681	1.9727	0.1931
	4,000	10	0.8	28	1.7415	0.9277	1.3292	0.2590
		5	0.4	24	5.3454	0.8756	1.2382	0.2474
		7.5	0.8	25	1.6089	0.7737	1.5832	0.2426
		7.5	1	20	1.3946	0.7595	2.1200	0.2091
		7.5	0.8	24	1.9671	0.9531	1.4760	0.2694
	5,000	10	0.4	24	5.9739	1.9415	1.0705	0.5751
		5	1	32	1.0528	0.7830	1.4617	0.2210
		7.5	0.8	34	1.7340	1.5183	2.1178	0.4679
55	3,000	10	0.4	22	4.7011	0.6940	1.1908	0.1873
		5	1	36	1.5545	0.9016	1.7754	0.2500
		7.5	0.8	21	1.4189	0.7170	1.8366	0.1896
	4,000	10	1	19	1.2124	0.9490	2.0176	0.2807
		5	0.8	43	1.9031	1.4039	1.0605	0.3928
		7.5	0.4	24	2.3252	1.0336	1.4666	0.3007
	5,000	5	0.4	22	4.1341	1.4009	1.4164	0.3636
		7.5	1	30	1.9495	1.1678	1.4009	0.3452

SC period is dependent on particle size. When particles greater than 0.8 mm are used, the rate is directly proportional to temperature, whereas below 0.8 mm, the rate is inversely proportional to temperature. For 0.8-mm-sized particle, there is a linear relationship between temperature and mass transfer rate. The fluid-phase mass transfer coefficient is determined by Eq. 1. Temperature, particle size, interaction of temperature and pressure, and the interaction of flow rate and particle size are significant ($p = 0.0063$, $p = 0.0347$, $p = 0.0118$, and $p = 0.0014$).

Diffusion rate is presented as the total mass of the bioactive compounds collected during the DC per unit volume of particle bed over the extraction time. **Diffusivity** is calculated by Eq. 2. The significant factors for diffusion rate and the diffusivity are pressure, particle size, temperature–particle size interaction, and pressure–particle size interaction ($p < 0.0001$, $p = 0.0008$, $p = 0.0206$, and $p = 0.0462$ for diffusion rate; $p < 0.0001$, $p = 0.0004$, $p = 0.0041$, $p = 0.0251$ for diffusivity). The intra-particle diffusion rate is considerably lower than the transfer rate during SC period. As noted with the transfer rate in SC period, higher diffusion rate is also achieved at higher pressure and smaller particle size. Pressure is the dominant factor in diffusion. The enhancement of 120 % in diffusion rate is achieved by increasing the pressure from 3,000 psi to 5,000 psi. The effect of pressure on diffusion rate is related to particle size. The strongest interaction is found when particles of 0.4 mm diameter were used. The interaction effect decreases when particle size increases from 0.4 to 1.00 mm. The effect of temperature is also related to particle size. For particle size smaller than 0.65 mm, the diffusion rate is inversely proportional to temperature, while for the particle size greater than that, the diffusion rate is found to be inversely proportional to temperature. The diffusivity that describes the solid-phase transfer coefficient is found to be two orders of magnitude lower than the fluid-phase mass transfer coefficient.

Conclusion

The solubility of black pepper's bioactive compounds in supercritical CO_2 was investigated. The highest solubility achieved was 10.38 kg/m^3CO_2, obtained at 5,000 psi and 45 °C. Higher mass transfer rate was obtained at higher operating pressure and smaller particle size for both SC and DC. The shortest transition time was obtained when 4-mm pepper particles were used. Diffusivity was found to be two orders of magnitude lower than the fluid-phase mass transfer coefficient.

Acknowledgments The authors would like to thank Dr. Dominique Dodge-Wan for her constructive feedback and the financial support from Malaysia Pepper Board.

References

Bruner, G. (2005). Supercritical fluids: Technology and application to food processing. *Reviews in Journal of Food Engineering*, *67*, 21–33.

Ferreira, S. R. S., Nikolov, L., Doraiswamy, L. K., Meireles, M. A. A., Petenatezivko, A. J. (1999). Supercritical fluid extraction of black pepper (Piper nigrum L.) essential oil. *Reviews in Journal of Supercritical Fluids*, 235–245.

Kumoro, A. C., Singh, H., & Hasan, M. (2009). Solubility of piperine in supercritical and near critical carbon dioxide. *Reviews in Thermodynamics and Chemical Engineering Data*, *67*, 1014–1020.

Li, Z. Y., Liu, X. W., Chen, S. H., Zhang, X. D., Xia, Y. J., Xiao, F. (2006). An experimental and simulating study of supercritical CO_2 extraction for pepper oil. *Reviews in Chemical Engineering and Processing*, 264–267.

Perakis, C., Louli, V., & Magoulas, K. (2005). Supercritical fluid extraction of black pepper oil. *Reviews in Journal of Food Engineering*, *71*, 386–393.

Reverchon, E., Daghero, J., Marrone, C., Mattea, M., & Poletto, M. (1999). Supercritical fractional extraction of fennel seed oil and essential oil: Experiments and mathematical modeling. *Reviews in Industrial and Engineering Chemistry Research*, 3069–3075.

Salgin, U., Döker, O., & Calimli, A. (2006). Extraction of sunflower oil with supercritical CO_2. *Reviews in Journal of Supercritical Fluids, 38*, 326–331.

Shi, J., Zhou, X. (2007). Solubility of food components and product recovery in the supercritical fluid separation process. In Shi John *functional food ingredients and nutraceuticas: Processing and technologies*. Boca Raton: Taylor & Francis Group, LLC.

Sovilj, M. N., Nikolovski, B. G., Spasojević, M. Đ. (2010). Critical review of supercritic al fluid extraction of selected spice plant materials. *Reviews in Macedonian Journal of Chemistry and Chemical Engineering*, 197–220.

Sovova, H. (2007). Modeling of supercritical fluid extraction of bioactives from plant materials. In Shi John *functional food ingredients and nutraceuticas*. Boca Raton: Taylor & Francis Group, LLC.

Srinivasan, K. (2009). Black pepper (Piper nigrum) and its bioactive compound, piperine. In B. B. Aggarwal, *Molecular targets and therapeutic uses of spices: Modern uses for ancient medicine*. Singapore: World Scientific Publishing Co. Pte. Ltd.

Comparative Study of Cellulose Extraction Processes from Palm Kernel Cake

Y. Y. Farm, S M. Anisuzzaman, D. Krishnaiah and A. Bono

Abstract Palm kernel cake (PKC) is one of palm oil industry by-products which are rich in edible cellulose, which can be used as potential source in food and pharmaceutical industry. The extraction of cellulose from PKC involved essentially delignification and hemicellulose removal processes. In this study, three delignification and two hemicellulose removal techniques were comparatively investigated. Response surface methodology (RSM) with D-optimal design was used for the analysis. In this analysis, delignification techniques, hemicellulose removal techniques, and hemicellulose removal time (HRT) were chosen as process factors, whereas quantity of hemicellulose removal, cellulose yield, and cellulose purity were chosen as process responses. The comparative result obtained in this study shows that the combination of liquid phase oxidation (LPO) of delignification technique and alkali treatment of hemicellulose removal was the best method of cellulose extraction from palm kernel. The result of FTIR spectrum analysis of the cellulose produced in this study was similar to those produced commercially validating the cellulose structure. The optimum cellulose extraction method in this study shows that hemicellulose can be removed up to 24 % with cellulose yield 70 % and purity 77 %.

Keywords Response surface methodology · Delignification · Hemicellulose · Process selection

Y. Y. Farm · SM. Anisuzzaman · D. Krishnaiah · A. Bono (✉)
School of Engineering and Information Technology, University Malaysia Sabah,
88400 Kota Kinabalu, Sabah, Malaysia
e-mail: bono@ums.edu.my

R. Pogaku et al. (eds.), *Developments in Sustainable Chemical and Bioprocess Technology*, DOI: 10.1007/978-1-4614-6208-8_33,

Introduction

In the recent years, the cellulose demand is growing in the food and chemical industrial applications. Cellulose is biocompatible, renewable, biodegradable, reproducible, and recyclable (Isogai 2001). The unique properties have prompted the development of cellulose in various applications such as smart material in electrical field (Kim et al. 2006), foods, fuel (Kammes et al. 2011), chemicals (Zhang et al. 2011), textile, membrane (Wu and Yuan 2002), and pharmaceuticals (Hoenich 2006). This has incited the research interest on sustainable agricultural residues which are rich in cellulose as alternative resources such as sugar beet pulp, wheat straw, and many other alternative sources.

Palm kernel cake (PKC) is one of the abundant by-products obtained from oil palm industries. In Malaysia, about 2.3 million tons of PKC was produced in 2011 (Chu 2011). The abundant and the reasonable content of cellulose are significant to consider PKC as potential resources for cellulose production (Farm et al. 2009). However, separation of cellulose from agriculture resource still constitutes as one of the major obstacles in efficient cellulose utilization in industries. To recover pure cellulose from resources, cross-linkage among cellulose, hemicelluloses, and lignin should be broken down. Hence, in this research, the cellulose extraction was divided into two stages which are hemicellulose removal and followed by delignification.

Hot water treatment and alkali treatment are common techniques to remove hemicelluloses (Xu et al. 2007; Liu et al. 2006; Zhang et al. 2007) from vegetation residues. High temperature could break down the linkage of hemicelluloses from cellulose and lignin. According to Dusterhoft et al. (1992), 66 % of hemicelluloses of PKC could be solubilized by with alkali and sodium chloride (Sundu et al. 2006). Additional delignification process is needed in order to enhance the cellulose extraction purity. Acidified sodium chlorite is commonly used to isolate cellulose from wood (Sun et al. 2004a). However, environmental concerns associated with chlorinating agents have led various green approaches to prevent ecological damage. Organosolv processes have been applied as one of the alternative cellulose extraction processes due to its environmental amity. Various researches have been studied on this approach (Astimar et al. 2002; Sun et al. 2004b; Sarwar et al. 2007; Feng et al. 2006; Xua et al. 2007). Environmental-friendly oxidation reagent is another potential green extraction process. Mae et al. (2000) and Farm et al. (2009) have studied the cellulose extraction by liquid phase oxidation (LPO) with hydrogen peroxide under mild conditions. Besides, Farm et al. (2009) have also found that the process is able to produce substantial quantities of useful by-products such as organic acids.

Hence, the aim of this study is to identify an efficient cellulose extraction method from PKC by comparing three different delignification techniques and two different hemicellulose removal techniques. The work was also extended to study the optimization of cellulose extraction methods from PKC.

Methodology

Materials

To maintain the consistency of raw material, fresh palm kernel (*Elaeisguineensis var. tenera*) was used to prepare PKC. The composition of oil-free PKC fiber is shown in Table 1. Prior to the cellulose extraction process, the kernel was washed, dried, ground, and made free from oil. After being washed, palm kernel was then dried at 60 °C in an oven for 16 h, and the palm kernel was ground about 50 g to pass through 0.30–0.425-mm screen. Oil-free palm kernel was produced using double-step solid–liquid extraction technique using *isopropanol* as the solvent for at least 8 h each step. All analysis was based on the composition of the oil-removed PKC fiber.

Analysis of Cellulose Content and Hemicellulose Released

Chemical analysis was carried out to determine cellulose content of PKC and cellulose purity. The method of cellulose content and purity determination was based on Japanese Standard JIS8101. The purity of cellulose was determined by the insoluble cellulose sample in 17.5 % NaOH at 20 °C as described in JIS 8101 which represented cellulose. The validation of cellulose produced in the process was conducted by comparing the Fourier transform infrared (FTIR) spectra with the spectra of commercial cellulose. The quantity of hemicellulose removed and the yield of extraction process were determined by gravimetry. The quantity of hemicellulose removed in the process was measured by comparing the weight of dry hemicellulose removed with the oil-free PKC used. Similarly, the yield of extraction process was quantified by comparing dry weight of extracted product with initial weight of oil-free PKC used.

Table 1 Analysis of oil-free PKC fiber (PKC)

Testing	Percentage
pH	5.38
Hot water soluble[a]	12.57
Cold water soluble[a]	13.36
Cellulose[b]	22.58
Holocellulose[c]	73.20
Lignin[d]	5.10
Protein	19.33
Ash content[e]	3.65

[a] TAPPI standard T 207 OS-75, water solubles in wood and pulp; [b] JIS 8101; [c] Determination of holocellulose from method of Wise et al. (1946); [d] TAPPI standard T 222 OS-74; [e] TAPPI standard T 15 OS-58. *Note* All percentages are on an oven-dried fiber basis

Hemicellulose Removal Process

Two techniques of hemicellulose removal process were selected for comparison purposes. Hot water treatment was conducted based on the technique of Mae et al. (2000). The technique consisted in the heating of oil-free PKC with water in a 2-L pressurized extractor (Parr Instruments, USA; range 0–140 bar) at 10 bar, 180 °C with liquid–solid ratio of 5:1 (w/w). The duration of the treatment experiment run was set in the range of 30–120 min. At the end of hot water treatment, reactor was cooled down and the content was filtered and then washed with solution of 95 % ethanol in water.

The second technique was alkali treatment. Oil-free PKC was mixed with NaOH solution of 17.5 % (w/w) at the liquid–solid ratio of 4:1 (w/w). The mixture was left stirred at room temperature, and the treatment duration time was set as hot water treatment experimental run. At the end of the alkali treatment, the mixture was then filtered and washed. The washing was conducted with distilled water and repeated until neutral filtrate was achieved.

Delignification Process

The delignification process was conducted using three different techniques: Organosolv, LPO with iron catalyst, and acidic sodium chlorite. Organosolv technique consisted in the mixing of 4-g hemicellulose-free PKC fiber with 40 g solution of formic acid and acetic acid in water at volume ratio of 30:60:10. HCl 0.1 % (w/w) was added into the mixture as catalyst at the ratio of 20:1 (mL/g). The mixture was left stirred at 85 °C for 2 h (Sun et al. 2004b; Feng et al. 2006).

For the acidic sodium chlorite delignification technique, 4-g hemicellulose-free PKC fiber was mixed with 40 g solution of 1.3 % (w/w) $NaClO_2$ at pH 3.5. The pH of the mixture was maintained at constant level by adjustment with 10 % (w/w) acetic acid. The process was carried out in heated open flask at 75 °C for 2 h as given by Sun et al. (2004a).

Hydrogen peroxide in the presence of Fe was used in LPO delignification process. It has been known that iron catalyzes the decomposition of hydrogen peroxide. The LPO technique was adopted from previous research work (Mae et al. 2000; Farm et al. 2009). In this technique, 4-g hemicellulose-free PKC fiber was mixed with 40 g solution of 30 % (w/w) H_2O_2 and 13.6 mg of $FeSO_4$. The reaction was left at constant temperature of 80 °C for 2 h.

At the end of each delignification process, the mixture was cooled to room temperature. For fast cooling, cold distilled water was used as a chiller. The product of each technique of the delignification process was obtained by washing the residue with distilled water and followed by 95 % ethanol. The product was then oven-dried at 60 °C for 16 h before it was analyzed for the yield and purity.

Table 2 Range and level of factors for the experimental design

Factors	Range and levels		
	−1		1
HRT (min)	30		120
Hemicellulose removal techniques	Hot water		Alkali
	−1	0	1
Delignification techniques	Organosolv	LPO	Acidic sodium chlorite

Experimental Design

The experimental work was designed to enable the comparison of the techniques with the selection of the best extraction process. For this purpose, response surface methodology (RSM) with D-optimal design via Design Expert software was employed (Duduku et al. 2011, 2012; Bono et al. 2006, 2008a, b). For the requirement of RSM, three factors and three responses were selected. Hemicellulose removal and delignification techniques were set as categorical factors, whereas hemicellulose removal time (HRT) was set as numerical factor. The range of each factors are shown in Table 2. The responses are cellulose yield, the quantity of hemicellulose removed, and the purity of cellulose. The number of experimental run and the condition of each run generated by Design Expert software based on the range and level of factors are shown in Table 3. There are some replications of experimental runs generated in Table 3. The replication was used to secure more accurate estimate of the experimental error, to decrease the experimental error, and thereby to increase precision, which is a measure of the variability of the experimental error.

Results and Discussion

Hemicellulose Removal Processes

Regression analyses were performed to fit the responses with the experimental data. The range of experimental conditions and methods employed in this research are shown in Table 4 for hemicellulose removal and cellulose yield. It is observed that the hemicellulose removal and cellulose yield depend on the time (HRT), treatment, and delignification methods.

As shown in Fig. 1, there are some hemicellulose was remaining using both these hemicellulose removal techniques. The oil-free PKC-removed hemicelluloses from via alkali treatment were in the range of 20–22.4 %. Biljanaet al. (2008) observed similar results during hemp fiber fractionation using 17.5 % sodium

Table 3 D-optimal design as composed by Design Expert® and the corresponding responses for selection of cellulose extraction process from PKC

Runs	Variables			Responses		
	HRT (min)	Treatment	Delignification	Hemicellulose removal (%)	Cellulose yield[b] (%)	Purity (%)
1	30	Hot water	Liquid phase oxidation	45.56	68.20	60.40
2	30	Alkali	Organosolv	20.07	70.27	72.20
3	30	Hot water	Acidic sodium chlorite	45.56	57.90	66.70
4	30	Hot water	Organosolv	44.44	66.17	72.80
5	30	Alkali	Organosolv	10.94	62.35	70.50
6	30	Hot water	Acidic sodium chlorite	45.00	56.80	67.80
7	50	Alkali	Acidic sodium chlorite	28.65	69.00	61.60
8	50	Alkali	Liquid phase oxidation	22.67	75.93	70.50
9	50	Hot water	Organosolv	50.83	62.97	78.10
10	75	Hot water	Acidic sodium chlorite	55.20	50.70	76.10
11	75	Hot water	Liquid phase oxidation	36.61	60.17	67.90
12	105	Alkali	Organosolv	21.11	50.65	75.60
13	105	Alkali	Organosolv	28.13	50.34	81.60
14	120	Hot water	Acidic sodium chlorite	56.07	66.90	78.00
15	120	Hot water	Organosolv	53.49	53.75	79.80
16	120	Alkali	Acidic sodium chlorite	21.12	67.92	65.50
17	120	Alkali	Liquid phase oxidation	21.49	58.18	79.80
18	120	Hot water	Liquid phase oxidation	53.98	54.20	70.80
19	120	Hot water	Organosolv	56.67	48.65	79.70

[a] Calculated on original PKC fiber based on oven-dried pretreated PKC fiber; [b] Calculated on pretreated PKC fiber based on oven-dried extracted PKC fiber, where t = HRT (min)

Table 4 Summary of fitted models for cellulose extraction process responses

Parameters	Hemicellulose removal (%)	Cellulose yield (%)	Cellulose purity (%)
Model	Quadratic	2FI	Quadratic
Prob>F	<0.0001	<0.0001	<0.0001
Lack of fit	0.2811	0.9252	0.0800
R-squared	0.9889	0.9608	0.9822
Adjusted R-squared	0.9858	0.9261	0.9508
Predicted R-squared	0.9784	0.8276	0.7842

hydroxide. Hot water was able to remove up to 55 % of hemicelluloses from PKC. This might be due to the stability of hemicellulose hydrogen bonding to cellulose fibrils in PKC. Hence, delignification is mandatory in order to recover the high purity cellulose.

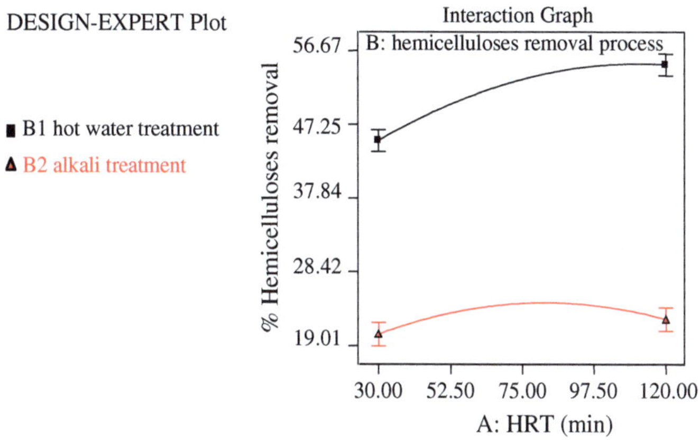

Fig. 1 Hemicellulose removal from PKC with hot water and alkali treatments

Delignification Processes

By using the Expert Design®, the experimental results for cellulose yield and purity were applied to fit the most compatible response surface models. Table 4 summarizes the results for delignification of PKC fiber and the adequacy of fitted models for all the responses. Quality of the fitted models quadratic and two-factor interaction (2FI) was evaluated by ANOVA to obtain the error. The models did not show any lack of fit and the determination coefficient (R-squared). The values obtained for hemicellulose removal, cellulose yield, and cellulose purity were 0.98, 0.96, and 0.98, respectively. Similar values of adjusted R-squared and predicted R-squared were found in respect of all the three responses.

Organosolv with Organic Acids

Delignification with Organosolv was carried out using formic acid, acetic acid, and water in the ratio of 30:60:10 (v/v), respectively, with 0.1 % (w/w) HCl. As shown in Fig. 2a, the cellulose yield of both pretreated PKC via Organosolv decreased with increase in HRT. The longer the HRT is, the yield of alkali-pretreated PKC is lower compared to hot water treated. The purity comparison is shown in Fig. 3a. The purity of cellulose with alkali-treated and hot water-treated PKC increased exponentially up to 90 min. Alkali-pretreated PKC with Organosolv produced maximum cellulose purity about 78.0 % at 90 min of HRT. However, a little degradation of hot water-pretreated PKC could be found besides delignification using Organosolv for pretreatment time longer than 90 min. Some of the cellulose fiber degraded and dissolved in solution. Hence, purity cellulose of Organosolv with hot water-pretreated PKC reduced when the HRT is more than 90 min.

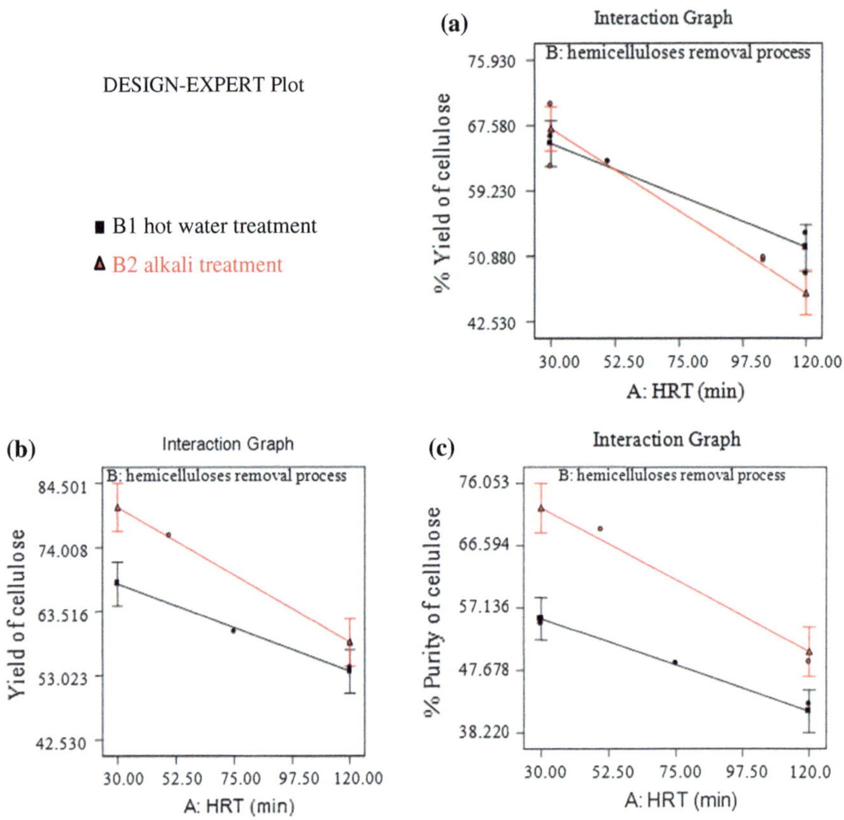

Fig. 2 Yield of cellulose using **a** Organosolv, **b** LPO, and **c** acidic sodium chlorite delignification technique

Acidic Sodium Chlorite Treatment

As shown in Figs. 2c and 3c, it is interesting to observe that cellulose yield of acidic sodium chlorite with alkali-pretreated PKC produced higher yield but lower purity of cellulose, while PKC treated using acidic sodium chlorite with hot water pretreatment produced lower yield but higher purity of cellulose compared to alkali-pretreated PKC cellulose. However, it was also found that purity cellulose slightly decreased for hot water-pretreated PKC more than pretreatment time of 98 min. This can be explained by the depolymerization or degradation of cellulose caused by the oxidizing agent at high concentration and long reaction times. However, to prevent the loss of yield due to degradation of cellulose, combination of longer pretreatment time and delignification time should be avoided (Asim 2006).

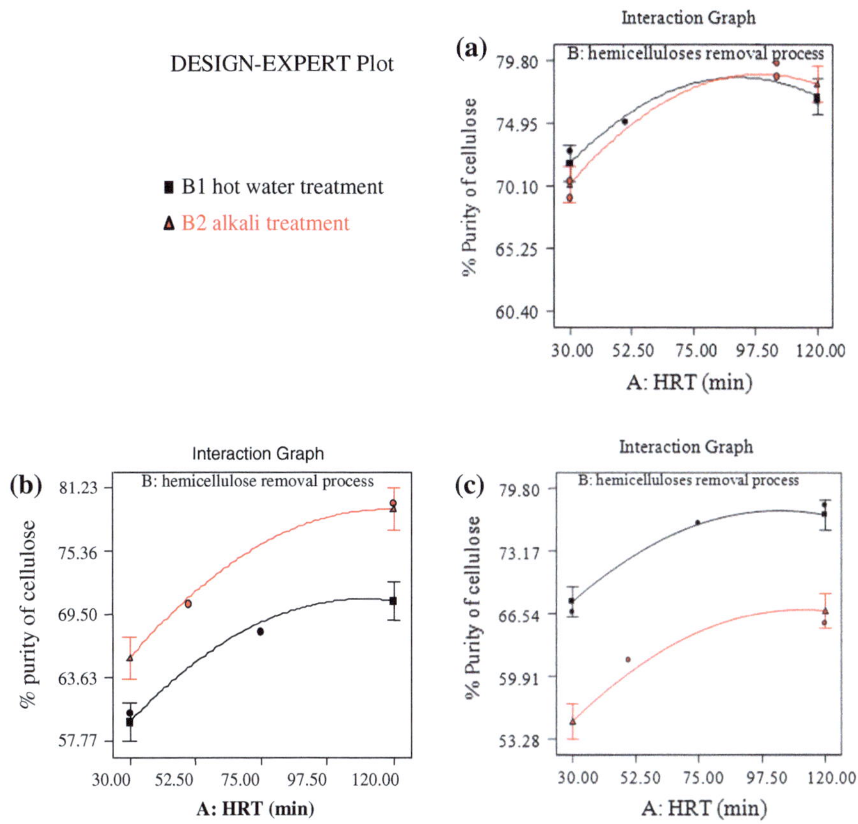

Fig. 3 Purity of cellulose using **a** Organosolv, **b** LPO, and **c** acidic sodium chlorite delignification technique

Liquid Phase Oxidation

The cellulose yield of LPO using Fenton's reagent obtained the highest yield compared to Organosolv and acidic sodium chlorite. As shown in Fig. 2b, the HRT increased with LPO, whereas cellulose yield continuously decreased. The delignification with alkali pretreatment obtained higher cellulose yield compared to hot water pretreatment as the longer pretreatment time is applied to PKC. The cellulose yield of LPO with alkali pretreatment decreased from 80.58 to 58.44 % with the increase in HRT from 30 to 120 min. Figure 3b shows the purity of cellulose increased significantly as the HRT time increased. In consequence, as in longer HRT, purity of cellulose PKC which is alkali treated and followed by LPO achieved higher than hot water treated. The maximum purity of cellulose PKC with alkali treated is about 79 % for 120 min HRT. Purity of cellulose with hot water treated in the process reached maximum purity about 71 % for 110 min

HRT. Purity of hot water-treated PKC cellulose has slightly decreased for HRT that is longer than 110 min.

Acidic Sodium Chlorite Treatment

As shown in Figs. 2c and 3c, it is interesting to observe that cellulose yield of acidic sodium chlorite with alkali-pretreated PKC produced higher yield but lower purity of cellulose, while PKC treated using acidic sodium chlorite with hot water pretreatment produced lower yield but higher purity of cellulose compared to alkali-pretreated PKC cellulose. However, it was also found that purity cellulose slightly decreased for hot water-pretreated PKC more than pretreatment time of 98 min. This can be explained by the depolymerization or degradation of cellulose caused by the oxidizing agent at high concentration and long reaction times. However, to prevent the loss of yield due to degradation of cellulose, combination of longer pretreatment time and delignification time should be avoided (Asim 2006).

FTIR Analysis of Extracted PKC Cellulose

Validation of extracted cellulose was conducted by comparison of the FTIR spectra with the spectra of commercial pure cellulose. Figure 4 illustrates the FTIR spectra of extracted cellulose from palm kernel fiber under different conditions: LPO with hot water treatment (C_{1a}), LPO with alkali treatment (C_{1b}), acidic sodium chlorite with hot water treatment (C_{2a}), acidic sodium chlorite with alkali treatment (C_{2b}), Organosolv with hot water treatment (C_{3a}), and Organosolv with alkali treatment (C_{3b}). Evidently, FTIR spectra of both extracted cellulose under different conditions indicate similar structure of residues. All the cellulose preparations clearly gave a typical signal pattern for the cellulose. The O–H stretch band (ca. 3,468 cm^{-1}) and stretch band 2,905 cm^{-1} correspond to CH stretching of CH_2 and CH_3 groups. A weak $C = 0$ band at 1,715 cm^{-1} for carboxylic group in FTIR spectroscopy is presented for cellulose extraction under Organosolv and acidic sodium chlorite delignification technique. In particular, the disappearance of principal band at 1,715 cm^{-1} is observed from cellulose treated under alkali treatment in LPO, which reveals that the LPO effectively cleaves the ester bond from hemicellulose and/or lignin.

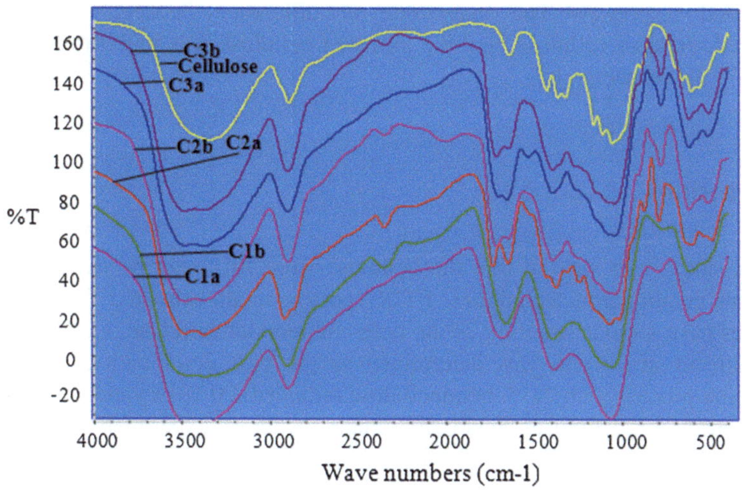

Fig. 4 FTIR spectra of cellulose obtained from PKC under different conditions

Optimization of Cellulose Extraction Methods

Central composite experimental design was used to establish an efficient cellulose extraction process. The optimization was done after each response has been analyzed to establish the appropriate model. In the range of yield and the maximum purity of cellulose compromise with respect to three responses acquired to determine efficient cellulose extraction process. Based on estimation models obtained, LPO with alkali treatment at 77 min HRT was determined as the best combination of cellulose extraction process. The selected process was able to remove 24 % of hemicellulose, 70 % of yield cellulose, and 77 % purity of cellulose. Optimized condition was then applied to the selected process with weight ratio of H_2O_2 to hemicellulose-free PKC at 10:30 (g/g), reaction time at 149 min, reaction temperature at 65 °C, and amount catalyst of $FeSO_4$ for 12.5 g. The outcome of the model was then compared with mean experimental values from duplicated experiments. As shown in Table 5, the validity of the responses of models was verified by small differences between the experimental values and the

Table 5 The experimental and calculated values of optimized cellulose extraction using LPO with alkali pretreatment

Response	Experimental value (%)	Calculated value (%)	Differenc[a]
Hemicellulose removed	26	24	7.70
Yield of cellulose	68	71	4.40
Purity cellulose	80	77	3.75

[a] Based on experimental value

calculated responses. The magnitude of difference between hemicellulose removed, yield of cellulose, and the purity of cellulose was found to be 7.7, 4.4, and 3.75 %, respectively.

Conclusions

In this study, it is proved that the cellulose from PKC can be extracted under different combination techniques. FTIR spectra of all extracted PKC cellulose samples have similar structure as the pure commercial cellulose. LPO with alkali pretreatment at 77 min was determined as the best combination of cellulose extraction process with 24 % hemicellulose removed, 70 % of yield cellulose, and 77 % purity of cellulose. Small differences between experimental values and calculated responses revealed that D-optimal design was verified as a model to study the process selection of cellulose extraction from PKC.

Acknowledgments Financial support for this study was provided by *Penyelidikan Siswazah Universiti* (PGD) 2006, MOSTI, Malaysia.

References

Alimon, A. R. (2004). The nutritive value of palm kernel cake for animal feed, Department of Animal Science, Faculty of Agriculture UPM. http://www.chgs.com.my/chinese/download/Palm%20Oil.pdf.

Asim, K. R. C. (2006). *Textile preparation and dyeing, chemistry of textile materials* (p. 11). New Hampshire: Science Publishers, 03784, USA.

Astimar, A. A., Husin, M., & Anis, M. (2002). Preparation of cellulose from oil palm empty fruit bunches via ethanol digestion: Effect of acid and alkali catalysts. *Journal of Oil Palm Research, 14*(1), 9–14.

Biljana, M. P., Mirjana, M. K., Petar, D. S., & Jovana, Z. P. (2008). The effects of hemicelluloses and lignin removal on water uptake behavior of hemp fibers. *Bioresource Technolnology 99*(15), 7152–7159.

Bono, A., Duduku K., Mariani R., & Nancy J. S. (2006). D-*optimal analysis of formula variation effect on Melamine-Urea Formaldehyde*. Studies in Surface Science and Catalysis (pp. 713–716). Elsevier BV. ISBN-10: 0-444-52751-6 159.

Bono, A., Rosalam, S., Mariani, R., & Duduku, K. (2008a). Effect of mixture components on the properties of MUF resin. *International Journal of Physical Sciences, 3*, 45–49.

Bono, A., Duduku, K., & Mariyani, R. (2008b). *Product and process optimization using response surface methodology*. Universiti Malaysia, Sabah. ISBN: 978-983-2369-80-6.

Chu, Y. M. (2011). *Overview of the Malaysian Oil Palm Industry 2010*, Malaysian Palm Oil Board, Malaysia.

Duduku, K., Rajesh, N., & Rosalam, S. (2011). Optimization of spray for drying Morinda citrifolia L. fruit extract. *Journal of Applied Sciences, 11*(13), 2276–2283.

Duduku, K., Anisuzzaman, S. M., Bono, A., & Rosalam S. (2012). Adsorption of 2, 4, 6-Trichlorophenol (TCP) onto activated carbon. *Journal of King Saud University-Science* (in press).

Dusterhoft, E. M., Posthumus, M. A., & Voragen, A. G. J. (1992). Non-starch polysaccharides from sunflower (*Helianthus annuus*) meal and palm kernel (*Elaeisguineensis*) meal preparation of cell wall material and extraction of polysaccharide fractions. *Journal of the Science of Food and Agriculture, 59*, 151–160.

Farm, Y., Duduku, K., Bono, A., & Mariani, R. (2009). Cellulose extraction from Palm kernel cake using Liquid phase oxidation. *Journal of Engineering Science and Technology, 4*(1), 57–68.

Feng, X., Jin-Xia, S., Runcang, S., Paul, F., & Mark, S. B. (2006). Comparative study of organosolv lignins from wheat straw. *Industrial Crops and Products, 23*, 180–193.

Hoenich, N. (2006). Cellulose for medical applications: past, present and future. *BioResources, 1*(2), 270–280.

Isogai, A. (2001). In D. N. S. Hon & N. Shiraishi (Eds.) *Chemical modification of cellulose in wood and cellulosic chemistry: Second Edition, revised and expanded* (pp. 599–625). New York: Mercel Dekker.

Kammes, K. L., Bals, B. D., Dale, B. E., & Allen, M. S. (2011). Grass leaf protein, a coproduct of cellulosic ethanol production, as a source of protein for livestock. *Animal Feed Science and Technology, 164*(1–2), 79–88.

Kim, J., Yun, S., & Ounaies, Z. (2006). Discovery of cellulose as a smart material. *Macromolecus, 39*, 4202–4206.

Liu, C. F., Fu, X., Sun, J. X., Ren, J. L., Curling, S. R., Sun, C., et al. (2006). Physicochemical characterization of cellulose from perennial ryegrass leaves (*Lolium perenne*). *Carbohydrate Research, 341*, 2677–2687.

Mae, K., Isao H., Noriaki S., & Kouichi M. (2000). A new conversion method for recovering valuable chemicals from oil palm shell wastes utilizing liquid phase oxidation with H_2O_2 under mild condition. *Energy & Fuels, 14*, 1212–1218.

Sarwar, J. M., Chowdhury, D. A. N., Khalidul, I. M., & Iqbal M. S. M. (2007). Characterization of Lignin Isolated from Some Nonwood Available in Bangladesh. *Bioresource Technology, 98*, 465–469.

Sun, J. X., Sun, X. F., Zhao, H., & Sun, R. C. (2004a). Isolation and characterization of cellulose from sugarcane bagasse. *Polymer Degradation and Stability, 84*, 331–339.

Sun, X. F., Sun, R. C., Fowler, P., & Baird, M. S. (2004b). Isolation and characterisation of cellulose obtained by a two-stage treatment with organosolv and cyanamide activated hydrogen peroxide from wheat straw. *Carbohydrate Polymers, 55*, 379–391.

Sundu, B., Kumar, A., & Dingle, J. (2006). Palm kernel cake in broiler diets: effect on chicken performance and health. *Worlds Poultry Science Journal, 62*, 316–325.

TAPPI T 15 os-58, Technical Association for the Pulp and Paper Industries. (1978). Ash in wood and pulp.TAPPI Test Method T 15 os-58. TAPPI PRESS, Atlanta.

TAPPI standard T 222 OS-74, TAPPI Standard Methods. (2009). Fibrous Materials and Pulp Testing. Technical Association of Pulp and Paper Industry, Atlanta.

TAPPI Test Method T 12 OS-75. (1978). Technical Association for the Pulp and Paper Industries Preparation of extractive free-wood. TAPPI PRESS, Atlanta.

Wise, L. E., Maxine, M., & D'Addieco, A. A. (1946). Chlorite holoeellulose, its fractionation and bearing on summative wood analysis and on studies on the hemicelluloses. *Paper Trade Journal, 122*, 35–43.

Wu, J., & Yuan, Q. (2002). Gas permeability of a novel cellulose membrane. *Journal of Membrane Science, 204*(1–2), 185–194.

Xu, F., Sun, J. X., Geng, Z. C., Liu, C. F., Ren, J. L., Sun, R. C., et al. (2007). Comparative study of water-soluble and alkali-soluble hemicelluloses from perennial ryegrass leaves (Loliump-eree). *Carbohydrate Polymers, 67*(1), 56–65.

Xua, Y., Kecheng, L., & Meiyun, Z. (2007). Lignin precipitation on the pulp fibers in the ethanol-based organosolv pulping. *Colloids and Surfaces A: Physicochemical and Engineering Aspects, 301*, 255–263.

Zhang, Y. H. P., Ding, S. Y., Mielenz, J. R., Cui, J. B., Elander, R. T., Laser, M., et al. (2007). Fractionating recalcitrant lignocellulose at modest reaction conditions. *Biotechnology and Bioengineering, 97*(2), 214–223.

Zhang, S. P., Jin, F. M., Hu, J. J., Huo, Z. B. (2011). Improvement of lactic acid production from cellulose with the addition of Zn/Ni/C under alkaline hydrothermal conditions. *Bioresource Technology, 102*(2), 1998–2003.

Screening and Separation of Industrially Useful Hydrolases from the Wasteful Skim Latex Serum of *Hevea Brasiliensis*

M. Nazhirah and Y. Faridah

Abstract Hydrolases extracted from *Hevea brasiliensis* are a source of very useful industrial enzymes. They play an important role in industries related to detergent, food, oil processing, fine chemicals, and agrochemicals. Rubber tree can be a source of phytochemicals which have the potential for use in making various nutraceutical and pharmaceutical products. Malaysia is one of the biggest producers of natural rubber in the world. However, Malaysia is still striving with the problems related to waste management and environmental pollution despite the high income from natural rubber production. Skim latex is a by-product and the most polluting waste in the rubber-processing industry. Therefore, it is very important to find new uses for this rubber waste. Protein constitutes about 2 % of dry mass of the total latex and has been reported to contain useful enzymes such as hydrolases. Compared to other enzymes, hydrolases are more frequently used in industrial applications. Over half of all detergents presently available contain hydrolases and their subclasses, including proteases, cellulases, amylases, and lipases. Malaysia has vast sources of hydrolases because it is one of the largest producers of natural rubber. Thus, it is unfortunate that the *Hevea brasiliensis* skim latex is as yet not fully utilized. This work proposes to systematically identify these potentially useful enzymes from the waste skim latex of *Hevea brasiliensis*. The work involved the identification of the hydrolases by screening for their presence using known in vitro enzymatic assays. This was followed by extraction and separation of the most active hydrolases identified via bioactivity-guided column chromatography techniques. From this study, lipid acyl hydrolase (LAH),

M. Nazhirah (✉)
Biomolecular and Bioprocess Engineering Research Unit (BPMERU),
Department of Biotechnology Engineering, Faculty of Engineering,
International Islamic University Malaysia, P.O. Box 10 50728 Kuala Lumpur, Malaysia
e-mail: nazhirah_mohd@yahoo.com

Y. Faridah
Department of Biotechnology Engineering, Faculty of Engineering,
International Islamic University Malaysia, P.O. Box 10 50728 Kuala Lumpur, Malaysia
e-mail: yfaridah@iium.edu.my

R. Pogaku et al. (eds.), *Developments in Sustainable Chemical and Bioprocess Technology*, DOI: 10.1007/978-1-4614-6208-8_34,

a subtype of lipase, was considered as the potential protein that can be recovered from the skim latex serum of *Hevea brasiliensis.* This is because its activity was found to be the highest (1.561 U/ml). Further separation by gel-filtration chromatography on Sephacryl™ S-200 gave the results for LAH activity (1.304 U/ml) and total activity (97.782 U). 10 ml of skim latex serum was successfully granted a purified LAH with specific activity of 0.593 μmol $min^{-1}mg^{-1}$. A running buffer solution of 100 mM Tris HCl (pH 7.0) was used in all the steps. The significances of this work are contribution toward the utilization of latex serum to produce value-added product, generation of additional income for the manufacturers, and development of environment-friendly techniques in latex processing.

Keywords Detergent · Industrial enzymes · Latex serum · Purification · Rubber waste

Introduction

According to (Yusof and Chow 2003), *Hevea brasiliensis,* the para rubber tree, is the most essential source of natural rubber (*cis*-1,4-polyisoprene). National Innovation Strategy (2012) reported that Malaysian rubber and rubber products are exported to more than 180 countries. Malaysia is currently the world's number one exporter of natural rubber latex gloves, nitrile gloves, latex threads, and rubber catheters. From the report, in 2010, close to 60 % of the world's consumption of gloves was met by the gloves manufacturing sector of Malaysia. According to Department of Statistics Malaysia (2011), the figures from October 2011 show a natural rubber production of 85,800 tones. However, even after the high income from natural rubber production, Malaysia still suffers from the rising costs for waste management and mounting environmental problems due to the pollution from skim latex.

Natural Rubber Processing

After tapping, the exuded latex is pooled and taken to the factories for processing. For the production of natural rubber, the collected latex is treated with ammonia, and after separation of the rubber particles, the aqueous layer is usually discharged into the environment. This milky white aqueous layer, which is the by-product of natural rubber latex concentrate, is known as skim latex. The main content of skim latex is its serum or the non-rubber aqueous part, while the dry rubber content is only 3–7 %. Skim latex is undeniably the most polluting waste in rubber industry. It has a distinct ammoniated odor that can affect the community living in the surrounding areas, as people who are repeatedly exposed to ammonia through inhalation may develop respiratory irritation. A membrane separation of skim latex yields skim rubber, which fetches low price, and serum. Usually, before being released into the main waterways, the pretreatment process of skim latex is done in the effluent treatment oxidation ponds. The used water which is discharged from the effluent treatment pond into the main waterways must meet the requirements of Malaysian Standard MS ISO/IEC 17025:2005. In the end, the rubber manufacturers spend a lot of money for the waste management and effluent treatment of skim latex.

Use of Hydrolases in Industries

Hydrolases belong to a family of enzymes which catalyzes the hydrolysis of a chemical bond. They are classified as EC3 and can be subdivided further into several subclasses based on the bonds they hydrolyze. Hydrolases are the most frequently used enzymes in organic chemistry due to several reasons, one of these being that they are easy to use and do not need cofactors. Also, compared to other enzymes, hydrolases are therefore more frequently used industrially. Since the mid-1960s, the use of enzymes in detergents has been their largest application. Over half of all detergents presently available contain hydrolases, of subclasses: proteases, cellulases, amylases, and lipases. These enzymes help in eliminating stains usually made up of starches, fats, and proteins from soiled garments and textiles. The natural task of hydrolases is to hydrolyze ester, amide, and glucoside bonds. They have different substrate specificity; however, many hydrolases can accept a wide range of substrates. In 1996 alone, (Tramper and Poulsen 2005), USD 414 million was spent globally on enzymes applied to detergents and it is projected that the average annual growth rate for enzymes in this sector is about 10 %. The second largest user of enzymes globally is the food industry (USD 214 million), followed by the textile industries (USD 32.4 million).

Types of Proteins in Skim Latex Serum

Hevea brasiliensis latex can be easily separated into three fractions: the upper layer rubber cream, the medium clear aqueous phase (the cytosol, called C-serum), and the sediment which consists mainly of vacuole-like organelles (lutoids) after high-speed centrifugation (Sunderasan et al. 2002). Proteins make up 2 % by weight of natural rubber latex (Perrella and Gaspari 2002). More than 75 % of the total soluble protein in latex is found in the serum fraction. According to Yusof and Chow (2003), latex is the milky sap that is discovered in several different plants and is generated by specialized cells known as laticifers. Wahler et al. (2009) state that some of the proteins and enzymes have been identified in the latices of laticiferous plants. These include trypsin inhibitor as well as chitinases and β-1,3-glucanase in the latex of the rubber tree *Hevea brasiliensis* (Martin 1991). Wahler et al. (2009) claimed that hydrolases is one of the major proteins that can be recovered from the skim latex.

Hevamine is one of the chitinases that has been isolated from latex (Lee et al. 2006). Hevamine is a member of one of the several families of plant chitinases and lysozymes which are important for a plant's defense against bacteria and fungi infection. A useful enzyme, named Hevea Cathepsin G, is one of the enzymes that can be recovered from the skim latex serum (Yusof et al. 2006). Superoxide dismutase (SOD) is another enzyme that has been detected and studied in natural rubber latex (Yusof and Abdullah 1997). SOD catalyzes the dismutation of superoxide ions into hydrogen peroxides and oxygen during the oxidative energy process (Chi et al. 2001). Another protein which has been studied extensively in natural rubber is hevein. This small molecular weight acidic protein of 14 kDa originates from the lutoid membrane, a vacuole-like structure in latex. It can be isolated by affinity chromatography on chitin column (Soedjanaatmadja et al. 1996) and is observed to be antifungal toward certain strains. Yusof et al. (1998) discovered an inhibitor to rubber biosynthesis. This proteinaceous inhibitor of rubber biosynthesis has a molecular weight of 43.7 kDa and the partial sequence revealed that this protein has regions of sequence similar to patatin in potato. Thus, this protein is named patatin-like-protein. Investigations also showed that like patatin, this protein demonstrates lipolytic acyl hydrolase activity.

Above are some examples of enzymes and proteins that have been discovered and studied in natural rubber latex. Looking at the electrophoretical analysis of protein profile, in previous studies, it is found that skim latex serum has hundreds of proteins or enzymes waiting to be discovered and identified. It has been predicted that with a systematic study and using sound experimental designs, more of these proteins and enzymes can be elucidated. Properly purified, these enzymes can be used in many industries including food, detergent, and medical. The utilization of the serum, by converting it into valuables substances, could be an effective and sustainable way to minimize the discharging of effluent into the main waterway. This would make latex processing an environment-friendly and even revenue-generating process for the concerned manufacturers.

Methodology

Sample Collection and Preparation

The skim latex (effluent) used in this project was collected from Mardec Industrial Latex Sdn. Bhd. in Tapah, Perak. The preparation of the sample, before being introduced into column chromatography steps, involved coagulation, centrifugation, precipitation, and dialysis. All processes were performed at 4 °C to preserve the stability of the proteins.

Hydrolases Extraction

After collection, the skim latex (approx. 17L, pH10) was immediately coagulated by adding acetic acid glacial to reach pH 5. Subsequently, the acidified sample was centrifuged at 10,000 x g, 4 °C for 30 min in an Avanti® J-E high-speed centrifuge to remove the coagulated rubber (skim rubber). The supernatant (skim latex serum) obtained after centrifugation was pooled (approx. 14.5 L, ~85 %). Solid $(NH_4)_2SO_4$ was added to the pooled sample up to 4 M, to concentrate the skim latex protein and it was left overnight under slow stirring at 4 °C. The precipitating 100 % saturation was collected by centrifugation at 5,000 x g for 15 min. The supernatant was discarded and the residue was dissolved in minimal phosphate buffered saline at pH 7.0 and dialyzed using 10,000 (MWCO) of Pierce Snake Skin™ Pleated Dialysis Tubing against 0.1 M Tris HCl buffer at pH 7.0 overnight.

Biochemical Analyses

Several biochemical analyses were conducted to measure the total protein concentration through the dye-binding method of Bradford's assay (Bradford 1976), with bovine serum albumin (BSA) as the standard; protein purity test by Sodium Dodecyl Sulfate–Polyacrylamide gel electrophoresis (SDS-PAGE)(Laemmli 1970); and several other enzyme activity assays.

Quantitation of Activity by Hydrolases Assay

Hydrolases activities were performed using the methods of calorimetric and spectrophotometric stop reactions, with some modification. All tests were carried out in triplicate.

Chromatography- Gel Filtration

The glass Econo® Column (Bio-Rad), with inner dimensions of 100 × 2.6 cm, was packed with 450 ml Sephacryl S-200 high resolution (GE Healthcare). The resin was washed with 3 column volume of degassed and filtered 20 % ethanol and was equilibrated with 2 column volume of degassed and filtered 0.1 M Tris HCl buffer at pH7. After the $(NH_4)_2SO_4$ precipitation, the dialysate was applied to the column. Elution was carried out with the same buffer. The flow rate was 1 ml/min, and fractions of 5 ml were collected. Absorbance measures of 280 nm were read from the AKTA-purifier™.

Results and Discussions

Hydrolases Extraction

When the skim latex of the rubber was coagulated with acid, the coagulated rubber was about three quarters of the total volume. After centrifugation, this coagulated rubber was less than a quarter (~ 15 %) of the total volume; the rest was comprised of the serum. This result agreed with the claim of Blackley (1997) in his book that skim latex contained around 2.5–10 % of rubber while most of it was comprised of the aqueous serum. After centrifugation, a light yellow supernatant was obtained, which became dark brown after precipitation with $(NH_4)_2SO_4$ and dissolved in minimal PBS buffer until it became a clear solution. This solution was dialyzed overnight. However, the $(NH_4)_2SO_4$ present was only partly removed by this dialysis. The next step in the separation procedure, gel filtration with Sephacryl S-200, removed the remaining $(NH_4)_2SO_4$.

Hydrolases Screening

The choice of the hydrolase enzyme for this study was made based on its importance in the industry. The enzyme activities of selected hydrolases are compiled in Fig. 1.

Based on the screening results in Fig. 1, lipid acyl hydrolase (LAH), a subtype of lipase, was considered as the potential protein that can be recovered from the skim latex serum of *Hevea brasiliensis*. This is due to its activity which was found to be the highest (1.561 U/ml). Very small activity was obtained for β-glucosidase.

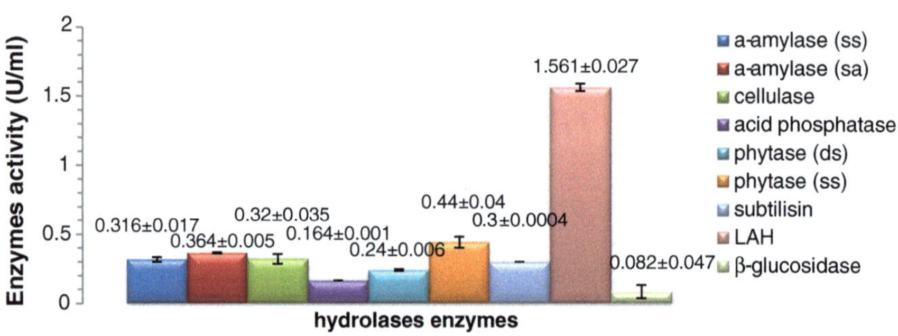

Fig. 1 Screening for hydrolases activity in skim latex serum of *Hevea brasiliensis*. * α-amylase (ss) indicates α-amylase assay using soluble starch as substrate. α-amylase (sa) indicates α-amylase assay using starch azure as substrate. phytase (ds) indicates phytase assay using phytic acid (Dedocasodium salt) as substrate. phytase (ss) indicates phytase assay using phytic acid (sodium salt) as substrate

Almost similar results for activity between 0.316 U/ml and 0.364 U/ml were obtained for α-amylase and cellulase. Similar work has been done by Abdullah et al. (2009) where the lyophilized crude sample of skim latex serum yielded 5.5 U/ml of LAH activity, while 2.4 U/ml, 1.4 U/ml, and 0.3 U/ml resulted from purification by ion exchange, gel filtration, and hydrophobic interaction chromatography, respectively.

Separation of Hydrolases by Sephacryl^{TM} S-200 Gel Filtration

Gel-filtration chromatography (GFC) was applied for partial purification of hydrolases protein from skim latex serum. Commonly, gel-filtration method is used in the initial stage for identification of unknown protein since this method does not allow removal of any proteins (all proteins were recovered). Yusof et al. (1998) did purify the patatin-like protein from C-serum of fresh latex obtained from *Hevea brasiliensis* by employing the Sephadex G-150 gel-filtration method followed by DEAE cellulose ion exchange and Phenyl Sepharose hydrophobic interaction chromatographies. In this study, the GFC step utilizes the Tris–HCl (pH 7.0) as the running buffer. The elutions of the fractions collected from the Sephacryl S-200 column are shown in Fig. 2.

From this chromatogram (Fig. 2), the elution volume was between 155 and 400 ml (fractions 31 and 80 under the protein peak). Thus, only these fractions were collected and subjected to analysis for total protein content, enzyme activity, and SDS-PAGE to visualize the separation of proteins. At 35 % of the column volume, fraction 31, the first separated sample was eluted. At this fraction, the size (molecular weight) of the protein is the highest while the last peak represents the lowest molecular weight of protein.

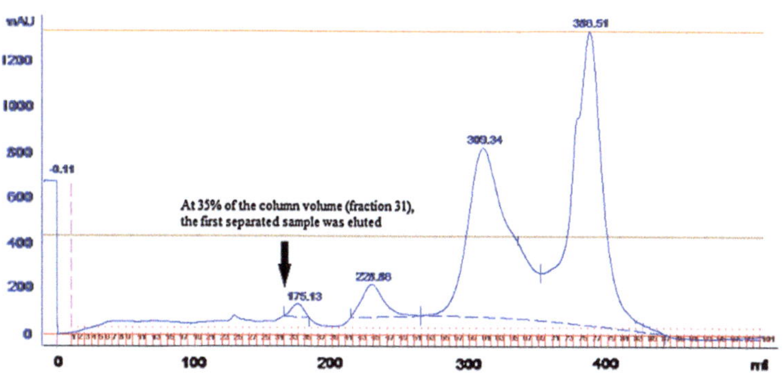

Fig. 2 Elution pattern of Sephacryl S-200 column of 85 × 2.6 cm. Absorbance was read at 280 nm using AKTA Purifier. Fractions 31–79 were collected and tested for the protein analysis

Biochemical Analysis

Qualitation of Purity by SDS-PAGE

The presence of target protein was detected (through SDS-PAGE analysis) in the early peak (which is estimated to be around 45 kDa). Similar result was obtained by Yusof et al. (1998) and Abdullah et al. (2009) in which this protein was eluted in the early peak of the protein profile from gel column chromatography using Sephadex G-150 and Sephacryl S-200 resin, respectively. Yusof et al. (1998) discovered a proteinaceous inhibitor of rubber biosynthesis that has a molecular weight of 43.7 kDa and has lipolytic acyl hydrolase (LAH) activity. From the observation, fractions between 43 and 57 showed higher band intensity at 45 kDa which indicates the presence of LAH enzyme at this fraction compared to the other fractions. It is suggested that these fractions be pooled before being processed in the next chromatography step.

Based on SDS-PAGE results, there were unidentified proteins with molecular sizes of around 21.5 and 5 kDa. In this study, the only known identified protein was a protein of size 30 kDa (Cathepsin G), a skim latex Hevea protease which is a homologue to cathepsin G found in the human body and is believed to be useful in skin regeneration mechanism (Yusof et al. 2006).

Quantitation of Protein by Bradford's Protein Assay

Bradford's protein assay was incorporated in the study to quantify the total protein purified at the separation process. BSA was used as the protein standard. The calculation of protein content was based on the equation:

$$y = -0.0042x^2 + 0.1106x + 0.0689 \tag{1}$$

where, y represents the absorbance reading of protein at 595 nm and x represents the amount of protein content (mg/ml). 5 ml elution was collected per fraction.

Specific activity is a measure of the amount of a particular protein among all the other contaminating proteins. It is calculated by dividing the activity by the total protein. The higher this value, the higher is the purity. Based on the calculated total protein, the specific activity of LAH is increasing after separation by GFC from 0.077 to 0.593 μmol of ρ-nitrophenyl palmitate hydrolyzed per milligram of protein per minute. A study done by Yusof et al. (1998) shows that 1.5 g of lyophilized C-serum was successfully granted a purified patatin-like-protein consistency with specific activity of 0.6 μmol $\mathrm{min}^{-1}\mathrm{mg}^{-1}$.

Conclusion

LAH, a subtype of lipase, was considered as the protein that can be potentially recovered from the skim latex serum of *Hevea brasiliensis*. This decision was based on the fact that its activity was found to be the highest (1.561 U/ml) and it was confirmed to be a LAH, by detection on SDS-PAGE. Further, GFC on Sephacryl^{TM} S-200 was used to separate hydrolases from crude skim latex serum to yield the results of LAH activity (1.304 U/ml) and total activity (97.782 U). 0.593 μmol of *p*-nitrophenyl palmitate was hydrolyzed per milligram of protein, per minute. A running buffer solution of pH 7.0 composed of 100 mM Tris HCl was used for all the steps to preserve consistency.

Summary

Extraction and separation of the hydrolase from the skim latex rubber *(Hevea brasiliensis)* resulted in a protein which demonstrated LAH activities. Future work will focus on continuing the purification steps on pooled fractions (43–57) using other steps of column chromatography, such as ion exchange for concentrating the LAH protein and hydrophobic interaction chromatography for polishing and achieving high purity by removal of any remaining traces of impurities or closely related substances.

Acknowledgments The authors would like to thank the International Islamic University Malaysia and the Ministry of Higher Education, Malaysia, for funding this research work.

References

Abdullah, N. A. H., Yusof, F. & Nor, Z. M. (2009). Development of patatin-like protein purification method from natural rubber industrial waste. In *Second International Conference and Workshops on Basic and Applied Science and Regional Annual Fundamental Science Seminar*, 2–4 June, Johor Bahru, Malaysia.

Blackley, D. C. (1997). *Polymer latices*. London: Chapman & Hall.

Bradford, M. M. (1976). A rapid and sensitive method for the quantitation of microgram quantities of protein utilizing the principle of protein-dye binding. *Analytical Biochemistry, 72*, 254–284.

Chi, T. S., Tung, L. L., & Jong, C. S. (2001). Purification and characterization of block porgy muscle Cu/Zn superoxide dismutase. *Zoological Studies, 40*(2), 84–90.

Department of Statistics Malaysia. (2011). *Monthly sector report: Rubber* (http://www.statistics.gov.my). Accessed 10 September 2012.

Laemmli, U. K. (1970). Cleavage of structural proteins during the assembly of the head of bacteriophage T4. *Nature, 227*, 680–685.

Lee, M. F., Chen, Y. H., Lin, H. C., Wang, H. L., Hwang, G. Y., & Wu, C. H. (2006). Identification of Hevamine and Hev b1 as Major Latex Allergens in Taiwan. *International Archives of Allergy and Immunology, 139*, 38–44.
Martin, N. (1991). The latex of *Hevea brasiliensis* contains both high chitinase and lysozyme activity. *Journal of Plant Physiology, 95*(2), 469–476.
National Innovation Strategy. (2012). *Rubber product innovation drafts.* (www.innovation.my). Accessed 10 September 2012.
Perrella, F. W., & Gaspari, A. A. (2002). Natural rubber latex protein reduction with an emphasis on enzyme treatment. *Methods, 27*(1), 77–86.
Soedjanaatmadja, U. M., Hofsteenge, J., Jeronimus, S. C. M., Bruins, A. P. & Beintema, J. J. (1996). *Journal of Biochemistry Biophysics*, 144–149.
Sunderasan, E., Ward, M. A., & Yeang, H. Y. (2002). Isolation and characterization of latex cyanogenic glucosidase in *Hevea brasiliensis. Journal of Rubber Research, 5*, 244–252.
Tramper, J., & Poulsen, P. (2005). Enzymes as processing aid and final products. In A. J. J. Strraathof & P. Adlercreutz (Eds.), *Applied Biocatalysis.* The Netherland: Taylor and Francis.
Wahler, D., Gronover, C. S., Richter, C., Foucu, F., Twyman, R. M., Moerschbacher, B. M., et al. (2009). Polyphenoloxidase silencing affects latex coagulation in taraxacum species. *Plant Physol, 151*, 334–346.
Yusof, F., & Abdullah, L. (1997). Purification and characterization of superoxide dismutase from Hevea brasiliensis latex. In *Proceedings of the 9th National Biotechnology Seminar*, 23–26 November, Pulau Pinang, Malaysia.
Yusof, F., & Chow, K. S. (2003). *The biosynthesis of rubber Hevea brasiliensis (MRB monograph no.4 ed.),* Malaysia, Malaysian Rubber Board.
Yusof, F., Ward, M. A., & Walker, J. M. (1998). Purification and characterization of an inhibitor of rubber biosynthesis from C-serum of *Hevea brasiliensis* latex. *Journal of Rubber Research, 1*(2), 95–110.
Yusof, F., Amid, A., Jimat, D. N., Nayan, M. Y., Osman, N. O., & Abdullah, N. A. H. (2006). Recovery of useful protease, cathepsin g, from the waste skim latex serum of hevea brasiliensis. In *Kulliyyah of engineering research and innovation exhibition (KERIE 2006)*, 5–6 December, Kuala Lumpur, Malaysia.

Study of H_2S Removal Efficiency of Virgin Zeolite in POME Biogas Desulfurization at Ambient Temperature and Pressure

H. Pourzolfaghar and M. H. S. Ismail

Abstract H_2S removal efficiency of a virgin zeolite is surveyed in this chapter by an adsorption unit fabricated in pilot scale and biogas produced directly from an anaerobic digester in ambient temperature and atmospheric pressure. The regeneration study and medium life-time were also surveyed by using it in several cycles of adsorption–desorption process. The H_2S removing efficiency of more than 90 % could be reached during the experiments. The virgin zeolite was analyzed with EDX and XRD equipments. Also the instinct regeneration ability and CH_4/CO_2 ratio-upgrading capability of zeolite are studied on this chapter. Instinct regeneration ability of the virgin zeolite showed an outstanding characteristic which makes it a promising adsorbent in industrial scale. From the results it is concluded that fresh zeolite is a suitable adsorbent for removing H_2S from biogas at ambient temperature and pressure in an economical view.

Introduction

Growing world demands for fossil fuels and its limited reserves led to the discovery of an alternative fuel source which is cheap, available, abundant, and environmental friendly. Biogas is a renewable fuel which is produced from fermentation of organic compositions (e.g.,) animal waste, crop residues, sewage sludge, household wastes, wastewater, and landfill (Wang et al. 2008a, b). It is clean, cheap, and environmentally friendly (Pipatmanomai and Vitidsant 2009). Physical and chemical characteristics of biogas are similar to those of natural gas; hence, it is a suitable alternative for LPG, fuel for heating, lighting, producing electricity or fuel for driving equipment (Wang et al. 2008a, b). Biogas composition depends on the

H. Pourzolfaghar (✉) · M. H. S. Ismail
Chemical and environmental Engineering Programme, School of Engineering,
Universiti Putra Malaysia (UPM), 43400 Selangor, Malaysia
e-mail: H_pourzolfaghar@yahoo.com

R. Pogaku et al. (eds.), *Developments in Sustainable Chemical and Bioprocess Technology*, DOI: 10.1007/978-1-4614-6208-8_35,

organic substrate. Biogas obtained from manure digestion generally consists of approximately (55–80 %) methane, carbon dioxide (20–45 %), nitrogen (0–10 %) and trace amounts of H_2S, and other impurities (Xuan et al. 2009; Wang et al. 2008a, b). H_2S concentration in biogas ranges from 10–30 to 3,000–6,000 ppm (Stepova et al. 2009).

For many reasons, H_2S, which is a persistent problem linked with biogas has to be removed to use as an alternative for fossil fuels. Concerns about H_2S uptake at low temperature are growing because it is a by-product of many industrial processes, such as the Claus process, anaerobic digestion, and natural gas sweetening (Pipatmanomai and Vitidsant 2009; Xiao et al. 2008). H_2S is an extremely toxic, odorous, and corrosive gas which can cause corrosion to the engine and pipeline and decrease plant lifetime (Xiao et al. 2008). Also, this toxic gas can cause health problems even at low concentrations (Gabriel and Deshusses 2003). Last but not least, sulfur compounds produced by H_2S oxidation are considered as a source of acid rains. And strict environmental regulations derived by developed countries' governments caused a growing concern over removing H_2S from gas streams (Seredych and Bandosz 2007). Current technologies for biogas desulfurization fall into one of the following: (1) biological conversion of sulfur compounds to elemental sulfur; (2) absorption into a caustic solution or a liquid; (3) adsorption by solid materials, for instance, activated carbons, impregnated activated carbons, zeolites, and oxide-based metals; and (4) selective catalytic oxidation (Cheah et al. 2009; Duan et al. 2007; Seredych and Bandosz 2007; Yasyerli et al. 2004; Gabriel and Deshusses 2003).

Each technology has its own advantages and disadvantages. Commercial biological processes for H_2S removal are also available and claimed to effectively reduce operating, chemical, and energy costs, but require higher capital costs than dry-based processes. Liquid-based and membrane processes require significantly higher capital, energy, and media costs, although regeneration is possible. Absorption by water or basic solution will cause wastewater problem, if nonregenerable or not treated well (Teresa 2002).

Among the various methods used to remove low-concentration H_2S from gases in low temperature, dry adsorption is considered as an efficient, environmental friendly, and economical approach (Bagreev et al. 2001; Bagreev and bandosz 2001; Teresa 1999). Materials used in dry adsorption process by different studies can be categorized into four groups: (1) metal-based materials, (2) carbon-based materials, (3) porous materials, and (4) polymeric materials (Khudiash and Al-Hinai 2006; Teresa 2002; Bagreev et al. 2001; Bagreev and bandosz 2001; Teresa 1999). Several studies have addressed the removal of sulfurous compounds in low temperatures using molecular sieves and (porous) adsorbents (Xuan et al. 2009; Khudaish and Al-Hinai 2006; Melo et al. 2004; Bae et al. 2003). These materials represent a series of valuable properties and characteristics to adsorption processes. Zeolites are a kind of molecular sieve, and because of their inherent ability to adsorb polar compounds, they have long been considered as excellent candidate materials for separation and purification of gases (Alonso-Vicario et al. 2010). In this context, application of a virgin zeolite to remove H_2S from biogas stream in an ambient temperature is surveyed. This study also investigated the characteristics of adsorption process by a pilot-scale adsorption column and biogas derived directly from an anaerobic digester.

Methodology

Material

In this study, a commercially available zeolite in three different particle sizes 1–3, 3–5, and 5–8 mm was utilized. Biogas used here was also derived directly from palm oil mill effluent (POME) anaerobic digester.

Characterization of Adsorbents

The adsorbents before, after 5, and after 7 cycles of applying biogas have been characterized by XRD and EDX in the laboratories of Chemical Engineering Department of the University of Putra, Malaysia. Crystalline structures of zeolite have been surveyed by X-ray diffractometer in a SIEMENS D-5000 diffractometer by scanning 2θ from 2 to 80 in a continuous scan with scan speed of 2 deg/min. EDX was performed to analyze the chemical composition of samples by HITACHI S-3400.

H_2S Adsorption Test

The adsorption properties of fresh material and also the influence of adsorption procedure variations were studied in the designed unit. The study was carried out on-site at Tennamaram Palm Oil Mill in Selangor province located about 50 km west of Kuala Lumpur, Malaysia. The H_2S removal unit, fabricated in this study, connected to the existing biogas line. Piping was established so that a small portion of biogas could be diverted to the test columns. The flow rate of the biogas was set up by the Rota flow meter for each column. The amount of the H_2S was varied during the test, so the inlet and outlet gas were analyzed with the portable gas analyzer in each sampling which was done every 10 min. For all experiments, columns were packed with 1,500 g of adsorbent, and biogas was applied to the column from down to up with flow rate 996 ml/min. The tests were repeated for 17 days and each took three hours.

Table 1 Chemical compositions of the zeolite

Element	C	O	Na	Mg	Al	Si	K	Ca	Fe
Atom (%)	6.26	1.60	0.89	0.73	10.93	65.47	6.08	4.29	3.73

Results and Discussion

Chemical composition of the zeolite is shown in Table 1. The adsorption characteristics of any zeolite are dependent upon the detailed chemical/structural makeup of the adsorbent. The Si/Al ratio and the cation type, number, and location are particularly influential in adsorption. The Si/Al ratio of the natural zeolite used in this study is 5:99. The higher Si/Al ratio may result in higher adsorption efficiency. The Si/Al ratio of natural zeolites cannot be modified except by acid treatment to dealuminate the structure.

Adsorption of H_2S by the Adsorbents

The time evolution of the ratio between H_2S concentration at the outlet and the inlet of the bed for fresh zeolite is illustrated in Fig. 1. As shown in Fig. 1, the H_2S concentration decreased to almost zero in some cases. This reveals that an excellent separation of H_2S can be obtained from biogas with fresh zeolite (Yasyerli et al. 2002). The results can be divided into two main categories, high efficiency and low efficiency. From Fig. 1, it can be seen that curves 9, 11, 12, and 16 show high efficiency of adsorbing H_2S and others show lower efficiency. This is because of time duration between experiments (TDBE) which is the days between experiments that the medium is not used and just exposed to the fresh air through up and down sides of the column. The TDBE is shown in Table 2.

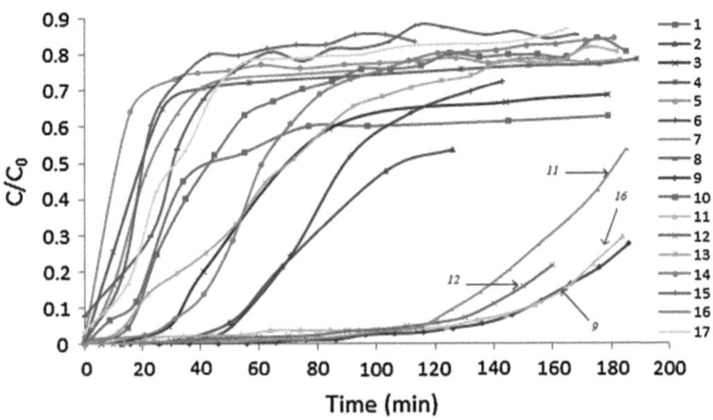

Fig. 1 Time evolution of the ratio between H_2S concentration at the *outlet* and the *inlet* of the bed for fresh zeolite

Table 2 Time duration between experiments (TDBE)

Experiment	1	2	3	4	5	6	7	8	9	10	11	12	13	14	15	16	17
TDBE (Day)	0	3	1	0	0	2	0	0	4	0	6	5	0	1	0	4	0
Efficiency (%)	62	76	69	40	28	64	38	27	94	38	93	94	47	43	26	88	30

Time duration between curves 9, 11, 12, and 16 are 4, 5, 6, and 4 days, respectively. It can be concluded from results that time duration of 4 days or more is a limit factor for high efficiency, and it is not related to the number of materials used as adsorbent. It can be concluded from the results that the virgin zeolite can be regenerated by itself without any extra regeneration process. This phenomenon is reported by other researchers (Yasyerli et al. 2002). Saturated molecular sieve totally recovers its activity and selectivity after this period of time. However, an unexpected behavior is observed for zeolite because it not only recovers its activity but also recovers its adsorption capacity for H_2S if time duration takes more than four days. From Table 2, it can be seen (for example experiments 13 and 14) that after 1 day of TDBE, the efficiency of saturated zeolite is recovered, and as the time duration increases, the efficiency of adsorbing H_2S molecules from biogas also increases. Repeating high-efficiency results prove that by exposing specific amount of oxygen to the saturated material, it can be recovered and even be improved. The exposure time for recovery of the material is not discussed in this chapter and needs more experiments and investigations.

Figure 2 shows the breakthrough sulfur capacity of fresh zeolite for 17 cycles of applying biogas. It can be seen from Fig. 2 that the highest breakthrough time is

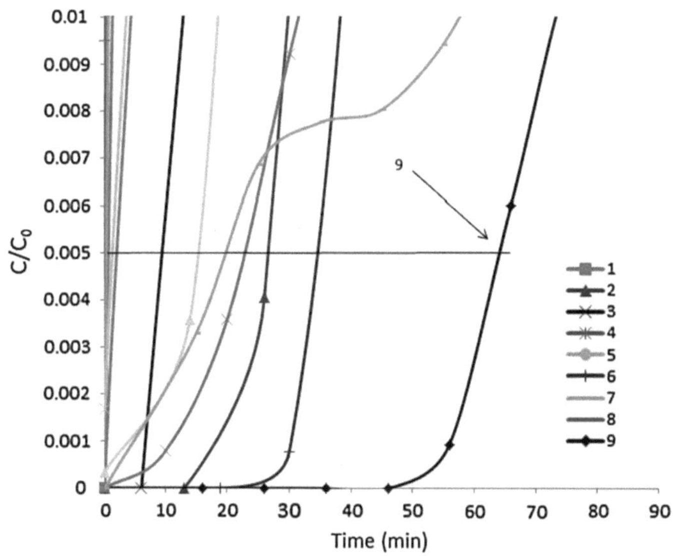

Fig. 2 Breakthrough sulfur capacity of fresh zeolite for 17 cycles of applying biogas

related to curve 9. Following curves 6, 2, 12, 16, and 11 have higher breakthrough time successively, compared to other curves. From the breakthrough curves, it can be concluded that although curves 9, 11, 12, and 16 have high efficiency, their breakthrough time is not the same. This might be because of filling the microporous spaces on the material by applying gas in different cycles. From Fig. 2, it can also be concluded that after cycle 9, the breakthrough time decreases. Breakthrough time of curve 9 is around 64 min.

Conclusions

According to the experiment results, fresh zeolite has capability of removing H_2S from biogas more than 94 % without any activation process provided that it is exposed to sufficient amount of oxygen. Although adsorption capacity of this material is lower compared to activated ones, it has attractive characteristics from economical view to be used in an industrial scale like no need to activation, improving the adsorption efficiency during cycles, and most importantly instinct regeneration. There are different criteria for a material to be used in an industrial scale like material, labor, preprocess, and regeneration costs and also kinetic of reaction, adsorption capacity, availability, and environmental friendly characteristics of material. As it can be seen, most of the criteria of an industrial adsorbent are related to economical. The fresh zeolite because of its characteristics mentioned before is a promising material to be used in gas desulfurization processes.

References

Alonso-Vicario, A., Ochoa-Gómez, J. R., Gil-Río, S., Gómez-Jiménez-aberasturi, O., Ramírez-López, C. A., & Torrecilla-Soria, J. (2010). Purification and upgrading of biogas by pressure swing adsorption on synthetic and natural zeolites. *Microporous and Mesoporous Materials, 134*(1–3), 100–107.

Bae, M. N., Song, M. K., Kim, Y., & Seff, K. (2003). Crystal structure of Mn 46Si 100Al 92O 384 · 89H 2S, a hydrogen sulfide sorption complex of fully dehydrated Mn 2+ exchanged zeolite X. *Microporous and Mesoporous Materials, 63*(1–3), 21–31.

Bagreev, A., & Bandosz, T. J. (2001). H_2S adsorption/oxidation on unmodified activated carbons: Importance of prehumidification. *Carbon, 39*(15), 2303–2311.

Bagreev, A., Rahman, H., & Bandosz, T. J. (2001). Thermal regeneration of a spent activated carbon previously used as hydrogen sulfide adsorbent. *Carbon, 39*(9), 1319–1326.

Cheah, S., Carpenter, D. L., & Magrini-Bair, K. A. (2009). Review of mid- to high-temperature sulfur sorbents for desulfurization of biomass and coal-derived syngas. *Energy and Fuels, 23*(11), 5291–5307.

Duan, H., Yan, R., Koe, L. C., & Wang, X. (2007). Combined effect of adsorption and biodegradation of biological activated carbon on H_2S biotrickling filtration. *Chemosphere, 66*(9), 1684–1691.

Gabriel, D., & Deshusses, M. A. (2003). Retrofitting existing chemical scrubbers to biotrickling filters for H_2S emission control. *Proceedings of the National Academy of Sciences, 100*(11), 6308–6312.

Khudaish, E. A., & Al-Hinai, A. T. (2006). The catalytic activity of vanadium pentoxide film modified electrode on the electrochemical oxidation of hydrogen sulfide in alkaline solutions. *Electroanalytical Chemistry, 587*(1), 108–114.

Melo, D. M. A., de Souza, J. R., Melo, M. A. F., Martinelli, A. E., Cachima, G. H. B., & Cunha, J. D. (2004). Evaluation of the zinox and zeolite materials as adsorbents to remove H_2S from natural gas. *Colloids and Surfaces A: Physicochemical and Engineering Aspects, 272*(1–2), 32–36.

Pipatmanomai, K. S., & Vitidsant, T. (2009). Economic assessment of biogas-to-electricity generation system with H_2S removal by activated carbon in small pig farm. *Applied Energy, 86*(5), 669–674.

Seredych, M., & Bandosz, T. J. (2007). Sewage sludge as a single precursor for development of composite adsorbents/catalysts. *Chemical Engineering, 128*(1), 59–67.

Stepova, K. V., Maquarrie, D. J., & Krip, I. M. (2009). Modified bentonites as adsorbents of hydrogen sulfide gases. *Applied Clay Science, 42*(3–4), 625–628.

Teresa, J. B. (1999). Effect of pore structure and surface chemistry of virgin activated carbons on removal of hydrogen sulfide. *Carbon, 37*(3), 483–491.

Teresa, J. B. (2002). On the adsorption/oxidation of hydrogen sulfide on activated carbons at ambient temperatures. *Colloid and Interface Science, 246*(1), 1–20.

Wang, X., Ma, X., & Xu, X. (2008a). Mesoporous-molecular-sieve-supported polymer sorbents for removing H_2S from hydrogen gas streams. *Topics in Catalysis, 49*(1), 108–117.

Wang, X., Sun, T., Yang, J., Zhao, L., & Jia, J. (2008b). Low-temperature H_2S removal from gas streams with SBA-15 supported ZnO nanoparticles. *Chemical Engineering, 142*(1), 48–55.

Xiao, Y., Wang, S., Wu, D., & Yuan, Q. (2008). Experimental and simulation study of hydrogen sulfide adsorption on impregnated activated carbon under anaerobic conditions. *Hazardous Materials, 153*(3), 1193–1200.

Xuan, J., Leung, M. K. H., Leung, D. Y. C., & Ni, M. (2009). A review of biomass-derived fuel processors for fuel cell systems. *Renewable and Sustainable Energy Reviews, 13*(6–7), 1301–1313.

Yasyerli, S., Ar, I., Dogu, G., & Dogu, T. (2002). Removal of hydrogen sulfide by Clinoptilolite in a fixed bed adsorber. *Chemical Engineering and Processing: Process Intensification, 41*(9), 785–792.

Yasyerli, S., Dogu, G., Ar, I., & Dogu, T. (2004). Dynamic analysis of removal and selective oxidation of H_2S to elemental sulfur over Cu–V and Cu–V–Mo mixed oxides in a fixed bed reactor. *Chemical Engineering Science, 59*(19), 4001–4009.

Investigation into Alternative for Bleaching Earth in Palm Oil Processing

L. N. Ngu and F. A. A. Twaiq

Abstract In this study, hexagonal mesoporous molecular sieves (HMS) were synthesized to investigate their potential to be used as alternative for bleaching earth in palm oil processing. This study also investigated their ability to be regenerated for reuse. HMS synthesized were found to have BET surface area ranging between 900 and 1,200 m^2/g and total pore volume of 0.9 cm^3/g. It was proven that synthesized HMS can successfully perform the bleaching of the palm oil and, in addition, can be regenerated for at least nine times to be reused for bleaching. It was found that the dosage of phosphoric acid added for degumming and the ratio of HMS to oil used during bleaching play important role in ensuring the success of the bleaching. The ratio of HMS to palm oil for successful bleaching is found to be 1:11.

Introduction

Bleaching is one of the critical stages in the crude palm oil refining process. In the bleaching process, bleaching adsorbents are used to remove the selective impurities and colour pigments. The colour pigments can adversely affect the taste of the oil and is generally disliked by consumers (Rohani 2006; Hassan and El-Shall 2004). Hence, the removal of the colour pigments is essential to improve the quality of the palm oil. Currently, bleaching clays have been most commonly and extensively used in the industry for bleaching of edible oil (Sabah 2007). According to a report by

L. N. Ngu (✉)
Department of Chemical Engineering, School of Engineering and Science, Curtin University, CDT 250 98009 Miri, Sarawak, Malaysia
e-mail: hannahngu@curtin.edu.my

F. A. A. Twaiq
Department of Chemical Engineering, College of Engineering and Architecture, University of Nizwa, P.O.Box 33616 Brikat Al Mouz, Nizwa, Oman
e-mail: farouq.twaiq@gmail.com

R. Pogaku et al. (eds.), *Developments in Sustainable Chemical and Bioprocess Technology*, DOI: 10.1007/978-1-4614-6208-8_36,

Malaysian Palm Oil Board (2012), crude palm oil production in Malaysia reached 19 million tons in 2011. With the high production rate of crude palm oil, there are significant quantities of spent bleaching earth generated in the refining of crude palm oil. Besides, spent bleaching earth usually contains retained oil of 20–40wt % after bleaching process (Yahia and Abdulrahim 1999), which can present a fire hazard. Also, as it is usually disposed into landfill without pretreatment, it could cause pollution problems, which includes polluting underground water. Therefore, a suitable alternative bleaching material that can be regenerated for repeated use would be both economically and ecologically advantageous.

In this research study, it is aimed to investigate an alternative bleaching agent that can be used to replace bleaching earth and can also be regenerated multiple times for reuse as a means of reducing the cost of refining. According to Patterson (1992), cited in Rohani (2006), for effective adsorption in bleaching process, porous adsorbents with high specific surface area are required. Mesoporous silica materials are known to have high surface area and pore volume. They have been used for industrial applications such as adsorbents for the removal of volatile organic compounds, catalysts, electronic applications, polymerization and similar applications (Beck and Vartuli 1996). However, their application as adsorbents for bleaching is yet to be explored.

Hexagonal mesoporous molecular sieves (HMS) silica has the potential to be used as bleaching agents due to their high specific surface area, high pore volume and, specifically, superior thermal stability (Peter and Thomas 1996). According to Peter and Thomas (1996), owing to more extensively cross-linked frameworks and thicker framework walls, HMS silica has superior thermal stability to calcination in air. Also, the complementary textural mesoporosity provides for better access to the framework-confined mesopores in adsorption by guest molecules. The aim of this paper is to synthesize, characterize HMS silica and assess its performance in bleaching crude palm oil.

Methodology

Synthesis of HMS

HMS were synthesized using the sol–gel technique (Peter and Thomas 1996) with deionized water, absolute ethanol, tetraethylorthosilicate (TEOS) and neutral surfactant template dodecylamine (DDA). Reactants were mixed with the molar ratio of 1.0 TEOS: 0.27 DDA: 9.09 EtOH: 29.6 H_2O. Ethanol and DDA were first added to deionized water and stirred for 30 min to allow the mixture to homogenize. Thereafter, TEOS was added as the silica source to the mixture while stirring. The mixture was then left at ambient temperatures for 18 h under vigorous stirring to form HMS. HMS products were recovered through filtration, followed by air-drying in ambient air. The products were finally calcinated in a muffle furnace at 630 °C for 4–8 h for template removal.

Characterization

Fresh HMS were characterized in terms of surface area and other porous characteristics using the Brunauer–Emmett–Teller (BET) approach. The ex situ XRD analyses were also performed on HMS using a D8 advance X-Ray diffractometer—Bruker AXS. To observe the morphology of HMS, scanning electron microscopy analysis was performed. Colour tests were performed on the bleached oil using Lovibond Tintometer Model F 1" cell.

Bleaching and Regeneration (Solvent Extraction and Calcination)

Prior to bleaching, crude palm oil was heated up to 100 °C and 0.2 or 0.3 mL of phosphoric acid was added for degumming. The HMS were then added to the crude palm oil while stirring, and the mixture was left at 150 °C for one hour for bleaching. The ratio of HMS to crude palm oil varied between 1:11 and 1:20. Spent HMS were filtered to obtain the bleached oil. Solvent extraction and calcinations were performed on spent HMS for regeneration. In solvent extraction, the excess oil retained on the surface or inside HMS was removed with the use of Soxhlet extractor utilizing hexane as solvent. To reactivate the spent HMS, it was calcinated in muffle furnace at 400 °C for 4 h.

Results and Discussion

Characterization

Figure 1 the SEM image of calcined HMS. The close-up view of the particles showed aggregates of fine particles, which are similar to those synthesized by Peter and Thomas (1996). It was observed from XRD patterns that the HMS prepared exhibited a single *d*100 reflection and retained the crystallinity after calcination in air at 630 °C for 8 h. This implies that HMS prepared had high thermal stability to calcination. This is an essential property for bleaching agent regeneration via calcination in order for it to be reused.

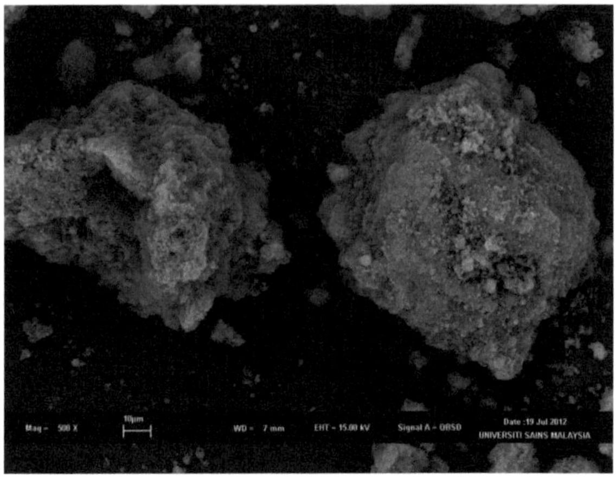

Fig. 1 SEM images of synthesized HMS

BET Analysis

Table 1 the porous characteristics of the HMS synthesized and the comparison with those of bleaching earths (Tsai et al. 2002; Syaiful 2006). It may be observed that the two samples of synthesized HMS have very high BET surface area (in the range of 900–1,200 m^2/g) compared to other bleaching earths that have commonly been used to bleach palm oil. Furthermore, the synthesized HMS samples also have much higher total pore volume compared to the bleaching earths (approximately two and three times higher). According to Rohani (2006) and Hassan and El-Shall (2004), adsorption process of bleaching earth is influenced by surface area, pore volume, pore size, and adsorbent polarity. Very porous adsorbents with large surface and high specific surface area are required for efficient adsorption (Patterson, cited in Rohani 2006, p. 60). Therefore, it is concluded that synthesized HMS samples possess the desired properties of high specific surface area and total pore volume, which are advantageous for efficient bleaching of palm oil. Thus, it is expected that HMS would perform better than bleaching earths.

Table 1 Porous characteristics of HMS synthesized and bleaching earth

Sample	HMS synthesized with poured reagent	HMS synthesized with reagent added dropwise slowly	Montmorillonite activated with acid treatment (Tsai et al. 2002)	Bleaching earth (Syaiful 2006)
BET surface area (m^2/g)	920	1,174	268	157
Total pore volume (cm^3/g)	0.9196	0.8994	0.3585	0.29
Average pore diameter ($\dot{A}$)	39.97	30.64	53.5	–

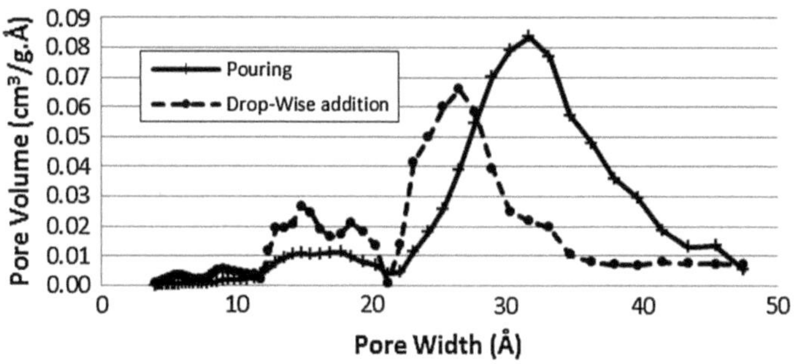

Fig. 2 DFT differential pore volume distribution for HMS synthesized. Effect of phosphoric acid dosage and bleaching material to oil ratio on bleaching efficiency

Table 2 Effect of dosage of acid and ratio of bleaching material to oil on bleaching efficiency

Batch	Ratio of HMS to oil (g:g)	Amount of phosphoric acid added (mL)	Speed of addition of TEOS to reactants	Success/failure of bleaching
1	1:20	0.2	Dropwise slowly	Fail
2	1:15	0.2	Dropwise at faster rate	Fail
3[a]	1:11	0.3	Pour	Success
4	1:11	0.3	Dropwise slowly	Success

[a] Regenerated HMS

Figure 2 the DFT differential pore volume distribution for HMS synthesized using the pouring and dropwise method, respectively. The variations in the methods were the speed of addition of TEOS to other chemical mixtures, namely added rapidly by pouring or slowly dropwise. Figure 2 indicates that HMS synthesized using the pouring method have a broad distribution of mesoporous type and compact distribution of microporous population. BET assessment shows that HMS express good surface area values, the majority of which are contributed by mesoporosities type and of moderate micropores groups. On the other hand, the features of HMS synthesized using the dropwise approach are quite similar; the only difference is micropores distribution, where they give better features of pore depth and broad porosity distribution region.

Table 2 the effect of phosphoric acid dosage and bleaching material to oil ratio on bleaching ability of HMS.

As shown in Table 2, the bleaching was not successful for the first two batches of experiments. With the increase in dosage of bleaching material and phosphoric acid, success in the bleaching of palm oil using HMS was achieved. A previous study (Kaynak et al. 2004) found that increasing the clay dosage improves the bleaching by improved colour reduction and chlorophyll removal. This is because with increasing dosage of bleaching material, there will be higher availability of adsorption sites for impurities and carotene, which tend to remove the colour. Also, as supported by Rohani (2006), sufficient dosage of phosphoric acid is essential to ensure success in degumming prior to bleaching. The finding in this

Table 3 Lovibond colour index of bleached palm oil using HMS (1" Lovibond cell)

Sample used for bleaching	Lovibond colour index (R)
None (crude palm oil)	23.0
Regenerated HMS (1st cycle)	7.6
Regenerated HMS (2nd cycle)	5.5
Regenerated HMS (3rd cycle)	4.0
Regenerated HMS (4th cycle)	4.1
Regenerated HMS (5th cycle)	6.7
Regenerated HMS (6th cycle)	4.9
Regenerated HMS (7th cycle)	5.5

study also shows that the bleaching of palm oil using synthesized HMS as bleaching material is affected by these two determining factors. On the other hand, as may be seen in Table 2, the speed of addition of TEOS to the reactant mixture has no significant effect on the bleaching efficiency.

Regenerability of HMS

To test the regenerability of HMS, bleaching was performed on repeatedly regenerated HMS with the ratio of HMS to oil fixed at 1:11, and 0.3 mL of phosphoric acid is added for degumming. Table 3 the Lovibond colour index of crude palm oil and bleached palm oil using regenerated HMS measured using a 1" Lovibond cell. It is the requirement of the industry that the colour index of bleached palm oil is below the range of 16–20 R with 1" cell. As observed in Table 3, the colour index of bleached palm oil using regenerated HMS was between 4 and 7.6 R, which indicated success in bleaching, regardless of the number of regenerations. Furthermore, it was found in this study that HMS can be regenerated for reuse for at least nine times.

The colour indexes for the 8th and 9th cycles were not available as the samples were not colour tested. However, visual observation indicted that bleaching had been successful for 8th and 9th cycles. Also, it can be seen from the colour test that the bleaching efficiency was maintained after several cycles. There was no significant change (deterioration or improvement) in the bleaching capability with increasing number of regeneration. This implies that the regeneration process and the increasing number of regenerations did not affect the capability of HMS to perform bleaching.

These findings have proven that HMS synthesized in this study have high regenerability. In fact, the number of regenerations could extend beyond the nine experimental runs that were conducted. The high regenerability of HMS could be due to its high specific surface area. As mentioned in the previous section, HMS have BET surface area ranging between 900 and 1,200 m^2/g, with total pore volume of approximately 0.9 cm^3/g. This is much higher than natural bleaching earth (clay) that has BET surface area of roughly 75–160 m^2/g (Kalam and Joshi 1988; Syaiful 2006) or much higher than acid-activated bleaching clay with BET surface of 270 m^2/g (Tsai et al. 2002) and total pore volume of 0.3 cm^3/g (Syaiful 2006).

On the other hand, it is also proven in this study that the bleaching approach (heating at 150 °C for 1 h) and regeneration method developed (solvent extraction using Soxhlet extractor followed by calcinations in furnace at 400 °C for 4 h) are effective for regeneration of used HMS. According to Al-Zahrani and Daous (2000), calcinations of bleaching earth will result in combustion of organic compounds, colour pigments and oil adsorbed on the surface or entrapped in the voids of the bleaching earth. It is therefore expected that during calcinations of HMS, these compounds were removed, and hence, the sites and surface area required for bleaching were recovered. These are significant findings that will definitely contribute to the industry. The current practice is to dump bleaching earth after once-off use. With the findings from this study, the cost incurred from continual purchase of bleaching earth can be reduced by replacement with HMS, given its much greater bleaching ability and regenerability. The use of HMS would also prevent the fire hazard and contamination risks that result from the dumping of bleaching earth with its inherent high oil content.

Conclusions

In this study, a bleaching agent suitable for the bleaching of palm oil has been successfully synthesized with a BET surface area of 900–1,200 m^2/g and a total pore volume of 0.9 cm^3/g. The dosage of HMS and phosphoric acid added to crude palm oil was found to play important role in ensuring successful bleaching. HMS not only has the ability to perform bleaching, but can also be regenerated repeatedly for at least nine cycles.

References

Al-Zahrani, A. A., & Daous, M. A. (2000). Recycling of spent bleaching clay and oil recovery. *Trans IChemE part B, 78*, 224–228.

Beck, J. S., & Vartuli, J. C. (1996). Recent advances in the synthesis, characterization and applications of mesoporous molecular sieves. *Current Opinion in Solid State and Materials Science, 1*, 76–87.

Hassan, M., & El-Shall, H. (2004). Glauconitic clay of El Gidida, Egypt: evaluation and surface modification. *Applied Clay Science, 27*, 219–222.

Kalam, A., & Joshi, J. B. (1988). Regeneration of spent Earth in aqueous medium. *Journal of the American Oil Chemists' Society, 65*, 1917–1921.

Kaynak, G., Ersoz, M., & Kara, H. (2004). Investigation of the properties of oil at the bleaching unit of an oil refinery. *Journal of Colloid and Interface Science, 280*, 131–138.

Patterson, H. W. B. (1992). Bleaching and purifying fats and oils. *Theory and Practice*. AOCS Press.

Peter, T. T., & Thomas, J. P. (1996). Mesoporous silica molecular sieves prepared by ionic and neutral surfactant templating: A comparison of physical properties. *Chemistry of Materials, 8*, 2068–2079.

Rohani, B. M. Z. (2006). Process design in degumming and bleaching of palm oil. *Centre of Lipids Engineering and Applied Research.* Universiti Teknologi Malaysia, Malaysia.

Sabah, E. (2007). Decolorization of vegetable oils: Chlorophyll-a adsorption by acid-activated sepiolite. *Journal of Colloid and Interface Science, 310*, 1–7.

Syaiful, N. H. (2006). Recovery of spent bleaching clay for reuse in water treatment. *Faculty of Chemical and Natural Resources Engineering.* University Teknologi Malaysia.

Tsai, W. T., Chen, H. P., Hsieh, M. F., Sun, H. F., & Chien, S. F. (2002). Regeneration of spent bleaching earth by pyrolysis in a rotary furnace. *Journal of Analytical and Applied Pyrolysis, 63*, 157–170.

Yahia, A. S. A., & Abdulrahim, A. A.-Z. (1999). Techno-economical evaluation of oil recovery and regeneration of spent bleaching clay. *Journal of King Abdulaziz University: Engineering Sciences, 11*, 115–126.

Part IV
Advanced Materials

A Brief Review on Photoanode, Electrolyte, and Photocathode Materials for Dye-Sensitized Solar Cell Based on Natural Dye Photosensitizers

R. F. Mansa, A. R. A. Yugis, K. S. Liow, S. T. L. Chai, M. C. Ung, J. Dayou and C. S. Sipaut

Abstract The dye-sensitized solar cell is a promising alternative for a new generation of photovoltaic devices due to its lightweight, flexibility, low cost, environmentally friendly materials. One important aspect of the DSSC is the potential of using dyes found in flowers, leaves, and fruits to be used as they are cheap and easily attained. However, the photovoltaic performances of dye-sensitized solar cells are greatly dependent on the CR which can be attributed to dye molecular size, semiconductor nanostructures properties, interaction between the semiconductor and dye, semiconductor/dye/electrolyte resistance, and dye aggregation (Narayan 2012). This paper briefly reviews recent developments in DSSC using natural dye photosensitizers, photoanode materials, various phases of the electrolytes, and nonmetal photocathode materials.

Introduction

O'Regan and Gratzel invented the dye-sensitized solar cell (DSSC) in 1991. Since then, the DSSC had attracted tremendous attention because the DSSC manufacturing process is less expensive, abundant source of materials and produce less environmental effect compared to commercial silicon-based solar cell (Senthil et al. 2011). The DSSC uses photosensitizers (i.e., dye sensitizers) adsorbed onto the surface of the photoanode, a mesoporous nanocrystalline-oxide semiconductor (e.g. TiO_2) to harvest energy from sunlight. The DSSC system is made of three

R. F. Mansa (✉) · A. R. A. Yugis · K. S. Liow · S. T. L. Chai · M. C. Ung · C. S. Sipaut
Energy and Materials Research Group, School of Engineering and Information Technology, Universiti Malaysia Sabah, Jalan UMS 88400 Kota Kinabalu, Sabah, Malaysia
e-mail: rfmansa@ums.edu.my

J. Dayou
Energy, Vibration and Sound Research Group (e-VIBS), School of Science and Technology, Universiti Malaysia Sabah, Jalan UMS 88400 Kota Kinabalu, Sabah, Malaysia

R. Pogaku et al. (eds.), *Developments in Sustainable Chemical and Bioprocess Technology*, DOI: 10.1007/978-1-4614-6208-8_37,

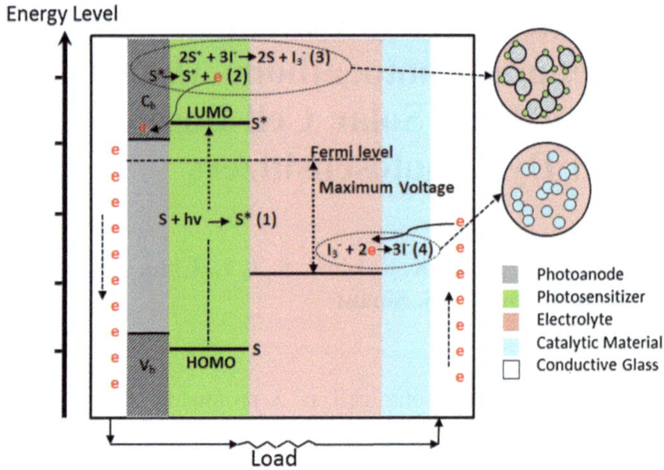

Fig. 1 Simplified mechanism of a dye-sensitized solar cell (*DSSC*). C_b Conduction band; V_b Valance band; *LUMO* Lowest unoccupied molecular orbital; *HOMO* Highest occupied molecular orbital; *hv* Photon, *S* Dye; *S** Excited dye

layers consisting of a dyed semiconductor layer, an electrolyte layer, and a counter electrode sandwiched between two conductive glass substrates.

Figure 1 shows a dye-sensitized TiO_2 layer receiving energy from the sun to excite electrons which then flows through the photoanode into the outer electrical circuit. The counter electrode completes the circuit and feeds the electron into the electrolyte layer which contains the iodine/triiodide (I^-/I_3^-) redox couple. Next, the electrolyte regenerates the dye by donating electrons to the oxidized dye (O'Regan and Gratzel 1991). The cell will continue generating current for as long as the visible light energy is provided and the cycle is not hindered. The main limitations of DSSCs based on natural dyes are charge recombination (CR), whereby the electrons fail to travel via the outer electrical circuit but instead recombine with an electron-hole (Narayan 2012). This paper aims to briefly review recent developments in DSSC research especially on photoanode, electrolytes, and counter electrodes materials suitable for DSSCs based on natural photosensitizers.

Components of the Dye-Sensitized Solar Cell

Photosensitizer

Photosensitizers are chemicals which absorb energy from visible light, which then produces excited electrons to be injected to the wide band gap semiconductor (Jayaweera et al. 2008). Table 1 presents a short list of anthocyanin- and betalain-type dye photosensitizers extracted from natural materials such as flowers and fruits.

Table 1 Photoelectrochemical parameters of anthocyanin- and betalain-based DSSCs

Dye	Source	C (nm)	D (h)	pH	η (%)	Reference(s)
A	Punica granatum peel	300–400; 540	–	4.0	1.86	Hernandez-Martinez et al. (2012)
A	Rosella	520	<3	1.0	0.70	Wongcharee et al. (2007)
B	Beet root	492; 500	>8	–	2.71	Sandquist and McHale (2011)
B	Red turnip	480; 540	–	1.0	1.70	Calogero et al. (2010)
B	Wild Sicilian prickly pear	450; 470	>24	1.0	1.19	Calogero et al. (2010)

(*A* Anthocyanin; *B* Betalain; *C* UV–Vis Spectra; *D* Photoelectrochemical stability)

Anthocyanins and betalains were chosen for this brief review because both were reported to have higher DSSC efficiency (Narayan 2012; Hernandez-Martinez et al. 2012; Sandquist and McHale 2011). Sandquist and McHale (2011) managed to improve the efficiency of the betanin (a betalain pigment) to η of 2.71 % by increasing betanin purity and reducing CR. The CR reduction methods are explained in the next section.

Wongcharee et al. (2007) reported that acidic conditions may improve cell efficiency due to increased light absorption. However, Calogero et al. (2010) stated that the acidic dye adsorption environment may result in low V_{oc} which lowers the DSSC efficiency. Furthermore, Sandquist and McHale (2011) also reported that the molecular size and location of the HOMO and LUMO on the molecule may affect the electron transfer kinetics.

Photoanode

The photoanode functions as a bridge and a charge separator (Kelly and Meyer 2001). It consists of a mesoporous metal-oxide semiconductors nanoparticles layer on the conductive surface of a transparent conducting electrode. Nanocrystalline TiO_2 had been extensively studied over other semiconductors due to its large surface area, safety, and matched energy band structure (Chen et al. 2012). But, studies on photoanode materials which can reduce the CR reactions between the photoinjected electrons with either oxidized dye molecules or electron-accepting species in the electrolyte may lead to a higher efficiency (Zhang et al. 2011). In Table 2, (Sandquist and McHale 2011) reported the highest DSSC efficiency (2.71 %) by reducing CR. CR was reduced by a dense TiO_2 blocking layer between the TiO_2 and mesoporous and substrate surface. This prevented shunting via the connection of the electrolyte to the surface of the substrate. Combining this with $TiCl_4$ treatment greatly improved the cell performance due to reduced defects on the mesoporous TiO_2 surface and improving the interparticle connectivity which facilitated electron transfer.

Table 2 Performance of the DSSC with different thin films and thickness

Photoanode material (size)/modification	Thickness (1m)	Dye	η (%)	Reference(s)
TiO_2 (<50 nm)	–	Pomegranate juice	1.076	(Jasim 2012)
TiO_2	0.9	Red cabbage	0.10	(Saelim et al. 2011)
TiO_2/SiO_2	1.5		0.09	
TiO_2/blocking layer	10	Beet root	2.71	(Sandquist and McHale, 2011)
$TiO_2/TiCl_4$ film treatment	10	Beet root	2.27	(Sandquist and McHale 2011)
TiO_2/Unmodified	10	Beet root	2.04	(Sandquist and McHale 2011)

(η Electric power conversion efficiency)

Electrolyte

An electrolyte is a substance that conducts electricity via the movement of ions. Several essential characteristics for high electrolyte efficiency are that the redox potential of the electrolyte should be negative compared to oxidation potential of the dye sensitizer and have high conductivity ($\sim 10^{-3}$ S.cm^{-1}), sufficient interfacial contact with the dye layer, long-term physical and chemical stability, and minimum absorption of light in visible range (Wu et al. 2008; Jena et al. 2012). Table 3 shows that the highest efficiency (7.10–7.90 %) was achieved by using an iodine redox couple liquid electrolyte. However, leakage and volatilization of organic solvents, possible desorption, and photodegradation of the attached dyes limited its long-term performance. Hence, researchers had worked on replacing liquid electrolytes with solid-state or quasi-solid-state electrolyte. Solid-state electrolytes overcame the disadvantages of fluidity and volatility of liquid electrolytes, but provided poor interface contact and lower efficiency. Thermoplastic gel electrolyte using poly(ethylene glycol) (PEG) achieved the highest cell efficiency, 7.22 % for a nonliquid electrolyte (Wu et al. 2008).

Photocathode

The photocathode completes the electrical circuit of the DSSC and reduces the triiodide to iodide. The highest cell efficiency (ŋ) was found to be 8.49 % using platinum as a CM due to the better electrocatalytic activity of platinum to I^-/I_3^- redox couple (Zhang et al. 2011). Lee et al. (2009) explained that the activated platinum nanoclusters reflected a portion of the unabsorbed incident light back to the TiO_2 electrode which resulted in higher amounts of excited electrons and higher solar energy conversion. In Table 4, most of the works on photocathodes were on nonnatural dye-sensitizer-based DSSCs. For a metal-free CE, the highest cell efficiency was 8.03 %. It used double-wall carbon nanotubes (DWCNT) as a CM. Zhang et al. (2011) reported that the high efficiency observed may be due to

Table 3 Performance achieved by different types of the electrolyte

Physical state	Type	Materials	η (%)	Reference(s)
Liquid		LiI/I_2	7.10 –7.90	O'Regan and Gratzel (1991)
Solid-state	HTMs	Spiro-OMeTAD	2.10	Sellinger et al. (2012)
		AS44	2.30	Sellinger et al. (2012)
		CuI	3.80	Meng et al. (2003)
	Containing redox couple	Poly(ethylene oxide)/titania polymer, iodide/triiodide redox couple	4.20	Wang et al. (2004)
Quasi-solid	TPGEs	PEG, PC, KI, I_2	7.22	Wu et al. (2008)
	TSGEs	PAA–PEG,iodide/triiodide redox couple	6.10	Wu et al. (2008)

(*HTMs* Hole transport materials; *TPGEs* Polymer-thermo-reversible/thermoplastic; *TSGEs* Polymer-thermo-irreversible/thermosetting gel; *PEG* Poly(ethylene glycol); *PC* Propylene carbonate; *PAA–PEG* Poly(acrylic acid)–poly(ethylene glycol)

the large inner surface area and porous 3D structures provided by DWCNT. However, the carbon-nanotubes (CNTs)-based cells have lower cell efficiency compared to platinum-based cells.

Table 4 Type of counter electrode for nonnatural-based DSSCs

Photocathode		Electrolyte	J_{sc}, (mA/cm^2)	V_{oc}, (mV)	FF (%)	η (%)	Reference(s)
CM	Substrate						
PEDOT+ 15 % CB	MM/PEN	I^-/I_3^-	12.60	739	58.00	5.50	Huang et al. (2012)
PEDOT+ 15 % CB	ITO/PEN	I^-/I_3^-	13.30	756	60.00	6.00	
Pt	FTO	I^-/I_3^-	16.08	810	65.00	8.49	Zhang et al. (2011)
DWCNTs	FTO	I^-/I_3^-	15.43	800	65.20	8.03	
SWCNTs	FTO	I^-/I_3^-	14.94	800	64.00	7.61	
MWCNTs	FTO	I^-/I_3^-	15.25	800	56.40	7.06	
G	FTO	I^-/I_3^-	16.99	750	53.60	6.82	
Gt–CB	FTO	I^-/I_3^-	11.34	830	71.20	6.67	
Gt	Copper	MPN-100	2.10	360	70.00	5.29	Bazargan et al. (2010)
Gt	ITO	MPN-100	1.40	345	70.00	3.38	
CoS	FTO	I^-/I_3^-	14.17	730	63.00	6.01	Lin et al. (2010)

(*FTO* Fluorine-doped tin oxide glass; *ITO* Indium-doped Tin Oxide; *Pt* Platinum; *G* Graphenes; *CB* Carbon black; *PEDOT* Poly (3,4ethylenedioxythiophen); *MM* Metal mesh; *PEN* Polyethylene naphthalate; *SWCNT* Single-wall carbon nanotubes; *DWCNT* Double-wall carbon nanotubes; *MWCNT* Multiple-wall carbon nanotubes; *CoS* Cobalt sulfide; *Gt* Graphite)

Conclusions

DSSC has been known to be a potential solar energy conversion device. It has received a huge amount of attention in the last decade. It was mainly due to its potential for low-cost and high solar energy conversion efficiency. However, the DSSC has a long list of limitations. For instance, the competition between generation and recombination of the photoexcited electrons within the photocathode and electrolyte layers can affect the DSSC efficiency. This is may be improved by studying the methods in which CR can be reduced. Additionally, the electrolyte and the CE will also affect the DSSC efficiency. Hence, further researches on materials are vital to the improvement of the DSSC efficiency.

Acknowledgments The authors would like to thank the Universiti Malaysia Sabah and the Ministry of Higher Education, Malaysia, for supporting this research under the grant numbers SBK0025-TK-2012, ERGS0019-TK-1/2012, and ERGS0022-TK-1-2012.

References

Bazargan, M. H., Byranvand, M. A., Kharat, A. N. (2010). A new counter electrode based on copper sheet for flexible dye sensitized solar cells. *Chalcogenide Letters, 7*(8), 515–519.

Calogero, G., di Marco, G., Cazzanti, S., Caramori, S., Argazzi, R., di Carlo, A., et al. (2010). Efficient dye-sensitized solar cells using red turnip and purple wild sicilian prickly pear fruits. *International Journal of Molecular Sciences, 11*, 254–267.

Chen, H.-Y., Kuang, D.-B., & Su, C.-Y. (2012). Hierarchically micro/nanostructured photoanode materials for dye-sensitized solar cells. *Journal of Materials Chemistry, 22*, 15475–15489.

Hernandez-Martinez, A. R., Estevez, M., Vargas, S., Quintanilla, F., Rodriguez, R. (2012). Natural pigment-based dye sensitized solar cells, Vol. 10. In :First International Congress on Instrumentation and Applied Sciences, Oct 26–29, Cancun, Quintana Roo, Mexico.

Huang, W. C., Zhang, X. L., Huang, F. H., Zhang, Z. P., He, J. J., & Cheng, Y. B. (2012). An alternative flexible electrode for dye-sensitized solar cells. *Journal of Nanoparticle Research, 14*, 838. doi:10.1007/s11051-012-0838-0.

Jasim, K. E. (2012). Natural dye-sensitized solar cell based on nanocrystalline TiO_2. *Sains Malaysiana, 41*, 1011–1016.

Jayaweera, P. V. V., Perera, A. G. U., & Tennakone, K. (2008). Why Gratzel's cell works so well. *Inorganica Chimica Acta, 361*(3), 707–711.

Jena, A., Mohanty, S. P., Kumar, P., Naduvath, J., Gondane, V., Lekha, P., et al. (2012). Dye sensitized solar cells: a review. *Transactions of the Indian Ceramic Society, 71*(1), 1–16.

Kelly, C. A., & Meyer, G. J. (2001). Excited state processes at sensitized nanocrystalline thin film semiconductor interfaces. *Coordination Chemistry Reviews, 211*, 295–315.

Lee, W. J., Ramasamy, E., Lee, D. Y., & Song, J. S. (2009). Efficient dye-sensitized solar cells with catalytic multiwall carbon nanotube counter electrodes. *ACS applied materials & interfaces, 1*, 1145–1149.

Lin, J. Y., LIAO, J. H., & WEI, T. C. (2010). Honeycomb-like CoS counter electrodes for transparent dye-sensitized solar cells. *Electrochemical and Solid-State Letters, 14*(4), D41–D44.

Meng, Q. B., Takahashi, K., Zhang, X. T., Sutanto, I., Rao, T. N., Sato, O., et al. (2003). Fabrication of an efficient solid-state dye-sensitized solar cell. *Langmuir, 19*(9), 3572–3574.

Narayan, M. R. (2012). Review: dye sensitized solar cells based on natural photosensitizers. *Renewable and Sustainable Energy Reviews, 16*(1), 208–215.
O'Regan, B., & Gratzel, M. (1991). A low-cost, high efficiency solar cell based on dye- sensitized colloidal TiO_2 films. *Nature, 353*, 737–740.
Sandquist, C., & McHale, J. L. (2011). Improved efficiency of betanin-based dye- sensitized solar cells. *Journal of Photochemistry and Photobiology A: Chemistry, 221*, 90–97.
Saelim, N.-O., Magaraphan, R., & Sreethawong, T. (2011). Preparation of sol– gel TiO_2/purified Na-bentonite composites and their photovoltaic application for natural dye-sensitized solar cells. *Energy Conversion and Management, 52*, 2815–2818.
Sellinger, A., Leijthens, T., Ding, I. K., Giovenzana, T., Bloking, J. T., & Mcgehee, M. D. (2012). Hole transport materials with low glass transition temperature and high solubility for application in solid-state dye-sentitized solar cells. *ACS Nano, 6*(2), 1455–1462.
Senthil, T. S., Muthukumarasamy, N., Velauthapillai, D., Agilan, S., Thambidurai, M., & Balasundaraprabhu, R. (2011). Natural dye (cyanidin 3-O-glucoside) sensitized nanocrystalline TiO_2 solar cell fabricated using liquid electrolyte/quasi-solid-state polymer electrolyte. *Renewable Energy, 36*, 2484–2488.
Wang, P., Dai, Q., & Zakeeruddin, S. M. (2004). Ambient temperature plastic crystal electrolyte for efficient, all-solid-state dye-sensitized solar cell. *Journal of the American Chemical Society, 126*, 13590.
Wongcharee, K., Meeyoo, V., & Chavadej, S. (2007). Dye-sensitized solar cell using natural dyes extracted from rosella and blue pea flowers. *Solar Energy Materials and Solar Cells, 91*, 566–571.
Wu, J., Lan, Z., Hao, S., Li, P., Lin, J., Huang, M., Fang, L., & Huang, Y. (2008).Progress on the electrolyte for the dye sensitized solar cell. *Pure Appl.Chem. 80*(11), 2241–2258.
Zhang, D. W., Li, X. D., Chen, S., Tao, F., Sun, Z., Yin, X. J., et al. (2011). Fabrication of double-walled carbon nanotube counter electrodes for dye- sensitized solar cells. *Journal of Solid State Electrochemistry, 14*, 1541–1546.

The Influences of Dope Composition on Gas Permeance of Hollow Fiber Carbon Membrane

W. M. H. F. W. Harun, M. A. T. Jaya and M. A. Ahmad

Abstract In this work, a carbon membrane was produced from a polymer blend of polyetherimide (PEI)/polyvinylpyrrolidone (PVP). Four samples of the PEI-based carbon membrane were prepared at various concentrations of its blending counterpart, which were 0, 5, 9, and 13 wt.% of PVP. All samples were subjected under similar conditions where the stabilization of their polymeric precursors was performed at 300 °C for 1 h under a continuous flow of purified air. The stabilization was followed by pyrolysis at 400 °C for 1 min under purified nitrogen flow. Sample S3, which was prepared at initial dope solution of 9 wt.% PVP, demonstrated that the decomposition temperature has increased at highest degree compared with the others, and this has brought advantages to its morphology. Single gas permeability tests using CO_2 and CH_4 gases showed that the polymer-blend-precursor-based carbon membranes possessed molecular sieving properties. Sample with 9 wt.% PVP (S3) was found highest in the delivery of CO_2 permeance and CO_2/CH_4 permselectivity, which were 14.61 and 1.95 GPU, respectively.

Introduction

The emission of excessive greenhouse gases has been one of the main contributors for global warming. In order to minimize the problem, many countries have regulated the utilization of cleaner fuels such as natural gas for domestic use. In fact, the minimum level of harmful emission by the by-products of natural gas meets the specification implemented under the Environmental Quality Regulation 1996 by Malaysian Government.

W. M. H. F. W. Harun · M. A. T. Jaya · M. A. Ahmad (✉)
School of Chemical Engineering, Universiti Sains Malaysia, Engineering Campus, Seri Ampangan, 14300 Nibong Tebal, Penang, Malaysia
e-mail: chazmier@eng.usm.my

R. Pogaku et al. (eds.), *Developments in Sustainable Chemical and Bioprocess Technology*, DOI: 10.1007/978-1-4614-6208-8_38,

The benefits of the natural gas utilization cause this commodity to account for 23 % of the total energy consumed in the developing countries. In 2003 alone, 95 trillion cubic feet of natural gas was consumed worldwide, whereas the number was predicted to increase to 182 trillion cubic feet in 2030 (Xiao et al. 2009). In natural gas processing, the crude natural gas undergoes enrichment to maximize the percentage volume of methane while eliminating carbon dioxide as an impurity. The carbon dioxide content in the crude natural gas may vary from 4 to 50 vol.%, and the typical product after the enrichment is a methane-enriched residue stream containing less than 2 vol.% of carbon dioxide (Safari et al. 2009). The removal of carbon dioxide also increases the calorific value of the natural gas (Datta and Sen 2006). In addition, the elimination of acidic carbon dioxide will also prevent corrosion in the transport line and diminishes the unnecessary total volume of gas, which brings economic advantages for transportation and storage (Hao et al. 2002).

The use of membrane technology in gas processing offers many advantages over other conventional technologies in gas separation such as cryogenic, adsorption, and absorption. The membrane process provides simple and versatile mechanisms that allow smaller space occupancy for relatively larger processing capacity. Its operation requirement only involves a pressure gradient toward the membrane as a driving force for operation supplied by gas compressor or vacuum pump. Maximizing the industrial margin is achievable by optimizing the power to engage the separation process. Two-thirds of the total components cost is associated with the feed-crude natural gas compressor, while the remaining is associated with the membrane module. Therefore, most of the operating cost is contributed by the consumed energy to power the compressor (Baker 2002). It is possible to pair a highly selective membrane with smaller compressor to achieve greater saving in the operation cost. Therefore, finding new materials or extensive modifications of existed materials to enhance the membrane selectivity and permeability has become current focuses.

Unlike the polymeric membranes, carbon membranes possess highly porous structure that allows the gas separation to be performed at higher permeability which means the separation process can be completed at a higher production rate. Another method in accomplishing high production rate is by applying the hollow fiber as the membrane configuration. The hollow fiber provides higher surface area as well as higher packing density, which leads to large permeability of the product (Saufi and Ismail 2004). In addition, this membrane configuration is simple and possesses easy module assembly (He and Hagg 2011). These advantages have attracted attention to enhance the hollow fiber quality and envision the capability of hollow fiber toward the main membrane configuration in future moment.

The porous structure of the carbon membrane consists of pores that lie within the ultramicropore region. These pores are generated during pyrolysis where the evolved decomposition gases channel their way through to the surface leaving traces of opening or pores throughout the membrane structure. The formation of such ultramicropores would enable the carbon membrane to discriminate the diffusing gases according to their kinetic diameters at relatively high sensitivity (down to 0.1 Å) through molecular sieving mechanism (Jones and Koros 1995).

One of the prominent polymer precursors to produce the CO_2-selective carbon membrane is polyetherimide (PEI) due to its inexpensive cost compared with other types of polyimides. Salleh and Ismail (2011) have reported the performance of carbon membrane derived from PEI blended with PVP at various pyrolysis temperatures. It was found that the carbon membrane pyrolyzed at higher temperatures demonstrated a remarkable permselectivity compared to lower pyrolysis temperatures. However, no investigation was reported regarding the effect of varying the dope composition on carbon membrane performance. Thus, this chapter reports a study on the influence of dope composition on the morphology and performance of carbon membrane.

Methodology

Materials

Commercially available PEI and N-methyl-2-pyrrolidone (NMP) (99.5 % purity) were supplied by Sigma-Aldrich, USA. Polyvinylpyrrolidone (PVP) was purchased from Fluka, USA. Ethanol (96 % purity) and n-hexane (60 % purity) were purchased from QReC, Malaysia.

Hollow Fiber Production

Four types of dope solutions were prepared by varying the input amount of PEI and PVP powder into NMP solvent as shown in Table 1. The dope solutions were stirred using a magnetic stirrer at 110 °C for 24 h. The dope solution was then fed into a spinneret to generate hollow fibers using a well-known method of dry/wet spinning process. The polymeric precursor hollow fibers obtained were then immersed in tap water for 24 h to complete the coagulation process. After that, the polymeric hollow fiber was immersed in ethanol solution for 2 h, followed by another 2 h in n-hexane for solvent exchange procedure to accommodate slow removal of water. Finally, the fibers were dried under normal air atmosphere. The spinning conditions are depicted in Table 2.

Table 1 Dope composition

Sample notation	PEI (wt.%)	PVP (wt.%)
S1	21	0
S2	21	5
S3	21	9
S4	21	13

Table 2 Spinning conditions

Dope temperature	25 °C
Coagulation bath	Tap water
Coagulation bath temperature	25 °C
Bore fluid	Distilled water
Bore fluid flow rate	100 ml/h
Air gap	20 cm

Carbon Membrane Production

The high-temperature treatment was performed using Carbolite™ vertical furnace. The various hollow fibers were stabilized at 300 °C for 1 h under purified air flow. Subsequently, the hollow fibers were pyrolyzed by increasing the heating temperature to 400 °C at 3 °C/min with holding time of 1 min. Then, the hollow fibers were naturally cooled to room temperature. Purified nitrogen was supplied at 50 ml/min throughout the processes.

Membrane Characterization

Thermogravimetric analysis of the membranes was characterized via simultaneous thermal analyzer (Model Perkin Elmer STA 6000, USA). The membrane was heated under nitrogen flow from 30 to 500 °C at 3 °C/min. The heating was held for 15 min as the process reached top temperature. The temperature of decomposition (T_d) is measured by a tangent line across the curve of the thermal plot. The cross-sectional sample images were observed under scanning electron microscopy (Model FEI Quanta 450 FEG, USA). The membrane permeability toward CO_2 and CH_4 was evaluated using single gas permeability test method. The permeance, P, and permselectivity, α, were calculated using equations below:

Permeance, P (GPU):

$$\left(P/l\right)_i = \frac{Q_i}{\Delta P.A} = \frac{Q}{n\pi Dl\Delta P}, \tag{1}$$

$$1\,\text{GPU} = 1 \times 10^{-6} \frac{\text{cm}^3\,\text{STP}}{\text{cm}^2\,\text{s cm Hg}} \tag{2}$$

Selectivity, α:

$$\alpha_{CO_2/CH_4} = P_A/P_B \tag{3}$$

where P/l is the permeance of the hollow fiber, Q_i is the volumetric flow rate of gas i at standard temperature and pressure (cm^3(STP)/s), Δp is the pressure difference between the feed side and the permeation side of the membrane (cm Hg), A is the effective membrane surface area (cm^2), n is the number of fiber in the module, D is an outer diameter of hollow fiber (cm), and l is the effective length of hollow fiber (cm). The pressure difference of 4 bar was applied throughout the experiments, while the outer diameter of the carbon membranes were measured from the SEM images.

Results and Discussion

Scanning Electron Microscopy

The cross sections of the membranes are depicted in Fig. 1. Before thermal treatment, S1 exhibits the fingerlike pore structure, while the S2 possesses the spongelike pore structure. The cross-sectional images of S3 and S4 before the thermal treatment are unavailable since the hollow fibers have been too elastic and unbreakable. It is believed that the presence of PVP has triggered the formation of sponge-like pore. The increase in the PVP concentration would increase the amount of sponge-like pore and may strengthen the membrane matrix. The thermal treatment process toward S1 had caused severe destruction of the hollow fiber membrane structure. The S1 composition could not retain the original shape and melt as the thermal treatment proceeds higher than 300 °C. Thus, no image of S1 after thermal treatment is depicted in Fig. 1. Meanwhile, the S2-based membrane

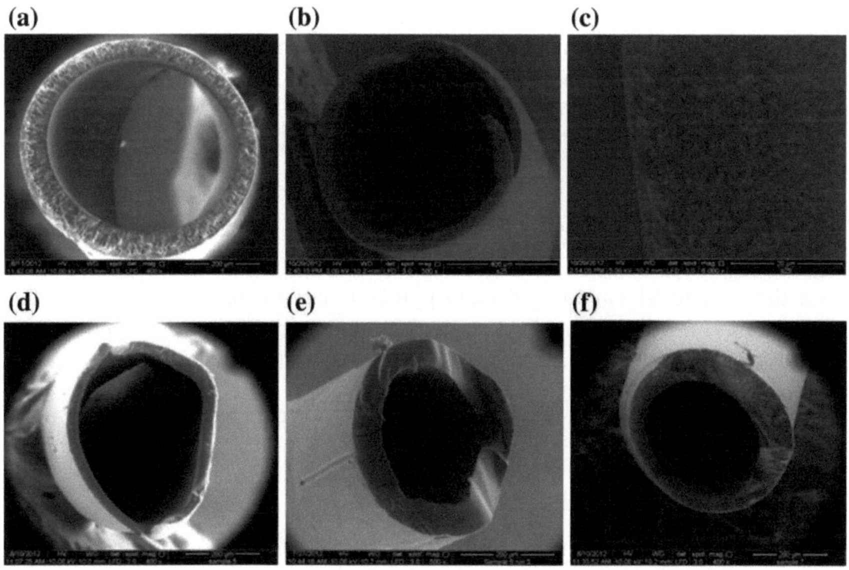

Fig. 1 SEM images of hollow fiber membranes; **a** S1 before thermal treatment. **b** S2 before thermal treatment. **c** S2 before thermal treatment under 6000X of magnification. **d** S2 after thermal treatment. **e** S3 after thermal treatment, and **f** S4 after thermal treatment

showed extra shrinkage, which was believed to have been caused by insufficient PVP, while S4 exhibited irregular structure of the membrane matrix. The irregularity of the structure possibly originated from the nature of the PVP itself. Excessive amount of PVP has increased the viscosity of dope solution and had caused the nascent fiber to be unstable and slightly non-uniform during spinning. S3 has lesser shrinkage impact on its structure, and the matrix is uniform throughout the membrane wall.

Thermogravimetric Analysis

Figure 2 shows the thermogravimetric analysis of each dope composition sample. The samples with blend polymer indicated two stages of weight loss, contributed by PEI and PVP subgroups release. Salleh and Ismail (2011) have categorized the PEI and PVP as the thermally stable and thermally labile polymer, respectively. Thus, the first weight loss of the thermal curve belonged to the PVP, while the later was belonged to the PEI.

The tangent of the thermal curve of each sample indicated the decomposition temperature (T_d) of PVP was slightly increased from S2 to S4. On the other hand, T_d of PEI in S2 was significantly higher than that of PEI in S1. The S3 possessed highest PEI T_d, while the T_d of PEI in S4 was slightly diminished. The increase in T_d values indicated that the presence of PVP might improve the thermal stability of the polymeric structure. The enhancement of thermal stability conserves the original shape of hollow fiber during thermal treatment. In fact, severe collapse of S1 during thermal treatment demonstrates the incapability of PEI to retain the original shape of hollow fiber without the presence of PVP. S3 was envisioned as the optimum composition for attaining the highest PEI's T_d. Further addition of PVP (S4) would cause the T_d of the PEI to deteriorate.

Single Gas Permeability Test

Since the sample S1 cannot withstand pyrolysis treatment, no permeability data were recorded for that particular sample. The permeabilities of pyrolyzed S2, S3,

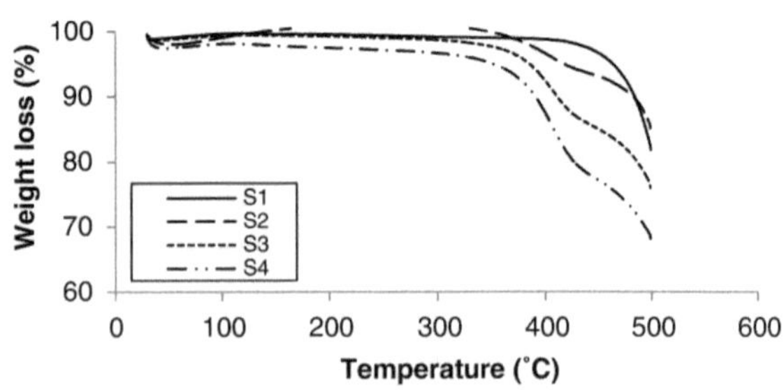

Fig. 2 Thermogravimetric analysis for membrane samples

Table 3 Permeance and permselectivity of samples after thermal treatment

Sample	Permeance (GPU)		CO_2/CH_4 permselectivity
	CO_2	CH_4	
S2	12.32	8.38	1.47
S3	14.61	7.47	1.95
S4	13.07	7.73	1.69

and S4 toward CO_2 and CH_4 gases are depicted in Table 3. The CO_2 permeance was larger than CH_4, which is in accord with their kinetic diameters. This phenomenon confirmed the molecular sieve mechanism might be involved in gas transport throughout the carbon membrane.

Based on Table 3, S3 possesses the highest CO_2 permeance and CO_2/CH_4 permselectivity. The micropore-type structure was believed to be well developed throughout the thermal treatment process. The PVP acted as the pillar to the polymer matrix that prevented the melting of PEI and retained the original shape of the hollow fiber after thermal treatment. The incapability of S1 to retain the original shape of the hollow fiber during thermal treatment has proven the requirement of PVP for structure reinforcement by the formation of sponge-like pore structure. The S2 exhibits the lowest CO_2 permeance and CO_2/CH_4 permselectivity. This was probably due to the collapsing of pores during thermal treatment. The insufficient amount of PVP has triggered the excessive shrinkage of carbon membrane and suppressed the pore development activities. On the other hand, the S4 exhibited low CO_2 permeance and CO_2/CH_4 permselectivity due to excessive amount of PVP in the polymer matrix. The thermally labile PVP would evolve earlier before the PEI. However, excessive amount of PVP has changed the properties of PEI/PVP blend system, which hinders complete decomposition during the pyrolysis. The remaining PVP in the membrane structure would disrupt the arrangement of carbon aggregation that originated from the PEI, which resulted in membrane with less porosity achieved after thermal treatment.

Conclusion

The thermal stability of the polymeric hollow fibers was found to be influenced by the amount of PVP presented in the dope solution. Sample S3 possesses uniform matrix structure with lesser shrinkage. The decomposition temperature of the membrane has increased for S3, indicating better thermal stability and structure preservation. Single gas permeability test has shown that S3 to be an optimum dope composition for production of PEI-/PVP-based hollow fiber carbon membrane.

Acknowledgments The authors wish to thank USM Membrane Cluster Grant and Ministry of Higher Education (MyPhD) for the financial supports.

References

Baker, R. W. (2002). Future direction of membrane gas separation technology. *Industrial and Engineering Chemistry Research, 41*, 1393–1411.

Datta, A. K., & Sen, P. K. (2006). Optimization of membrane unit for removing carbon dioxide from natural gas. *Journal of Membrane Science, 283*, 291–300.

Hao, J., Rice, P., & Stern, S. (2002). Upgrading low-quality natural gas with H_2S- and CO_2-selective polymer membranes: Part I. Process design and economics of membrane stages without recycle streams. *Journal of Membrane Science, 209*, 177–206.

He, X., & Hagg, M. G. (2011). Optimization of carbonization process for preparation of high performance hollow fiber carbon membranes. *Industrial and Engineering Chemistry Research, 50*, 8065–8072.

Jones, C. W., & Koros, W. J. (1995). Characterization of ultramicroporous carbon membranes with humidified feeds. *Industrial and Engineering Chemistry Research, 34*, 158–163.

Safari, M., Ghanizadeh, A., & Montazer-Rahmati, M. M. (2009). Optimization of membrane-based CO_2-removal from natural gas using models considering both pressure and temperature effects. *International Journal of Greenhouse Gas Control, 3*, 3–10.

Salleh, W. N. W., & Ismail, A. F. (2011). Carbon hollow fiber membrane derived from PEI/PVP for gas separation. *Separation and Purification Technology, 80*, 541–548.

Saufi, S., & Ismail, A. F. (2004). Fabrication of carbon membranes for gas separation: A review. *Carbon, 42*, 241–260.

Xiao, Y., Low, B. T., Hosseini, S. S., Chung, T. S., & Paul, D. R. (2009). The strategies of molecular architecture and modification of polyimide-based membranes for CO_2 removal from natural gas–a review. *Progress in Polymer Science, 34*, 561–580.

Effect of the Pyrolysis Soaking Time on CO_2 Separation of Polyetherimide/ Polyethylene Glycol-Based CMS Membranes

W. Z. Wan Nurul Huda and M. A. Ahmad

Abstract Carbon molecular sieve (CMS) membranes were successfully prepared from polymer blends consisting of polyetherimide (PEI) and polyethylene glycol (PEG). Pyrolysis process was carried out at 600 °C with the soaking time of 2, 4, 8, and 16 h. Gas permeation properties were tested using pure CH_4, CO_2, and N_2 and performed at ambient temperature. The permeability for examined gases decreased with increasing soaking time, whereas the selectivity of gas pairs of CO_2/CH_4 and CO_2/N_2 increased. CMS membrane prepared at 8 h of soaking time gave the highest selectivity of CO_2/CH_4 and CO_2/N_2 of 11.2 and 7.8, respectively. It was shown that pyrolysis soaking time played a crucial part and affected the separation capability of the final CMS membrane.

Introduction

Global warming has become one of the biggest challenges facing the world. Carbon dioxide (CO_2) emission is major contributor to this problem. CO_2 is inevitably created by fossil fuel and gasoline burning. The emissions of CO_2 have increased dramatically within the last 50 years and are still increasing by almost two percent each year (Olivier et al. 2011). The most mature CO_2 capture processes depend on the use of amine solvents to wash CO_2 out of a gas mixture. It is well known that amines in some applications can cause health risks. Gas separation using polymeric membranes is an alternative way to solve this problem. However, current polymeric membrane materials have reached a limit in their permeability/selectivity trade-off. To overcome this limitation, many researchers

W. Z. Wan Nurul Huda · M. A. Ahmad (✉)
School of Chemical Engineering, Engineering Campus, Universiti Sains Malaysia, Seri Ampangan 14300 Nibong Tebal, Seberang Perai Selatan, Penang, Malaysia
e-mail: chazmier@eng.usm.my

R. Pogaku et al. (eds.), *Developments in Sustainable Chemical and Bioprocess Technology*, DOI: 10.1007/978-1-4614-6208-8_39,

have investigated the potential of carbon molecular sieve (CMS) membranes as an advanced membrane material for CO_2 separation applications.

CMS membranes have become promising materials due to their excellent permeation and separation, and thermal and chemical stability under harsh conditions (Kim et al. 2004). Many studies have been done to optimize and improve the preparation technique in the fabrication of CMS membrane. Several attempts were made to control the pore structure such as modifying the precursor by polymer blending, fine-tuning the pyrolysis process parameters, pretreatment, as well as post-treatment. During pyrolysis, the microporous structure of the CMS membranes will be developed by the evolution of gaseous products channeling out from the carbon matrix. Many studies have shown that the pyrolysis conditions have a strong effect on the gas transport properties of CMS membrane. Studies by Centeno et al. (2004) and David and Ismail (2003) have shown that at some point of soaking time, the gas selectivity of CMS membranes will be enhanced. At this point, the microstructural rearrangement occurred and narrowed the existing pores, thus resulting in more selective CMS membranes (Vu et al. 2002). However, the structure of the final membranes is still dependent on the precursor, casting solution, and other pyrolysis conditions.

Centeno et al. (2004) have investigated the effect of pyrolysis soaking time on phenolic resin-derived carbon membranes. They found that at 700 °C, soaking time between 0 and 4 h resulted in poor gas selectivity. However, at 8 h of soaking time, the prepared CMS membrane showed the best performance for separation of CO_2/N_2 and CO_2/CH_4 of 28 and 103, respectively. Prolonged soaking time for 8 h had favored the rearrangement and subsequent alignment of carbonaceous aggregates, resulting in smaller microporous structure.

In this work, we report on a new approach in the preparation of PEI CMS membrane containing polymer blend. PEG is used as a pore-forming agent to introduce controllable mesoporosity into the carbon matrix. The effects of the pyrolysis soaking time on the permeation performances as well as morphology and structure properties of derived CMS membrane were investigated.

Methodology

Materials

PEI and PEG, used as precursors, were supplied by Sigma-Aldrich (M) Sdn. Bhd., while N-methyl-2-pyrrolidone (NMP), a solvent for PEI/PEG, was supplied by Acros Organics Ltd. Wellgas Sdn. Bhd. supplied purified N_2 (99.95 % purity), CO_2 (99.99 % purity), and CH_4 (99.99 % purity) gases.

Experimental

The polymer solution was prepared by dissolving 7 wt % of PEI in NMP. PEG was added into the solution with a weight ratio of 1:0.05 (PEI:PEG). The mixture was stirred 3 h to produce clear yellow solution and then degassed in an ultrasonic bath for 1 h. The PEI/PEG solution was deposited on the surface of the support using dip-coating technique. The pyrolysis process was carried out in a Carbolite vertical tube furnace. In the first step, the membranes were heated to 300 °C with a heating rate of 3 °C/min under air atmosphere. The temperature was held for 30 min before being heated further to 600 °C under nitrogen flow. At this stage, the holding time was varied from 2, 4, 8, and 16 h. Finally, the resulted CMS membranes were cooled gradually to room temperature under nitrogen flow. The gas transport properties of CMS membranes were studied by determining the permeation rates of three pure gases (CH_4, CO_2, and N_2) at ambient temperature. The morphological structure of carbon membranes was observed under SEM (Quanta FEG 450). ATR-IR spectra of PEI/PEG precursor and CMS membranes were recorded using IR Prestige-21, Shimadzu, over the wavenumber range of 4,000–650 cm^{-1}. The X-ray diffraction analysis was carried out using Philips PW 1729 X-ray generator with Philips PW 1820 diffractometer with diffraction angle 2θ from 10 to 50°.

Results and Discussion

Microstructure Properties

Table 1 gives a summary of ATR-IR results of PEI/PEG precursor. After pyrolysis, all peak intensities in CMS membranes prepared disappeared. This indicated that the polymer had sufficient time to break the polymer chain and remains the carbon element (David and Ismail 2003). This result confirmed the fact that polymer structures transformed to amorphous carbon structures during the process of pyrolysis.

XRD profile for CMS membrane obtained at 8 h of pyrolysis soaking time is illustrated in Fig. 1. The *d*-spacing calculated using Bragg's law indicated the distances between side-group atoms and skeletal atoms or distances between atoms in neighboring planes. Two peaks are relevant when examining polymer-based

Table 1 ATR-IR results of PEI/PEG precursor

Frequency (cm^{-1})	Functional group
1,172, 1,357	C–N
1,716, 1,774	C=O
1,597	C=C
1,273	C–O–C
744, 779	C–H

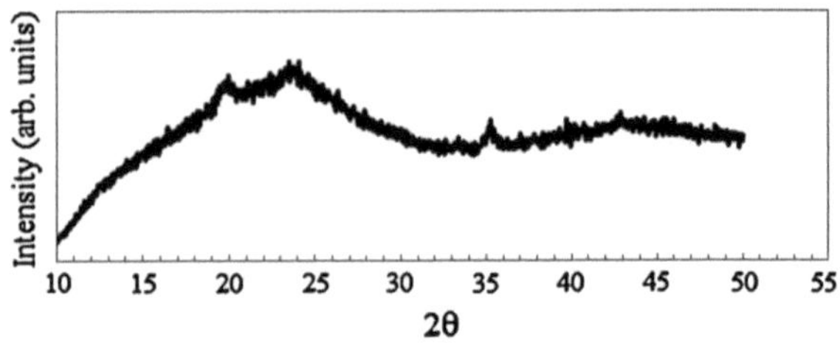

Fig. 1 XRD pattern of the PEI/PEG CMS membranes pyrolyzed at 8 h of soaking time

carbon, namely the (002) and (100) peaks, which, respectively, derived from the spacing between graphite sheets and distance between carbon atoms within a graphite sheet (Vu et al. 2002). The 002 and 100 peaks were presented at $2\theta = 24.09°$ and 42.87°, corresponding to average *d*-spacing of 0.369 nm and 0.211 nm, respectively. This behavior was due to the amorphous structure of the CMS membrane and disordered graphitic planes. This spectrum suggested a wide variation in interlayer spacing present in CMS membrane. Besides, this broadband indicated that the average microcrystallite size is very small since the width of the peak is inversely proportional to the crystallite size (Mariwala and Foley 1994). This result is in agreement with CMS membranes derived from polymer blends of PEI/PVP (Salleh and Ismail 2011).

Effect of Pyrolysis Soaking Time on Gas Permeation Properties

Table 2 shows the permeance of pure gases and selectivity for CMS membranes prepared at different pyrolysis soaking time. At 2 h of soaking time, all gases have almost the same value of permeance. The permeance of N_2 and CH_4 was decreased with the increase in soaking time from 2 to 8 h. The CO_2 permeance was slightly decreased by about 8 percent when the soaking time increased from 2 to 4 h. The drop of N_2 and CH_4 permeances in the CMS membranes showed that the micropore structure became narrower and restricted the larger molecular size gas to permeate. It showed that prolonged pyrolysis soaking time would lead to shrinkage of the pores to some extent, thereby resulting in decreasing permeances

Table 2 Permeability and selectivity for the PEI/PEG CMS membranes prepared at different pyrolysis soaking time

Pyrolysis soaking time (h)	Permeance (x 10^{-10} mol.m^{-2}.s^{-1}.Pa^{-1})			Selectivity	
	CO_2	N_2	CH_4	CO_2/CH_4	CO_2/N_2
2	898	886	851	1.06	1.02
4	829	467	435	1.90	1.70
8	1,071	137	95	11.20	7.80
16	1,669	469	192	8.70	3.60

(Su and Lua 2006). However, when the soaking time further increased from 8 to 16 h, the permeance for all gases began to rise. This was due to the collision of enlargement of microdomains, which resulted in more permeable membranes. The results are in agreement with the result of Steel and Koros (2005), which found that more compact and less permeable carbon membranes were produced at longer soaking time.

For gas selectivity, CMS membranes pyrolyzed at 8 h showed that the best performance with CO_2/CH_4 and CO_2/N_2 selectivities of 11.2, and 7.8, respectively. Longer soaking time will lead to the reduction in the pore sizes and thus resulted in the increase in selectivity. In fact, it has been observed that with prolonged soaking time, the porous structure would change closer to the graphitic and non-porous structure. Therefore, the difference in porosity of carbon membranes at different heating duration was attributed to the kinetic conversion of amorphous carbon to more crystalline graphitic carbon (Lafyatis et al. 1991).

Effect of Pyrolysis Soaking Time on Morphological Structure

Figure 2 shows the SEM micrographs of PEI/PEG CMS membranes prepared at different pyrolysis soaking time. Three distinct layers (carbon layer, intermediate layer, and porous alumina support) can be distinguished in CMS membranes as shown from the cross-section angle. The carbon film is a dense layer with a thickness of around 1–2 μm. It can be seen that, by extending the soaking time, the carbon layer became denser. During pyrolysis process, polymer precursor has undergone a plastic stage and turned into a more dense membrane surface (Salleh and Ismail 2011). For the top layer, a smooth and almost defect-free surface was obtained for all CMS membranes prepared at 4 and 8 h. For CMS membrane prepared at 2 and 16 h, a few defects spotted on the surface of the membrane can be observed. Defects on these two CMS membranes were not correlated with the soaking time but might be due to the uneven surface of the support.

Based on gas permeation results, it was shown that the most effective pyrolysis soaking time for PEI/PEG CMS membrane was 8 h. In this condition, the absence of crack and pinhole at the carbon layer surface indicated a suitable thermal expansion behavior of the coating in comparison with the support and appropriate support–membrane compatibility (Centeno and Fuertes 2001). The selectivities of CO_2/CH_4 and CO_2/N_2 were relatively high compared to the other membranes in this study which were 11.2 and 7.8, respectively.

Conclusion

The effects of pyrolysis soaking time on the microstructure properties, morphological structure, and gas transport properties of resulted CMS membranes have been studied. From the study, PEI/PEG CMS membrane prepared at 600 °C with

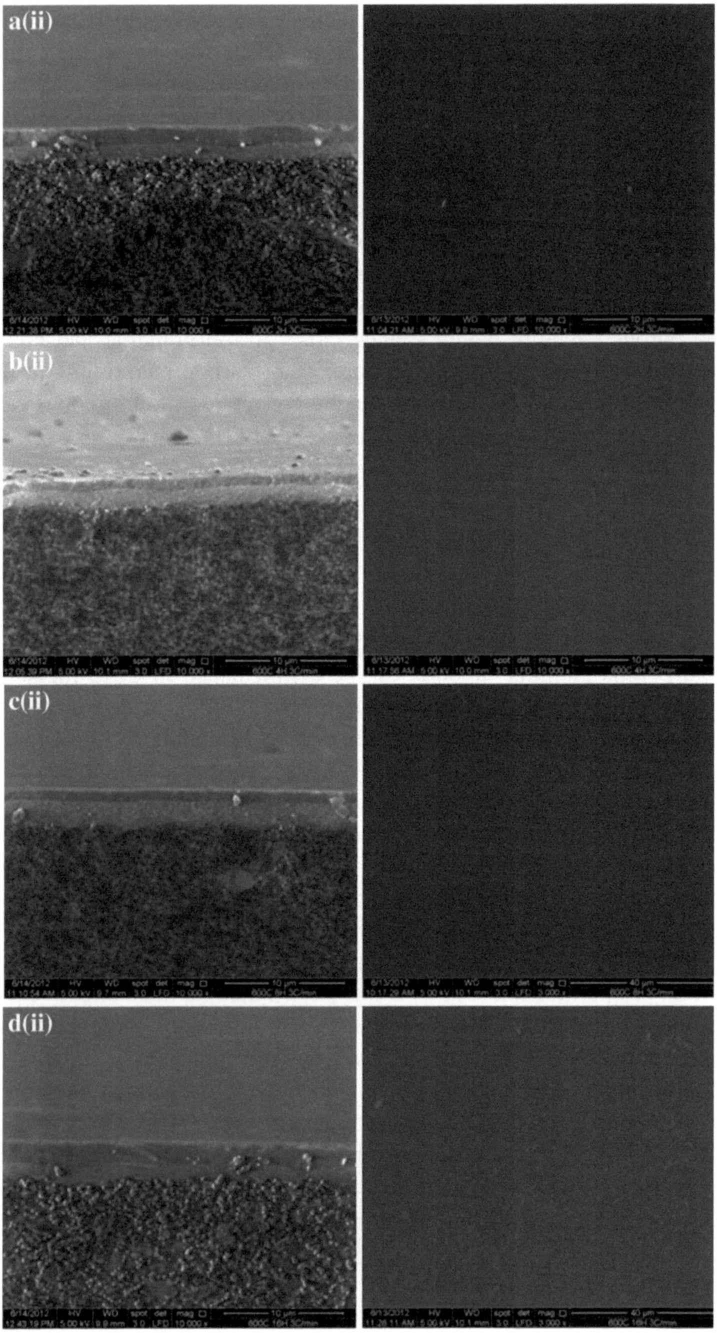

Fig. 2 SEM micrographs of (*1*) cross-section and (*2*) top layer of the CMS membrane prepared at (**a**) 2 h, (**b**) 4 h, (**c**) 8 h, and (**d**) 16 h

soaking time of 8 h showed the best performance in terms of selectivity. The selectivities of CO_2/CH_4 and CO_2/N_2 were 11.2 and 7.8, respectively. This research suggested that by fine-tuning the pyrolysis soaking time, it is able to produce carbon membranes that possess micropores of desired dimension for a different gas separation.

Acknowledgments The authors gratefully acknowledge the financial supports from USM Membrane Cluster Grant and Ministry of Higher Education (MyBrain).

References

Centeno, T. A., & Fuertes, A. B. (2001). Carbon molecular sieve membranes derived from a phenolic resin supported on porous ceramic tubes. *Separation and Purification Technology, 25*, 379–384.

Centeno, T. A., Vilas, J. L., & Fuertes, A. B. (2004). Effects of phenolic resin pyrolysis conditions on carbon membrane performance for gas separation. *Journal of Membrane Science, 228*, 45–54.

David, L. I. B., & Ismail, A. F. (2003). Influence of the thermastabilization process and soak time during pyrolysis process on the polyacrylonitrile carbon membranes for O_2/N_2 separation. *Journal of Membrane Science, 213*, 285–291.

Kim, Y. K., Park, H. B., & Lee, Y. M. (2004). Carbon molecular sieve membranes derived from thermally labile polymer containing blend polymers and their gas separation properties. *Journal of Membrane Science, 243*, 9–17.

Lafyatis, D. S., Tung, J., & Foley, H. C. (1991). Poly(furfuryl alcohol)-derived carbon molecular sieves: dependence of adsorptive properties on carbonization temperature, time, and poly(ethylene glycol) additives. *Industrial and Engineering Chemistry Research, 30*, 855–873.

Mariwala, R. K., & Foley, H. C. (1994). Evolution of ultramicroporous adsorptive structure in poly(furfuryl alcohol)-derived carbogenic molecular sieves. *Industrial and Engineering Chemistry Research, 33*, 607–615.

Olivier, J. G. J., Janssens-maenhout, G., Jeroen, A. H. W. P., & Wilson, J. (2011). Long-term trend in global CO_2 emissions 2011 report. (http://edgar.jrc.ec.europa.eu/news_docs/C02%20Mondiaal_%20webdef_19sept.pdf) Accessed 9 November 2012.

Salleh, W. N. W., & Ismail, A. F. (2011). Effects of carbonization heating rate on CO_2 separation of derived carbon membranes. *Separation and Purification Technology, 88*, 174–183.

Steel, K. M., & Koros, W. J. (2005). An investigation of the effects of pyrolysis parameters on gas separation properties of carbon materials. *Carbon, 43*, 1843–1856.

Su, J., & Lua, A. C. (2006). Influence of carbonization parameters on the transport properties of carbon membranes by statistical analysis. *Journal of Membrane Science, 278*, 335–343.

Vu, D. Q., Koros, W. J., & Miller, S. J. (2002). High pressure CO_2/CH_4 separation using carbon molecular sieve hollow fiber membranes. *Industrial and Engineering Chemistry Research, 41*, 367–380.

Performance of Particleboard with Palm Kernel Cake as Filler

A. Bono, S. M. Anisuzzaman, N. M. Ismail and R. Haziami

Abstract Particleboard is a common product of wood industry. Usually, the particleboard is used as a component of furniture manufacturing or part of building construction. In the production of particleboard, wood glue or binder is applied to form hard bonded particles. The usage of melamine–urea–formaldehyde (MUF) resin as a binder in particleboard production is widely practised. Due to low solid content and the cost of MUF resin, some particleboard manufacturers are adding filler to increase the solid content and to reduce the cost of binder. For example, industrial-grade wheat flour and palm shell powder are commonly used as filler. The performance of wood products is significantly affected by the type of filler used due to the compound composition. Therefore, a study conducted in this work is to investigate the performance of MUF resin-based particleboard produced with palm kernel cake powder as filler. Palm kernel cake (PKC) was chosen due to the protein content that might help in the binding strength and the reduction in free formaldehyde in the wood product. The investigation was conducted by the preparation of particleboard in laboratory with various curing time and quantity of MUF resin and filler used. The performance properties of particleboard such as water absorption (WA), thickness swelling (TS), modulus of rupture (MOR), and modulus of elasticity (MOE) were investigated. Response surface methodology (RSM) of design expert software was used for the experimental design and optimization. The results shown that there is some effect of filler on the performance properties particleboard produced. The optimized or targeted performance also can be identified via design expert software.

Keywords Melamine–urea–formaldehyde · Modulus of elasticity · Modulus of rupture · Wood binder · Wood glue

A. Bono (✉) · S. M. Anisuzzaman · N. M. Ismail · R. Haziami
School of Engineering and Information Technology, Universiti Malaysia Sabah,
88400 Kota Kinabalu, Sabah, Malaysia
e-mail: bono@ums.edu.my

R. Pogaku et al. (eds.), *Developments in Sustainable Chemical and Bioprocess Technology*, DOI: 10.1007/978-1-4614-6208-8_40,

Introduction

Particleboard can be considered as a common product of wood industry. Particleboard is widely used around the world in furniture manufacturing and house construction (Youngquist 1999). The production of particleboard involved compression of wood particles with other lignocelluloses materials and adhesive at high pressure. The adhesives used to bind the wood particles made from natural or synthetic products (Bono et al. 2010). At the present time, the majority of particleboard manufacturer employs formaldehyde-based adhesives such as phenol–formaldehyde (PF), urea–formaldehyde (UF), and melamine–urea–formaldehyde (MUF) as the main adhesives. The current global trend shows that the marketplace is moving in relation to use particleboards with reduced or no formaldehyde (Hashim et al. 2009; Bono et al. 2010). Formaldehyde-based adhesive accounted for 99 % of the total volume and 97 % of the total sale value, which indicated that formaldehyde-based adhesives played a dominant role in the wood adhesive market (Sellers 2001). However, formaldehyde-based adhesives are derived from nonrenewable petrochemicals and natural gas. In addition, the emission of formaldehyde, especially from the breakdown of formaldehyde-based resins in wood composites, poses a great hazard to human health because formaldehyde is carcinogenic to human body. Finite oil reserves, the expanding wood adhesive market, and hazardous issues associated with formaldehyde-based adhesives generate an urgent need for the development of environment-friendly alternative wood adhesives from renewable materials. In the year 1995, the International Agency for Research on Cancer (IARC) agreed that formaldehyde is a probable human carcinogen. However, in a reassessment of available data in June 2004, the IARC rearranged formaldehyde as a known human carcinogen (Hai 2005). Due to the cost of synthetic adhesive such as MUF resin, filler was used to reduce the production cost of bonded wood products.

Filler can be defined as additive added to material such as fine particle of agriculture by-products, plastics, composite materials, and concrete to decrease the utilization of expensive binder material or to improve the properties of the mixture material. The availability of palm kernel cake (PKC) around the world can be utilized as filler. PKC as agriculture by-product of palm oil industry can be similar or better than the current filler of agriculture by-products such as industrial-grade wheat flour or fine powder of palm shell. PKC contains protein content and some essential amino acid. Apart from that, PKC also contains 21–23 % of crude fibers which is rich in cellulose, hemicelluloses, and lignin (Farm et al. 2009; Bono et al. 2009; Boateng et al. 2008). The amino groups (-NH_2) in the crude protein of PKC may form hydrogen bonding between the melamine and urea in MUF resin also fibers (cellulose and hemicelluloses) to facilitate the bonding strength of MUF resin (Ong et al. 2011; Bono et al. 2011). The presence of small quantity of lipid in the PKC may decrease intrusion of water to disrupt the hydrogen bonding between MUF resin and the wood surface. Other than that, PKC acts as a formaldehyde scavenger which helps to reduce formaldehyde emission that is carcinogenic to human being. Utilization of PKC as filler may fill up the gaps or voids between the wood bonding surface and MUF resin to help promote the contact surface area and hence increase water resistance and the bond strength performance of MUF resin.

Therefore, a study of the effect of filler on the performance of particleboard with MUF resin was conducted. The main objective of this work was to investigate the performance properties of particleboard such as water absorption (WA), thickness swelling (TS), modulus of rapture (MOR), and modulus of elasticity (MOE) with the variation of MUF/filler ratio and the curing time. The experimental work was designed and optimized using response surface methodology (RSM) (Bono et al. 2008).

Methodology

The experimental works involved in this study were categorized into three parts: preparation of oil-free PKC powder, preparation of MUF resin, and production and characterization of particleboard.

Preparation of Palm Kernel Cake Powder

PKC was obtained from Borneo Samudera Lumadan Palm Oil Mill, Sabah, Malaysia. Fresh PKC was oven-dried at 80 °C for 12 h. The residue oil content was removed by Soxhlet extractor with hexane as the solvent for 8 h as suggested by Farm et al. (2009). Then, the PKC powder was prepared by grinding of oil-free PKC to less than 200 micron. The powder was kept at oven at 60 °C for at least 24 h prior to the usage for the particleboard production.

Melamine–Urea–Formaldehyde Resin Preparation

The MUF resin was prepared using industrial-grade melamine, industrial-grade urea, and 37 % formaldehyde as the raw materials. The method of synthesis of MUF resin was adopted from reflux process (Pizzi 1994; Bono et al. 2003, 2006). The adopted MUF resin synthesis process was divided into three stages: the first stage was the initial process, the second stage was the polymerization process, and the third stage was the post-refluxing process. The MUF resin produced was used for the production of particleboard within 3 h after it was synthesized.

Production and Characterization of Particleboard

The production process of particleboard involves mixing of wood particle with the adhesive and filler, molding and pressing at hot press with 120 °C. Acacia species was chosen as the wood particle due to the availability. Prior to the production process, the wood particles were dried approximately 5 % moisture content. The quantity of wood particle was measured to meet the target density of particleboard 600 kg/m^3, with the final target board equilibrium moisture content (EMC) of 12 %. The quantity of MUF resin and PKC powder were varied based on the experimental design. Board dimension was prepared as the capability of the hot press of 300 × 300 × 10 mm. The particleboard was then characterized for the WA, TS, MOR, and MOE.

WA test and TS were carried out based on Method B of ASTM D1037-93 method of soaking in water bath. The 50 × 50 mm piece of the particleboard was soaked in water at room temperature for 24 h. The thickness and weight were measured before and immediately after soaking. The board size and weight measured before and after soaking were used to calculate thickness swell and WA.

Modulus of rupture (MOR) and MOE following Japanese Industrial Standard (JIS 5908-1994) using GOTECH testing machine (Model AI-7000 LA10). The crosshead speeds were 10 mm/min. All of the test specimens were prepared with dimension of 250 × 50 mm.

Experimental Design

The experimental design and optimization were carried out using Design Expert software (Stat-Easy, Inc., Minneapolis, USA). Three independent process parameters, such as quantity of MUF, quantity of PKC powder, and curing time, were chosen as the factors, whereas the WA, TS, MOR, and MOE were selected as the process responses. The range of process parameters studied are shown in Table 1. The experimental runs were generated by the Design Expert Software according to RSM of D-optimal design.

Table 1 Range of process factors

Process factors	Limit	
	Low	High
MUF (%w/w)	5	15
PKC filler (%w/w)	10	20
Curing period (min)	8	16

Results and Discussion

Experimental run was conducted according to the experimental design with the parameters range set up as suggested in Table 1. The number of runs with the suggested parameters variation and measured responses is shown in Table 2.

The ANOVA of responses with variation of factors is represented by polynomial as Eq. (1) below:

$$\text{Response} = x_1 + x_2A + x_3B + x_4C + x_5AB + x_6AC + x_7BC + x_8A^2 + x_9B^2 + x_{10}C^2 \quad (1)$$

where A, B, and C are the actual factors of amount of PKC filler, amount of MUF resin, and curing time. The values of coefficients $x_1, x_2, x_3, x_4, x_5, x_6, x_7, x_8, x_9$and x_{10} for all responses are shown in Table 3. These values showed the significant of the single and combination of factor contribution toward the predicted responses. The responses were also analyzed graphically. The graphical analysis of responses is categorized into the effect of PKC filler on the WA, TS, MOE, and MOR of particleboard.

Table 2 The measured responses with various factors based on experimental design

Run	Experiment parameter			Responses			
	PKC filler (%)	MUF resin (%)	Curing period (min)	WA (%)	TS (%)	MOE (N/mm^2)	MOR (N/mm^2)
1	20.00	15.00	8.00	71.89	4.34	761.70	4.92
2	20.00	5.00	16.00	141.50	14.73	56.34	0.93
3	15.00	5.00	12.00	108.00	8.58	350.90	1.62
4	10.00	15.00	16.00	56.41	7.47	513.00	1.12
5	20.00	15.00	16.00	73.35	2.63	829.70	3.83
6	15.00	5.00	8.00	96.96	10.59	274.80	1.31
7	15.00	10.00	16.00	111.10	2.74	502.30	2.21
8	15.00	15.00	8.00	82.73	4.22	673.20	4.04
9	10.00	15.00	16.00	56.41	7.47	513.00	1.12
10	12.50	12.50	12.00	84.96	2.59	579.80	2.59
11	10.00	5.00	8.00	46.23	5.40	456.60	2.79
12	10.00	15.00	8.00	61.35	6.87	470.30	3.16
13	20.00	15.00	16.00	73.35	2.63	829.70	3.83
14	10.00	5.00	16.00	65.92	3.89	509.10	2.94
15	20.00	5.00	8.00	115.50	18.54	60.12	0.21
16	15.00	10.00	16.00	111.10	2.74	502.30	2.21
17	10.00	5.00	16.00	65.92	3.89	509.10	2.94
18	20.00	5.00	16.00	141.50	14.73	56.34	0.93
19	10.00	10.00	8.00	64.43	3.08	431.50	2.98
20	20.00	10.00	12.00	110.80	6.32	418.10	2.38

Table 3 ANOVA results

Coefficients	Responses			
	WA (%)	TS (%)	MOE (N/mm^2)	MOR (N/mm^2)
X_1	−243.22	9.63	624.87	7.29
X_2	28.58	0.67	−19.35	−0.58
X_3	18.36	−0.94	−98.04	−0.22
X_4	2.58	−1.05	57.46	0.05
X_5	−0.59	−0.16	7.42	0.05
X_6	0.08	−0.03	−0.18	9.74×10^{-3}
X_7	−0.31	0.03	0.22	−0.02
X_8	−0.65	0.06	−1.97	–
X_9	−0.43	0.12	1.11	–
X_{10}	0.03	0.04	−2.14	–

Effect of PKC Filler on Water Absorption of Prticleboard

The effect of PKC filler on WA of particleboard with variation of MUF composition at curing period of 8, 12, and 16 min is shown graphically in Fig. 1a, b, c. This figure shows that the WA is increased with the amount of PKC added and the longer curing period but decreases with the amount of MUF used. The increases in WA with the composition of PKC could be due to the increases in polysaccharide that is hydrophilic. The decreases in WA with the amount of MUF used in the formulation may be caused by the hydrophobic characteristic of MUF. Whereas longer pressing time may cause of over curing and may give severe effect on mechanical and physical properties of particleboard produced (Izran et al. 2009).

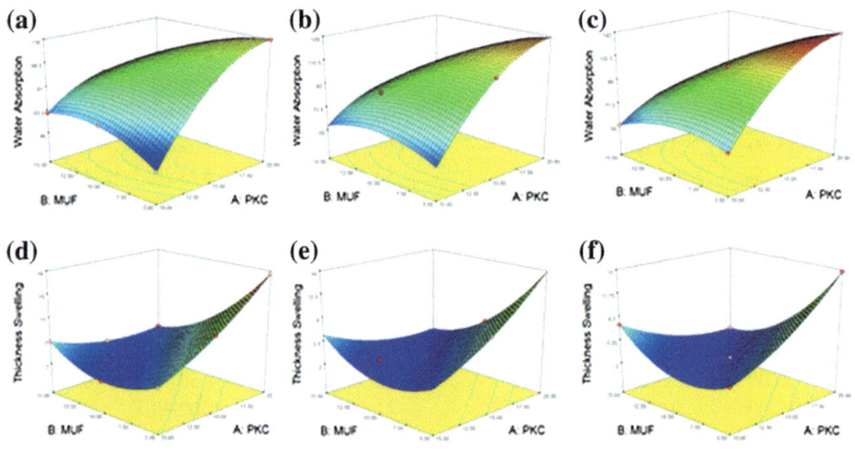

Fig. 1 Water absorption (%) and thickness swelling (%) of particleboard with various compositions of PKC (%) and amount of MUF (%) at curing period of (**a**) 8, (**b**) 12, and (**c**) 16 min

Effect of PKC Filler on Thickness Swelling of Particleboard

The TS of particleboard due to the PKC filler and the MUF composition at curing period of 8, 12 and 16 min are presented graphically in Fig. 1d, e, f. The magnitude of TS is increased with the quantity of PKC added but decreased with the curing period. Whereas, the quantity of MUF used in the production of particleboard showing a curving trend of the TS with a minimum point in the middle. The rise of TS of particleboard with the quantity increment of PKC may due to the increase in the fiber with ionic or hydrogen bonding inside PKC which is also related with the increase the WA. The changes of physical properties during curing period could be the main reason for causing the changes of TS behavior. The trend of TS shows with the variation of MUF composition shows that there is possible interaction between MUF and PKC. The curving shape of TS trend shows that the composition of PKC filler and MUF resin can be optimized to obtain desirable properties of particleboard.

Effect of PKC Filler on Static Bending Test (MOE and MOR) of Particleboard

The results of static bending test of particleboard, MOE are presented graphically in Fig. 2a, b, c and MOR in Fig. 2d, e, f respectively. Figure 2a, b, c is representing the MOE with various MUF resins and PKC filler at curing period of 8, 12

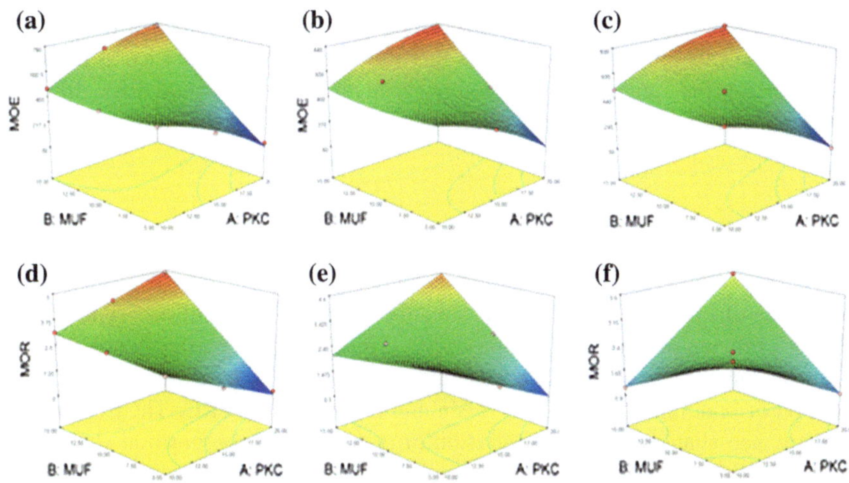

Fig. 2 Modulus of elasticity (N/mm^2) and modulus of rapture (N/mm^2) of particleboard with various compositions of PKC (%) and amount of MUF (%) at curing period of (**a**) 8, (**b**) 12, and (**c**) 16 min

Table 4 The optimization criteria for the process condition of the particleboard production

Responses	Target	Range	
		Low	High
WA (%)	Minimum	36.1	140.76
TS (%)	Minimum	1.52	19.78
MOE (N/mm^2)	Maximum	0.213	6.55
MOR (N/mm^2)	Maximum	45.134	874.91

and 16 min. Similarly, Fig. 2d, e, f is representing the MOR with various MUF resins and PKC filler at curing period of 8, 12 and 16 min. With respect to the rise of curing period, the magnitude of MOE increases, whereas MOR is decreased. At lower composition of MUF resin, the magnitude of MOE and MOR decrease as the quantity of PKC filler increases. In contrast, the magnitude of MOE and MOR increase as the quantity of PKC filler increases at higher composition of MUF resin. The lower values of MOE and MOR at lower composition of MUF may be due to lack of MUF to form interlock with the wood fiber.

Optimization of Process Factors with Properties of Particleboard with PKC Filler

The optimization of process factors conducted using the procedure included in Design Expert Software. The criteria selected were based on the range of factors which is similar to the range of factors selected for the experimental run as shown in Table 1, whereas the target value for the responses were selected as shown in Table 4.

The result of optimization showed that the maximum desirability of the optimization for the range of MUF composition of 12.5–15 % and PKC filler of 15–20 % at curing period 10.72 min.

Conclusions

The usage of PKC as a filler in the production of particleboard affects significantly the performance of particleboard produced. The rise of the particleboard strength with the increase in PKC filler used indicates that there is a possibility of the reaction or bonding between MUF resin and protein inside PKC. The trend of the performance changed with the variation of particleboard composition shows that the production process can be optimized to the desire properties of the product produced. Finally, the investigation in this work can be extended to the study of the other process parameters such effect of particle sizes, pressing temperature and the various types of fillers.

References

Boateng, M., Okai, D. B., Baah, J., & Donkoh, A. (2008). Palm Kernel Cake extraction and utilization in pig and poultry diets in ghana. *Livestock Research for Rural Development, 20*, 7.

Bono, A., Yeo, K. B., & Siambun, N. J. (2003). Melamine-Urea-Formaldehyde (MUF) resin: The effect of the number of reaction stages and mole ratio on resin properties. *Jurnal Teknologi. Universiti Teknologi Malaysia, 38*(F), 43–52.

Bono, A., Krishnaiah, D., Rajin, M., & Siambun, N. J. (2006). Variation of reaction stages and mole composition effect on melamine-urea-formaldehyde (MUF) resin properties, study in surface science and catalysis. *Elsivier, 159*, 713–716.

Bono, A., Krishnaiah, K., & Rajin, M. (2008). *Products and process optimization using response surface methodology*. Malaysia: Universiti Malaysia Sabah Press.

Bono, A., Ying, P. H., Farm, Y. Y., Chng, L. M., Sarbatly, R., Krishnaiah, D. (2009). Synthesis and characterized of Carboxylmethyl cellulose from palm kernel cake, *Advances in Natural and Applied Sciences, 3*(1), 5–11.

Bono, A., Farm, Y. Y., Krishnaiah, D., Ismail, N. M., Yee, L. Y. K., & Lee, L. M. (2010). Palm kernel based wood adhesive. *Journal of Applied Sciences, 10*(21), 2638–2642.

Bono, A., Nur, M., Anisuzzaman, S. M., Saalah, S., & Chiw, H. K. (2011). The performance of MUF resin with palm kernel as filler. *Advanced Materials Research, 3*, 233–235.

Farm, Y. Y., Krishnaiah, D., Rajin, M., & Bono, A. (2009). Cellulose extraction form palm kernel cake using liquid phase oxidation. *Journal of Engineering Science and Technology, 4*(1), 57–68.

Hai, B. (2005). WTO declares formaldehyde as human carcinogen. *China Wood-based panels*, 30.

Hashim, R., Hamid, S. H. A., Sulaiman, O., Ismail, M. H., & Jais, H. (2009). Extractable formaldehyde from waste medium density fibreboards. *Journal of Tropical Science, 21*, 25–33.

Izran, K., Abdul Malek A. R., Mohd. Yusof, M. N., Masseat, K., & Ashaari, K. (2009). Physical and mechanical properties of flame retardent treated Hibicus cannabinus particleboard. *Journal of Modern Applied Science, 3*, 2–8.

Ong, H. R., Prassad, R., Rahman Khan, M. M., & Kabir Chowdhury, M. N. (2011). Effect of palm kernel meal as MUF adhesive extender for plywood application: Using a FTIR spectroscopy study. *Advanced Mechanics and Materials, 493*, 121–126.

Pizzi, A. (1994). Melamine-formaldehyde adhesive. In A. Pizzi & K. L. Mitti (Eds.), *Handbook of adhesive technology* (393–403). New York: Marcel Dekker Inc.

Youngquist, J. A. (1999). Wood-based composites and panel products. In *Wood handbook: Wood as an engineering material, General Technical Report. FGL-GRT-113., USDA Forest Serv.* (pp. 1–31). Madison, WI: Forest Prod. Lab.

Starch-Based Biofilms for Green Packaging

R. R. Ali, W. A. W. A. Rahman, N. B. Ibrahim and R. M. Kasmani

Abstract The aim of this study is to develop degradable starch-based packaging film with enhanced mechanical properties. A series of low-density polyethylene (LDPE)/tapioca starch compounds with various tapioca starch contents were prepared by twin-screw extrusion with the addition of maleic anhydride-grafted polyethylene as compatibilizer. Palm cooking oil was used as processing aid to ease the blown film process; thus, degradable film can be processed via conventional blown film machine. Studies on their mechanical properties and biodegradation were carried out by tensile test and exposure to fungi environment, respectively. The presence of high starch contents had an adverse effect on the tensile properties of LDPE/tapioca starch blends. However, the addition of compatibilizer to the blends improved the interfacial adhesion between the two materials and hence improved the tensile properties of the films. High content of starch was also found to increase the rate of biodegradability of LDPE/tapioca starch films. It can be proved by exposure of the film to fungi environment. A growth of microbes colony can be seen on the surface of LDPE/tapioca starch film indicates that the granular starch present on the surface of the polymer film is attacked by microorganisms, until most of it is assimilated as a carbon source.

R. R. Ali (✉) · W. A. W. A. Rahman · N. B. Ibrahim · R. M. Kasmani
Polymer Engineering Department, Faculty of Chemical Engineering,
Universiti Teknologi Malaysia, 81310 Skudai, Johor, Malaysia
e-mail: roshafima@cheme.utm.my

N. B. Ibrahim · R. M. Kasmani
Gas Engineering Department, Faculty of Petroleum and Renewable Energy Engineering,
Universiti Teknologi Malaysia, 81310 Skudai, Johor, Malaysia

R. Pogaku et al. (eds.), *Developments in Sustainable Chemical and Bioprocess Technology*, DOI: 10.1007/978-1-4614-6208-8_41,

Introduction

Plastics are widely used economical materials characterized by excellent all-round properties, easy moulding, and manufacturing. Approximately 140 million tonnes of synthetic polymers are produced worldwide each year to replace more traditional materials, particularly in packaging. Over 60 % of post-consumer plastic waste is produced by households and most of it as single-use packaging (Zheng and Yanfu 2005). Plastics are manufactured and designed to resist the environmental degradation and also more economical than metal, woods, and glasses in terms of manufacturing costs and energy required. Due to these issues, plastic resins have become one of the most popular materials used in packaging. However, these plastics are made of petroleum-based materials that are not readily biodegradable. Synthetic plastics such as polyethylene and polypropylene have a very low water vapor transmission rate, which is a good characteristic of packaging materials. Furthermore, they are totally non-biodegradable and therefore lead to environmental pollution, which pose serious ecological problems.

Polyolefin is not degraded by microorganisms in the environment, which contributes to their long lifetime of hundreds of years. Plastic packaging has a cycle less than a year and continuously enters the waste stream on a short turnout of time. The continuous growth of plastic industries has led to the increased volume of plastic waste in the landfill. With the growing concern about environmental pollution, the accumulation of plastic waste needs immediate resolution. Biodegradable plastics have been intensively studied in recent years (Khabbaz et al. 1998; Erlandsson et al. 1997; Akaranta and Oku 1999; Arvanitoyannis et al. 1998; Manzur et al. 2004) and have been commercialized into various products such as garbage bags, composting yard-waste bags, grocery bags, and agriculture mulches.

Biodegradable plastics are defined as plastics with similar properties to conventional plastics, but it can be decomposed after disposal to the environment by the activity of microorganisms to produce end products of CO_2 and H_2O (Tharanathan 2003). It is also an alternative to the petroleum-based non-biodegradable polymers. Biodegradable plastics can be used in hygiene products, household goods, horticultural products, agriculture, medical products, and many more. It decreases the solid waste problems created by plastic waste. Biodegradable polymers can be divided into two main categories, which are naturally occurring biodegradable polymers and synthetic biodegradable polymers (Danjaji et al. 2002).

Research on biodegradable plastics based on starch began in the 1970s and continues today at various laboratories all over the world. Technologies have been developed for continues production of extrusion films and injection-moulded plastics containing 50 % or more of starch (Tharanathan 2003). Starch satisfies the requirements of adequate thermal stability, minimum interference with melt properties, and disturbance of product quality (Danjaji et al. 2002). Starch is an readily available inexpensive, renewable, and fully biodegradable natural raw material which has generated a renewed in its use as a component in plastic formulations (Shah et al. 1995). Incorporation of starch into the synthetic polymer will increase the biodegradability of synthetic polymer when starch is consumed by microorganisms. It is believed that under a rapid enzymatic hydrolysis, starch will be degraded leading to a void-containing matrix, reduced the mechanical properties of the plastics, and might promote the biodegradation of synthetic polymer due to the increased surface area available for interaction with microorganisms (Chandra and Rustgi 1998). In this present study, low-density polyethylene (LDPE)/tapioca starch biodegradable film was prepared by blown film extrusion with the addition of compatibilizer and palm oil-based glycerin as processing aids.

Methodology

Sample Preparation

Low-density polyethylene resin grade (TITANLENE® low-density polyethylene, 71009A, LDF 260GG) and tapioca starch (particle size of those starches ranged from 9.73 to 83 ηm with an average particle size of 32.97 ηm; moisture content of tapioca starch has average of 11.5 %) were oven-dried for 24 h at 80 °C before premixing and compounding to remove moisture. The compounding of LDPE/ tapioca starch at different blend ratios was done using twin-screw extruder with the addition of palm cooking oil (PCO) (5 wt % phr) as processing aids and polyethylene-grafted maleic anhydride and LLDPE-g-MA (5, 10, 15, and 20 wt % phr) as compatibilizer. The compounding process was carried out at a screw speed of 80 rpm, and the temperatures were set at 150 °C/150 °C/140 °C/140 °C from feeder to die. The strands were then cooled using water and then passed through the palletizer and palletized. The pallets were collected and used for film blowing. The prepared pallets of different composition, as indicated in Table 1, were put in the feed hopper with the temperature set at 115 °C/120 °C/130 °C/130 °C/120 °C/ 120 °C (from feeder to die). The films of LDPE/starch blends were blown at a drawer screw of 50 rpm and extruder screw speed of 120 rpm.

Table 1 The abbreviations of blends

Samples	LDPE (wt %)	TS (wt %)
LDPE	100	0
LDPE/TS:90/10	90	10
LDPE/TS:80/20	80	20
LDPE/TS:70/30	70	30
LDPE/TS:60/40	60	40
LDPE/TS:50/50	40	50

Mechanical Testing

The tensile strength and elongation at break of sample films were determined by using tensile testing machine, Lloyd. The films were cut into uniform shaped according to ASTM D882-02. The test was carried out at a crosshead speed of 50 mm/min. At least five samples were tested for each formulation.

Biodegradability Study

For biodegradation analysis, sample films were inoculated with *Aspergillus Niger (A. Niger)* on a medium and incubated at surrounding temperature (25–37 °C) for 21 days (Akaranta and Oku 1999; Chandra and Rutsgi 1998). Samples were cut (2.5 × 2.5 cm) and faced on the surface of mineral salt agar in a Petri dish containing no additional carbon source. Before faced the samples, agars surfaced were cultivated with *A. Niger* from the tapioca slices. Thereafter, the films were examined for evidence of colony growth.

Results and Discussion

Mechanical Test

The tensile strength, which is a measure of the resistance to direct pull, is one of the important plastic properties in machinability and packaging applications. Tensile properties of different LDPE/tapioca starch blend extruded films were measured and shown in Fig. 1a. It was observed from graph that the tensile strength of LDPE gradually decreased with the addition of 40 % tapioca starch in LDPE/tapioca starch blends. The decrease in tensile strength observed for the films containing the fillers as the addition level increases has been attributed to possible increasing filler–filler interactions at the expense of filler–polymer interactions (Liu et al. 2003). This is also due to incompatibility of hydrophilic tapioca starch with hydrophobic LDPE. In fact, addition of starch generally results in an increase in stiffness, which corresponds to decreasing tensile strength as well as elongation at break (Chandra and Rustgi 1998). Incorporation of starch had also reduced the elongation at break of all LDPE/tapioca starch blends as shown in Fig. 1b. Reduction in elongation at break of the LDPE/tapioca starch film was due to no chemical interaction between starch and LDPE, has been proven from the FTIR result. Starch incorporation produced discontinuity in the film matrix, resulting in lower elongation (Baldev et al. 2004).

In order to enhance the compatibility between two immiscible polymers, an increased interest has appeared in the use of polymers containing reactive groups,

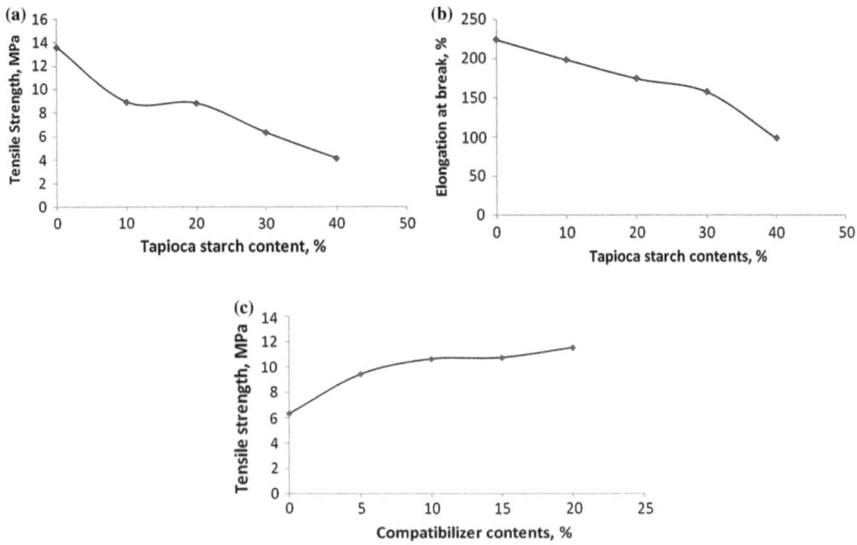

Fig. 1 (**a**) Tensile strength, (**b**) elongation at break, and (**c**) tensile strength of LDPE/tapioca starch-based films with various starch contents

such as polyethylene-grafted maleic anhydride as compatibilizer. Addition of PE-g-MA to this blend showed a significant increase in tensile strength of LDPE/tapioca starch blends (Fig. 1c). Tensile strength of LDPE/starch:70/30 (30 wt % of tapioca starch content in the blends) film increased as the compatibilizer contents rose. It was discovered that anhydride groups in polyethylene-grafted maleic anhydride could react with hydroxyl groups in starch to produce chemical bonding, thus improving the dispersion of starch, the interfacial adhesion, and subsequently the mechanical properties of the blends (Liu et al. 2003).

Biodegradability Study

There was no fungi growth on the surface of LDPE, due to microbial resistance behavior present in the film, as shown in microscopy image analyzed with magnification of 200x (Fig. 2). LDPE is formed by carbon–carbon (C–C) linkages in which these linkages are not susceptible to microbial attack (Shah et al. 1995). In contrast, after 21 days, the fungi growth was clearly visible for LDPE/starch films as the incorporation of starch has attracted fungi to attack the film. Figure 3 shows the fungal growth for various compositions of LDPE/starch blends. Starch loading up to 20 % shows the apparent growth of fungi, and as the starch contents increased, most of the film surfaces were covered by fungi growth. This indicated that the growth of *A. Niger* colony increases as the starch content is increased.

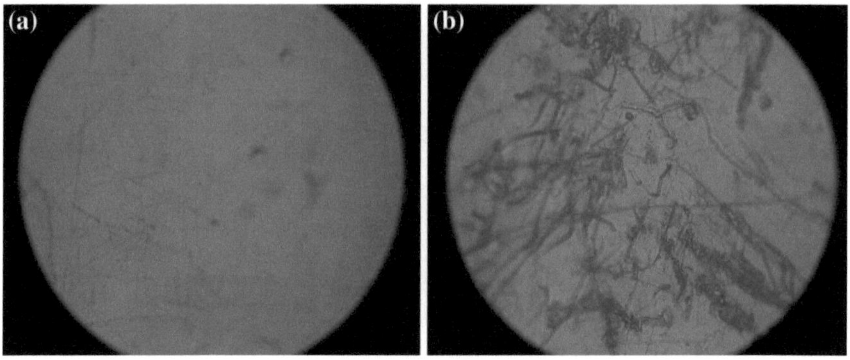

Fig. 2 Evidence of fungal growth (A. Niger) on surface of LDPE films with and without starch under microscope magnification 200X. **a** LDPE film. **b** LDPE/tapioca starch-based film

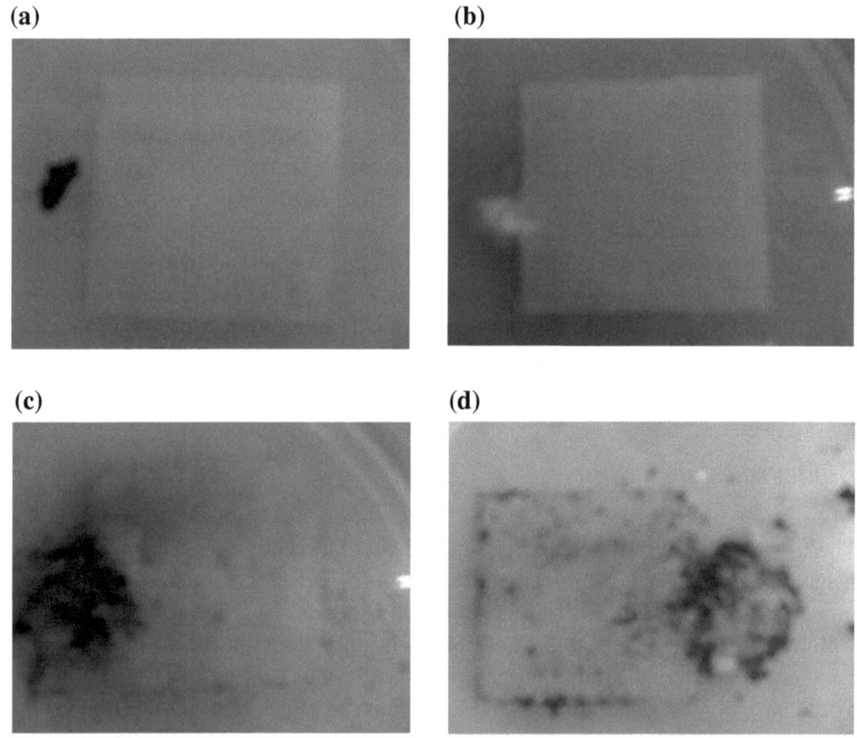

Fig. 3 Fungal growth for various compositions of LDPE/starch blends. **a** LDPE/tapioca starch:90/10 film. **b** LDPE/tapioca starch:80/20 film. **c** LDPE/tapioca starch:70/30 film. **d** LDPE/tapioca starch:60/40 film

The granular starch present on the surface of the polymer film is attacked by fungi. This weakens the polymer matrix and increases the surface volume ratio, hydrophilicity, and permeability of the film (Shah et al. 1995).

Conclusions

LDPE/tapioca starch-based biofilms were prepared and their mechanical properties and biodegradability were measured and correlated with their composition. The addition of palm cooking oil as processing aid improved process ability of the blends; thus, degradable film can be processed via conventional blown film machine. Incorporation of polymer containing PE-g-MA as compatibilizer improved the dispersion of starch, interfacial adhesion between starch and LDPE, and subsequently the mechanical properties of the resultant blends. Starch loading up to 20 % shows the apparent growth of fungi, and as the starch contents increased, most of the film surfaces were covered by fungi growth, as the incorporation of starch has attracted fungi to attack the film, which shows that this biofilm is degradable. This starch-based biofilm can be decomposed after disposal to the environment by the activity of fungi and microorganisms. Starch-based biofilm can be used as an alternative to the petroleum-based non-degradable polymers. It can be used as degradable plastics for food packaging, agricultural mulching films, and many more.

Acknowledgments The author would like to thank the government of Malaysia through Ministry of Higher Education (MOHE) for furnishing the grant to expedite the research work (Vot: Q.J130000.2544.03H26). Last, but not least the authors thank Universiti Teknologi Malaysia and research colleagues for all the assistance granted throughout the research period.

References

Akaranta, O., & Oku, G. E. (1999). Erratum to: Some properties of cassava mesocarp carbohydrates-low density polyethylene blends. *Carbohydrate Polymers, 34*, 403–405.

Arvanitoyannis, I., Biliaderis, C. G., Ogawa, H., & Kawasaki, N. (1998). Biodegradable films made from low density polyethylene (LDPE), rice starch and potato starch for food packaging application: Part I. *Carbohydrate Polymer, 36*, 89–104.

Baldev, R., Udaya, S. R., & Siddaramaiah, N. H. (2004). Low density polyethylene/starch blend films for food packaging application. *Advance in Polymer Technology, 42*, 32–45.

Chandra, R., & Rustgi, R. (1998). Biodegradable polymers. *Progress in Polymer Science, 23*, 1273–1335.

Danjaji, I. D., Nawang, R., Ishiaku, U. S., Ismail, S., & Mohd ishak, Z. A. (2002). Degradation studies and moisture uptake of sago-starch filled linear low-density polyethylene composites. *Polymer Test, 21*, 75–81.

Erlandsson, B., Karlsson, S., & Albertsson, A. C. (1997). The mode of action of corn starch and a prooxidant system in LDPE: Influence of thermooxidation and UV irradiation on the molecular weight changes. *Polymer Degradation and Stability, 55*, 237–245.

Khabbaz, F., Albertsson, A. C., & Karlsson, S. (1998). Trapping of volatile low molecular weight photoproducts in inert and enhanced degradable LDPE. *Polymer Degradation and Stability, 61*, 329–342.
Liu, W., Wang, Y. J., & Sun, Z. (2003). Effects of polyethylene-grafted-maleic anhydride (PE-g-MA) on thermal properties, morphology and tensile properties of low density polyethylene (LDPE) and corn starch blends. *Journal of Applied Polymer Science, 88*, 2904–2911.
Manzur, A., Limon-Gonzalez, M., & Favela-Torres, E. (2004). Biodegradation of physicochemically treated LDPE by a consortium of filamentous fungi. *Journal of Applied Polymer Science, 92*, 265–271.
Shah, P. B., Bandopadhyay, S., & Bellare, J. R. (1995). Environmentally degradable starch filled low density polyethylene. *Polymer Degradation and Stability, 47*, 165–173.
Tharanathan, R. N. (2003). Biodegradable films and composite coatings: past, present and future. *Trends in Food Science and Technology, 14*, 71–78.
Theubaud, S., Aburto, J., Alric, I., Borredon, E., Bikiaris, D., Prinos, J., et al. (1997). Properties of fatty acid estera of starch and their blends with LDPE. *Journal of Applied Polymer Science, 17*, 705–721.
Zheng, Y., & Yanful, E. K. (2005). A review of plastic waste biodegradation. *Critical Review in Biotechnology, 25*, 243–250.

Influence of Potassium Hydroxide Concentration on the Carrageenan Functional Group Composition

S. M. Anisuzzaman, A. Bono, S. Samiran, B. Ariffin and Y. Y. Farm

Abstract Potassium Hydroxide treatment of seaweed is one of the common processes in the production of semi-refined carrageenan (SRC) across Southeast region. However, potassium hydroxide will influence the properties of the product through the effect on the variation of functional group composition in the carrageenan molecule. Therefore, a study on the influence of potassium hydroxide (KOH) concentration on the functional group composition of kappa carrageenan extracted from seaweed (*Kappaphycus alvarezii*) was conducted. The functional groups investigated were 3,6-anhydrogalactose, ester sulphate, 3,6-anhydrogalactose-2-sulphate and galactose-4-sulphate. To investigate the effect of KOH concentration on the variation of functional groups composition, a set of KOH concentration of 0.01–0.1 M was employed. The SRC powder was produced by spray dryer of alkali-treated seaweed. The SRC powder produced was used for the Fourier transform infrared spectroscopy (FTIR) analysis for identification and determination of the composition of functional groups. The results of the data revealed that there is composition variation in all of the functional groups analysed due to the influence of KOH concentration.

Keywords *Kappaphycus alvarezii* · 3,6-anhydrogalactose · Ester sulphate · 3,6-anhydrogalactose-2-sulphate · Galactose-4-sulphate

S. M. Anisuzzaman (✉) · A. Bono · S. Samiran · Y. Y. Farm
School of Engineering and Information Technology, Universiti Malaysia Sabah, 88400 Kota Kinabalu, Sabah, Malaysia
e-mail: anis_zaman@ums.edu.my

B. Ariffin
Universiti Malaysia Terengganu, 21030 Kuala Terengganu, Terengganu, Malaysia

R. Pogaku et al. (eds.), *Developments in Sustainable Chemical and Bioprocess Technology*, DOI: 10.1007/978-1-4614-6208-8_42,

Introduction

Traditional exploitation of seaweeds in South-east region was started for many centuries. The exploitation was usually as a natural source of food, food ingredients or medicines. Nowadays, around six million tons of fresh algae are cultivated per year across Asian countries, which amounting to about 90 % of the commercial demands (Besada et al. 2009). Seaweeds can be classified into four categories based on their colour such as red (division Rhodophyta), green (division Clorophyta), blue-green (division Cyanophyta) and brown algae (division Phyaeophyta) (Phang 2006). The major seaweed genera are *Sargassum*, *Eucheuma*, *Caulerpa*, *Gracilaria*, *Hypnea*, *Padina* and *Hydroclathrus* are commonly found along the coastal area of South-east countries. Among these, *Gracilaria changgi* and *Eucheuma* mainly serve as a raw material for agar or carrageenan production. The extracted agar or carrageenan is used in many applications such as in the food industries or in the production of tissue culture media (Glickman 1987). Carrageenan classified into kappa, iota and lambda types (Campo et al. 2009), the difference between them is the number and position of ester sulphate groups as well as the composition of 3,6-anhydrogalactose. Higher composition levels of ester sulphate mean lower solubility temperature and lower gel strength. Kappa-type carrageenan has an ester sulphate composition of 25–30 % and a 3,6-anhydrogalactose of 28–35 %. Iota-type carrageenan has an ester sulphate composition of 28–30 % and a 33,6-anhydrogalactose content of 25–30 %. Lambda-type carrageenan has an ester sulphate content of about 32–39 % and no content of 3,6-anhydrogalactose. Thus, the variation functional groups composition will determine the properties of carrageenan. Kappa carrageenan will form strong and rigid gels when combine with water. Iota carrageenan can bind water, but it will form a dry, elastic gel, especially in the presence of calcium salts. This is due to the 2-sulphate group on the outside of the iota carrageenan molecule which not allows the helices to aggregate to the same extent as kappa carrageenan, but to form additional bonds through calcium interactions. Lambda carrageenan, it is highly sulphated, thus less likely to form a gel structure. The ester sulphate distribution of lambda carrageenan is randomly distributed on the molecule which prevents formation of gel and promotes viscous solutions.

In the production of semi-refined and refined carrageenan, seaweed was first subjected to alkali treatment at high temperature. The temperature of 70–130 °C is required to catalyse the formation of the 3,6 anhydrous bridges of k-carrageenan for the gel formation ability. The temperature inside the mass of the carrageenan should not exceeded 130 °C in the modification of alkali in order to prevent any degradation of the kappa carrageenan (Mishra et al. 2008).

The main purpose of alkali treatment is to swell the seaweed. The alkali is selected based on different salt type of carrageenan produced by process. It is crucial as the selection of alkali will affect the properties of extract, such as thickening and gel formation of carrageenan (Bono et al. 2012). In addition, kappa seaweed contains 6-sulphate ester groups and some of the 3,6-anhydro-D-galactose contains 2-sulphate ester groups. As 6-sulphate ester groups will reduce the gelling power, alkali treatment can be applied to transeliminate 6-sulphate groups, which results in formation of 3,6-anhydro-D-galactose. It helps to impart a higher degree of regularity to the molecule. Moreover, alkali treatment will promote an internal arrangement which modifies the polysaccharides backbone. For instance, kappa carrageenan can form firm and brittle gels as the bridge formed from rearrangement allows its adjacent chains to form helical structures after neutralization of exposed sulphate groups. However, iota carrageenan forms weak and elastic gels after alkali modification. Thus, it is crucial to know which type of raw seaweed is used before any process is started. As *Kappaphycus alvarezii* is used as raw material in this research, potassium hydroxide (KOH) is chosen as the alkali in alkali treatment (Rideout et al. 1998).

Therefore, a study on the influence of KOH concentration on the functional group composition during alkali treatment was conducted. Four functional groups, namely 3,6-anhydrogalactose (3,6-AG), ester sulphate, 3,6-anhydrogalactose-2-sulphate (3,6-AG-2-sulphate) and galactose-4-sulphate of carrageenan, were investigated. For the determination of functional groups compositions, Fourier transform infrared spectroscopy (FTIR) was employed (Pereira et al. 2003).

Methodology

The experimental work was divided into four sections; preparation of dry seaweed, alkali treatment of seaweed, production of semi-refined carrageenan (SRC) powder and determination of functional groups composition. Analytical grade of chemicals such as standard kappa carrageenan (Sigma I, C-1013), KOH and KBr were used.

Preparation of Dry Seaweed

Fresh seaweed was obtained from local farm and sun-dried immediately after collected to protect from any degradation by microorganism. Prior the usage of sun-dried seaweed, it was washed with distilled water in order to eliminate all epiphytes and encrusting material. Then, samples were chopped for about 10–20 mm in size. Seaweed sample was then oven-dried for 24 h maintaining temperature range 60–70 °C for avoiding the decomposition of volatile components.

Alkali Treatment of Seaweed

Alkaline treatment was conducted by adding the seaweed sample into KOH solution and cooked for 30 min at temperature of 80 °C. The solid loading to liquid solution was 25 g of seaweed sample to 400 ml of KOH solution. For the influence of KOH concentration variation, the KOH solution was prepared as 0.01–0.1 M. The alkali-treated seaweed sample was then washed and dried for the SRC powder preparation.

Production of SRC Powder

The powder of carrageenan was prepared by spray dryer of carrageenan solution. The solution was prepared by dissolving of 25 g SRC particle into 400 ml distilled water at 70 °C. The carrageenan solution was then fed into laboratory-scale spray drier at flow rate of 8 ml/h. The hot air flow rate and the temperature were set to 40 m^3/h, and 140 °C. The carrageenan powder produced was kept in airtight plastic container and stored in a desiccators containing silica gel before it was analysed.

Determination of Functional Group Composition

A quantity of carrageenan powder was mixed with the KBr salt as suggested in the standard preparation of thin pellet of FTIR analysis. All IR spectra were recorded at room temperature (26 ± 1 °C), and the infrared spectra of all formulations were recorded with FTIR. A commercial standard, Type I, predominantly kappa carrageenan (SIGMA) was used as comparison. In this analysis, 32 scans were signal-averaged for a single spectrum. The resolution is 4 cm^{-1} and the data spacing is 1.928 cm^{-1}. Each spectrum is displayed in terms of absorbance as calculated from the reflectance–absorbance spectrum using the OMNIC software. The recorded value of absorbance was translated into the percentage composition of functional groups.

Results and Discussion

FTIR Spectrum of Carrageenan and Confirmation of Functional Group

The FTIR spectrum of SRC produced with KOH concentration compared to standard carrageenan sample shown in Fig. 1. The presence of 3,6-AG, ester sulphate, 3,6-AG-2-sulphate and galactose-4-sulphate were shown by the absorption peaks at wave length number 930 cm^{-1}, 1,260 cm^{-1}, 803 cm^{-1}, 846 cm^{-1}, respectively. Figure 1 shows that the SRC produced in this work is similar to commercial standard.

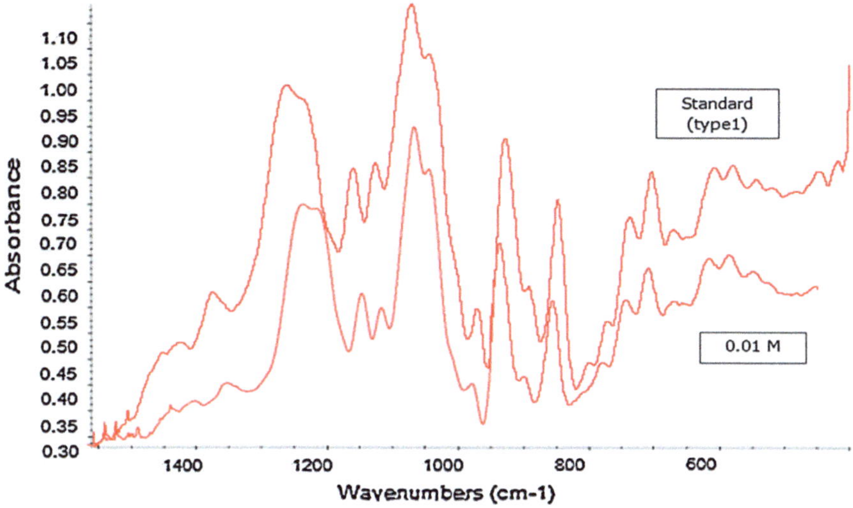

Fig. 1 FTIR spectrum of commercial standard carrageenan and semi-refined carrageenan produced at KOH concentration of 0.01 M

Complete FTIR spectrum of SRC produced at various KOH concentrations shown in Fig. 2. The spectrum shows the similarity to the commercial standard of carrageenan and proves the existence of functional groups. The magnitude of recorded absorbance is obviously varies with the variation of KOH concentration used in the alkali treatment.

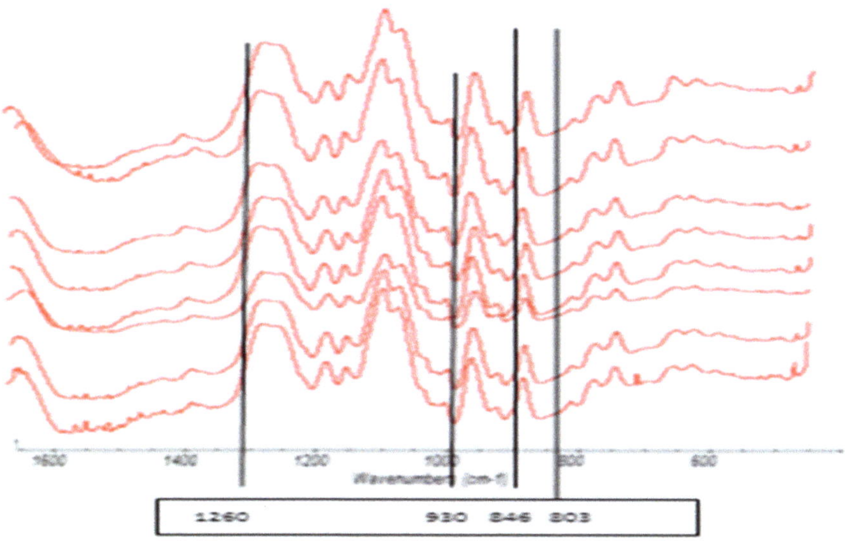

Fig. 2 FTIR spectrum of semi-refined carrageenan produced at various KOH concentrations

Based on the FTIR analysis result that shown in Fig. 2, it can be said that the absorbance band of 3,6-anhydrogalactose contents is higher in the powder sample which has been extracted and produced in the alkaline treatment of high KOH concentrations. High contents of 3,6-anhydrogalactose in the powder sample increase the strength of gel of carrageenan while alkaline treatment will reduce the sulphate contents. In summary, it can be said that increase in the gel strength will increase the 3,6-anhydrogalactose content while the sulphate content will be decreased (Buriyo and Kivaisi 2003). Therefore, it shows that the strength of gel is solely dependent on the content of 3,6-anhydrogalactose.

Analysis of Functional Groups Composition

The compositions of functional group were obtained from the magnitude of absorbance and tabulated in the Table 1.

Table 1 shows the two maximum contents of 3,6-anhydrogalactose occur at the concentration of KOH of 0.07 M and also at 0.1 M which was 15 % and 13.5 %, respectively. The content of 3,6-anhydrogalactose in the reference which was from the commercial carrageenan (Sigma Type I) was 13 %. This also shows that the content of this functional group was still in the range of standard type even though some of the sample has low content of 3,6-AG.

As mentioned previously, sulphate content reduced in high concentration of KOH. This is proved based on the result from FTIR that the absorbance for ester sulphate was decreased in order the K^+ ion concentration was increased. Generally, the degree of sulphation will determine the gel strength of extract of carrageenan where an increase in sulphation will results in decreasing of carrageenan's gel strength. It means that higher level of ester sulphate has a lower solubility temperature and also lower gel strength of kappa carrageenan. This is due to

Table 1 The variation of functional groups with various type of SRC produced at various KOH concentrations

KOH concentration (M)	Functional group			
	3,6-AG (%)	Ester sulphate (%)	3,6-AG-2-sulphate (%)	D-galactose-4-sulphate (%)
0.01	5.16	18.00	37.56	18.05
0.02	7.55	6.30	30.12	14.11
0.03	10.00	3.00	26.77	9.00
0.04	9.00	6.40	21.45	9.00
0.05	9.79	5.01	15.86	7.00
0.07	15.00	4.50	12.39	4.00
0.08	12.51	2.16	15.80	5.248
0.1	13.5	4.50	8.00	9.30
Type I standard	13.00	10.47	34.67	16.60

negatively charged gels that given by ester sulphate with potassium. It also believed that potassium ion, K^+, will neutralize the sulphate group in the kappa carrageenan. Kinks in the helical structure that responsible for formation of gel were caused by the sulphate units at point in the polysaccharide chains comprising the gel which result in the lower gel strength of carrageenan (Rees 1972). The maximum content of ester sulphate was found in concentration of KOH of 0.01 M, and the minimum content was in the KOH concentration of 0.08 M. For the reference or standard type, the ester sulphate content obtained was at 10.47 %. Another functional group that contains in the carrageenan extract is 3,6-anhydrogalactose-2-sulphate. The sulphate content was indicated in all the spectra by the very weak and poorly defined band in the region of 800–850 cm^{-1} where the primary and secondary equatorial and axial sulphate group absorbs but the content is relatively low. Based on the result shown in Table 1, the highest value of 3,6-AG-2-sulphate content was found in the sample extracted with KOH concentration of 0.01 M and the least was found in the kappa carrageenan extract with 0.1 M of KOH. The standard type of commercial carrageenan used has 34.67 % of 3,6-AG-2-sulphate content in it. From the table, the results that have been obtained still include in the range of the standard type. However, the values were getting lower as the concentration of KOH increase. As mentioned previously, the gel strength of carrageenan was decreased in the presence of high level of sulphate content. Furthermore, the gel strength of carrageenan was more sensitive towards the presence of 2-sulphate. Similar to the ester sulphate, the sulphation neutralize the sulphate group. Since the gel strength increases as the KOH concentration increases, the sulphation will be decreased which result in the lower value of 3,6-AG-2-sulphate content.

The KOH concentration has influenced the content of sulphate group in the carrageenan extract; thus, it also has effect on the content of galactose-4-sulphate as it has in the ester and 3,6-AG-2-sulphate. It appears that the physical and chemical properties of the carrageenan's extract were influenced by the KOH concentration (K^+ ion). In the previous section, it already stated that the content of 3,6-AG helps in increasing the gel strength. Different for D-galactose-4-sulphate, it has poor gelling ability as the content of it increases in the sample powder of carrageenan. Furthermore, after the alkaline treatment, the yield of carrageenan was reduced. It appears to be related to degradation of polysaccharide which was the galactose-4-sulphate itself during the alkaline treatment, also from the carrageenan loss by diffusion during the processing of SRC powder (Montano et al. 1999). That is why the content of this functional group was decreased in the high concentration of KOH. Based on the result shown in Table 1, the higher value of galactose-4-sulphate appear in the KOH concentration of 0.01 M and lower value was found in the sample extracted at 0.07 M of KOH. The content of this functional group was supposed to be decrease as the concentration of KOH increase, but as it can be seen in the results shown in Table 1, the content has increased in the concentration of KOH ranging from 0.07 to 0.1 M. This is due to stabilization and neutralization of galactose-4-sulphate by alkali in order to prevent the lowering of gelling ability of kappa carrageenan extract. Generally, it can be said that the alkaline treatment with KOH solution helps to increase the content of 3,6-AG while reducing the sulphate content in the extract of carrageenan. Chemical analysis shows an inverse relationship between sulphate and 3,6-anhydrogalactose content (Praiboon et al. 2006). The 3,6-AG is responsible in increasing the gel strength of carrageenan extract where the higher the gel strength, the more content of 3,6-AG in the sample. The presence of sulphate content in the extract of kappa carrageenan is appeared to have responsibility on the quality of the carrageenan where the more sulphated the carrageenan, the greater the tendency of the carrageenan's gel to turn into solution (Tsuneaki and Suzuki 1975). As shown in the result above, the alkali-treated carrageenan had lower sulphate content (Praiboon et al. 2006). Increasing in concentration of KOH actually influenced the 3,6-AG, ester sulphate, 3,6-AG-2-sulphate and also the D-galactose-4-sulphate of kappa carrageenan extract. Based on the results obtained, we can know that *Kappa Alvarezii* have a high content of ionizable groups (carboxyl groups from mannuronic and guluronic acid) on the cell wall of polysaccharides where it makes it very liable to the influence of the concentration of potassium ion.

Conclusion

The physical and chemical properties of the SRC produced by alkaline treatment are dependent on the functional group compositions. The composition of functional groups in the carrageenan was strongly influenced by the strength of KOH used during the production process. The influence of KOH concentration can be manipulated to control the composition of functional groups. FTIR can be considered as the suitable equipment in analysing the functional group of SRC extract. Increasing of KOH concentration increased the 3,6-anhydrogalactose functional group concentration. Contrary to this, all other functional groups ester sulphate, 3,6-anhydrogalactose-2-sulphate and galactose-4-sulphate concentration decreased with increase in KOH concentration.

Acknowledgments Authors would like to acknowledge the financial support from the Universiti Malaysia Sabah (UMS) for the UMS Fasa 1/2011 project (SLB0017-TK-1/2012).

References

Besada, V., Andrade, J. M., Schultze, F., & Gonzalez, J. J. (2009). Heavy metals in edible seaweeds commercialized for human consumption. *Journal of Marine Systems, 75*, 305–313.

Bono, A., Anisuzzaman, S. M., Ding, O.W. (2012). Effect of process conditions on the gel viscosity and gel strength of Semi-Refined Carrageenan (SRC) produced from seaweed (*Kappaphycus Alvarezii*). *Manuscript accepted in Journal of King Saud University: Engineering Sciences.*

Buriyo, A. S., & Kivaisi, A. K. (2003). Standing Stock, Agar yield and properties of Gracilaria salicomia harvested along the Tanzanian coast, Western Indian Ocean. *Journal of Marine Science, 2*, 171–178.

Campo, L. C., Kawano, D. F., Dilson, B. S. J., & Carvalho, I. (2009). Carrageenans: Biological properties, chemical modifications and structural analysis—a review. *Carbohydrate Polymers, 77*, 167–180.

Glickman, M. (1987). Utilization of seaweed hydrocolloids in the food industry. *Hydrobiology*, 151/152, 31–47.

Mishra, D. K., Tripathy, J., & Behari, K. (2008). Synthesis of graft copolymer (k-carrageenan-g-N, N-dimethylacrylamide) and studies of metal ion uptake, swelling capacity and flocculation properties. *Carbohydrate Polymers, 71*, 524–534.

Montano, N. E., Villanueva, R. D., & Romero, J. B. (1999). Chemical characteristics and gelling properties of agar from two Philippine Gracilaria spp. (Gracilaria, Rhodophyta). *Journal of Applied Phycol, 11*, 27–34.

Pereira, L., Sousa, A., Coelho, H., Amado, A. M., Paulo, J. A., & Ribeiro, C. (2003). Use of FTIR, FT-Raman and C-NMR spectroscopy for identification of some seaweed phycocolloids. *Biomolecular Engineering, 20*, 185–202.

Phang, S. M. (2006). Seaweed resources in Malaysia: Current status and future prospects. *Aquatic Ecosystem Health and Management, 9*, 185–202.

Praiboon, J., Chirapart, A., Akakabe, Y., Bhumibhamon, O., & Kajiwara, T. (2006). Physical and chemical characterization of agar polysaccharides extracted from the Thai and Japanese species of Gracilaria. *Science Asia, 32*(Supplement 1), 11–17.

Rees, D. A. (1972). Shapely polysaccharides: the eighth Colworth medal lecture. *Biochem Journal, 126*, 257–273.

Rideout, S. C., Hill, R., Bernabe, M. G., & Markham, (1998). Method for extracting semi refined carrageenan from seaweed. *United States patent, 5*(801), 240.

Tsuneaki, F., & Suzuki, T. (1975). Preparation and properties of agar sulphates. *Agriculture Biolological Chemistry, 39*, 119–126.

Effect of 3-Mercaptopropionic Acid on Polymerization of Thermo-Responsive Poly(N-Isopropylacrylamide)

D. Krishnaiah, S. M. Anisuzzaman, S. F. Shi and A. Bono

Abstract Poly(N-isopropylacrylamide) (PNIPA) is a thermally sensitive polymeric material. The temperature-sensitive nature of PNIPA makes it an attractive candidate for controlled drug delivery devices. A series of temperature-responsive PNIPA were produced by free radical polymerization using various amount of 3-mercaptopropionic acid (MPA) as chain transfer agent. The effect of chain transfer agent on the lower critical solution temperature (LCST) and chemical structure of PNIPA was characterized by using UV–vis spectrophotometer and Fourier transform infrared spectroscopy (FTIR), respectively. It was found that, with increasing MPA content, the typical band of the PNIPA in respective FTIR spectrum was shorter or getting narrower due to the decrease in the strength of respective bond in the PNIPA structure. Furthermore, the increase in MPA volume in the polymerization increased the LCST of PNIPA. The various amounts of 3-MPA ratios to monomer preparation have shown that hydrophobic modification of carboxylated PNIPA.

Keywords Poly N-isopropylacrylamide · 3-Mercaptopropionic acid · Fourier transform infrared spectometer · UV–vis spectrophotometer · Lowe critical solution temperature

Introduction

Water-soluble polymers and hydrogels are responsive or sensitive to external stimuli like temperature, pH, ionic strength, solutes or solvent composition. These smart materials represent an important and active research area due to many

D. Krishnaiah (✉) · S. M. Anisuzzaman · S. F. Shi · A. Bono
Chemical Engineering Programme, School of Engineering and Information Technology, University Malaysia Sabah, 88400 Kota Kinabalu, Sabah, Malaysia
e-mail: krishna@ums.edu.my

R. Pogaku et al. (eds.), *Developments in Sustainable Chemical and Bioprocess Technology*, DOI: 10.1007/978-1-4614-6208-8_43,

possible applications in diverse technological fields (Pagonis and Bokias 2007). Due to the specific intelligent nature, "smart" polymers have attracted increasing interest and considerable research attention in recent years. Further researches into these polymer synthesis, behavior, and application show that these polymers have great potential to become the "next generation" of separation media for cost-effective and environment-friendly extraction and purification of high value biomolecules from agricultural food and other raw materials (Maharjan et al. 2008). Poly(N-isopropylacrylamide) (PNIPA) is a typical thermo-sensitive polymeric material which demonstrates a transition temperature (T_{tr}) or lower critical solution temperature (LCST) at ~32 °C in aqueous solution (Zhang et al. 2002; Zhang and Zhuo 2000a). PNIPA has been studied extensively for the past 20 years due to its useful applications in many fields such as artificial muscle, on–off switches, immobilization of enzymes, thermo-sensitive absorbents, drug release, and others applications (Zhang and Zhuo 2000b). In the process to synthesis the PNIPA, free radical polymerization is a more preferable method to be used. In the previous years, researches show that free radical polymerization method done on PNIPA can lead to the formation of stable nanospheres instead of linear chains if the solutions are incubated at temperatures well above the LCST of PNIPA (Gao and Frisken 2003).

Telomerization is free radical polymerization in the presence of a chain transfer agent (telogens) which is to produce low average molecular weight products called telomere (Chung 2000). Using chain transfer agent which aids the shift of the radical from the growing polymer chain to another molecular species, in turn can reinitiate chain growth, is a convenient way to control the molar mass and the polydispersity of polymers produced by free radical polymerization. Many compounds can be used as chain transfer agents (telogens) in free radical polymerization. Materials such as halogens and mercaptans are the most popular among all. Besides having the capacity to react as chain transfer agents, these chain transfer agents also supply a well-defined end group to the final polymer (telomer) (Costioli et al. 2005). The chain transfer polymerization of N-isopropylacrylamide (NIPAAm) is increasingly used for the synthesis of defined telomers bearing a particular functional (ester, carboxylic acid, amino) end group that can be conveniently employed for the construction of bioconjugates. For such applications, the ability to control the size and heterogeneity of the polymer is the most important aspect besides the presence of a reactive end group which is known as degree of polymerization. Therefore, the molecular mass of the telomer can be controlled via the initial ratio of the telogen and monomer concentration (Costioli et al. 2005). Furthermore, the molecular average weight of telomer is easily affected by the variation of the concentration of chain transfer agent.

In particular, researches on the molecular weight changing behavior in the different concentration of chain transfer agent applied in telomerization have been studied worldwide (Chung et al. 1998). However, a detailed study of composition and structure changing behavior of PNIPA in effect of various concentration of chain transfer agent is still missing.

In this research, free radical polymerization method is used and 3-mercaptopropionic acid (MPA) is chosen as chain transfer agent. MPA is an isolated polymer that contains a single specific functional end group and is often used as a chain transfer agent due to its high transfer reactivity. The AIBN is used as initiator to transform the thiol (R–S–H) into a radical. By applying various amount of MPA, the effect of the MPA on the characteristic, structure, and compositions of PNIPA is investigated. The analysis is carried out by using UV–vis spectrophotometer and Fourier transform infrared spectroscopy (FTIR). FTIR is used to identify the quality or consistency and amount of components in an unknown sample. Like a fingerprint, no two unique molecular structures produce the same infrared spectrum. This special characteristic makes infrared spectroscopy very useful for several types of analysis. UV–vis spectrophotometer is used to determine the absorption or transmission of UV/VIS light (180–820 nm) by a sample. It is used to measure concentrations of absorbing materials based on developed calibration curves of the material. UV–vis spectrophotometer is routinely used in the quantitative determination of solutions of transition metal ions and highly conjugated organic compounds.

Methodology

Materials

All the analytical-grade reagents, such as 3-MPA, ethanol, diethyl ether, were purchased from Sigma–Aldrich.

Synthesis of Poly(N-isopropylacrylamide)

The synthesis of PNIPA was carried out as follows: First, 10 g of NIPAAm was dissolved in 20 ml of ethanol which acts as a solvent. Theoretically, any solvent should be suitable for polymerization as long as the monomer, and the chain transfer agent are well dissolved and, more importantly, as long as the solvent does not act itself as chain transfer agent (Costioli et al. 2005; Choi et al. 2007). The solution is then added with 0.1 g of AIBN and various concentrations (0, 12, 24, 36, 60 g/l) of MPA which acts as initiator and chain transfer agent, respectively (Table 1 shows the composition for all the samples prepared in this study).

The mixture is incubated at 60 °C for 20 h. The resulting precipitate was isolated by precipitation into 20 ml of diethyl ether and recover by rotary evaporator. After evaporation process for an hour, the final product was collected and drying in the oven at temperature 80 °C for 16 h or more to achieve complete dry up.

Table 1 Characterization of the PNIPA

Sample	NIPAAm (g)	AIBN (ml)	MPA (g/l)	Ethanol (ml)
A	10	0.1	0	20
B	10	0.1	12	20
C	10	0.1	24	20
D	10	0.1	36	20
E	10	0.1	60	20

Fourier Transform Infrared Spectroscopy

The FTIR measurements were conducted on a Perkin-Elmer Spectrum 100 Series FTIR Spectrometer at room temperature (25 °C) in the 4,000–600 cm^{-1} range.

Measurement of LCST of Poly(N-isopropylacrylamide)

About 0.2 g of PNIPA was poured into distilled water (100 ml) at room temperature. Afterward, the samples were measured by UV–vis spectrophotometer. The LCST of PNIPA could be determined from the curve of relationship between UV light absorbance and temperature.

Results and Discussion

Effect of MPA on FTIR Spectroscopy

The FTIR spectra of PNIPA with different ratio of cross-linking agent to monomer are shown in Fig. 1 which confirms the presence of the 3-MPA in the polymer formed. The typical amide I band (1,680–1,600 cm^{-1}) consisting of $C = O$ stretch of PNIPA and the amide II band (1,548–1,459 cm^{-1}), including N–H vibration were evident in every spectrum. Every spectrum showed a broadband in the range of 3,600–3,200 cm^{-1}, which belongs to N–H and O–H stretching vibration of the PNIPA. Another typical band that consists of the poly(NIPAM) spectrum is in the wavelength (2,973–2,855 cm^{-1} and 1,389–1,368 cm^{-1}) and contains CH, CH_3, and CH_2 stretches of PNIPA. Furthermore, the clear increase in the relative intensity of the characteristic bands due to carboxyl bond (1,749–1,739 cm^{-1}) indicates an increase in the 3-MPA in the final polymer.

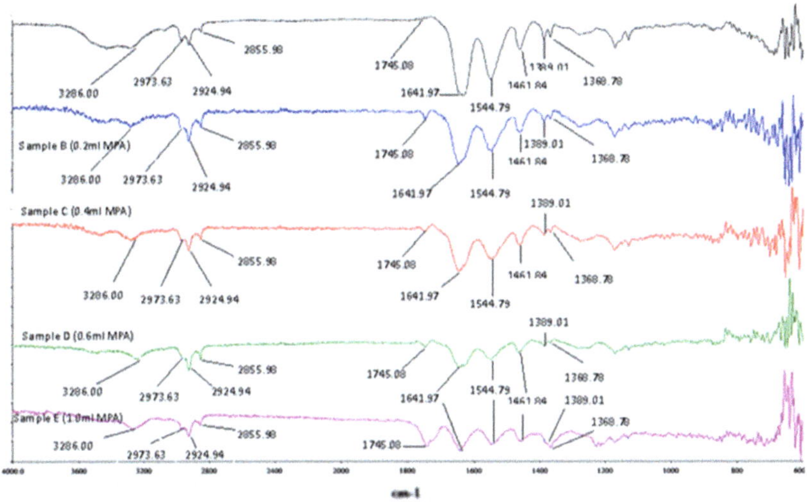

Fig. 1 FTIR spectra of various prepared poly(NIPAM) by chemical process

Effect of MPA on the LCST of Poly(N-isopropylacrylamide)

Figure 2 shows the LCST of the PNIPA with various concentration of MPA. The increase in MPA concentration increases the LCST of PNIPA. These phenomena can be explained by the structure of the NIPAM monomer. In NIPAM structure, there is a hydrophilic/hydrophobic balance of the side chain. MPA is a chemical where it contains a carboxyl group, and it is a hydrophilic agent. An addition of MPA to the end of the monomer increases the LCST of the polymer form due to the higher fraction mole of hydrophilic group than hydrophobic group in PNIPA. Terminal incorporation of hydrophilic or hydrophobic groups dramatically influences the phase transition of PNIPA because of the critical role of the polymer chain ends in initiation of the phase transition. So, the addition of MPA (hydrophilic) increases the LCST of the PNIPA.

Fig. 2 Lower critical solution temperature versus different volume of MPA

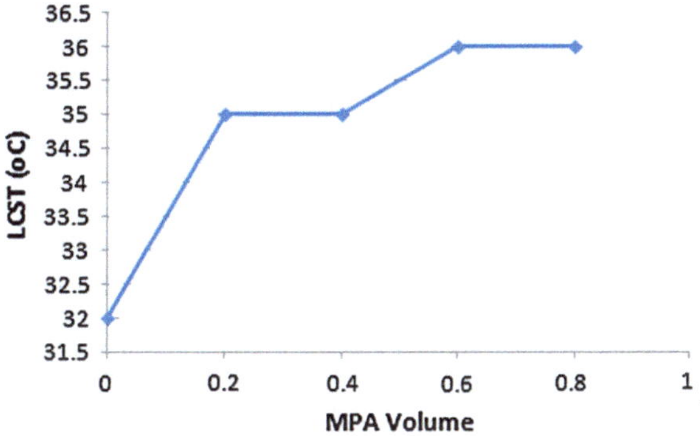

Conclusions

A thermo-sensitive polymer that can be either soluble or insoluble in water was synthesized successful by chain transfer polymerization with copolymerizing NIPAAm and 3-MPA. All the PNIPA were characterized by Fourier transform infrared spectrometer and UV–vis spectrophotometer. In the analysis of lower critical solution temperaturc, LCST of PNIPA increased with the increasing volume of the cross-linking agent (MPA). The importance of the LCST is due to the field of the polymer applied. Different field required the different value of LCST, so the factor which may increase or decrease the thermo-sensitive polymer is need to be further investigated to expand the utilities of the PNIPA in different field of usage. In the FTIR analysis, the result shows the carboxylated groups that appear in PNIPA. The typical bands of the PNIPA appear in each spectrum for each sample. This shows that the final product which matches with the PNIPA was produced successfully. The peak intensity of the carboxyl bond in the wavelength 1,749 cm^{-1} increased gradually with the increasing MPA. The other typical band becomes shorter or narrower when increasing MPA content due to the shorter polymer.

References

Choi, C., Jang, M. K., & Nah, J. W. (2007). Preparation and characterization of nanoparticles using poly(N-isopropylacrylamide)-poly(ε-caprolactone) and poly(ethylene glycol)-poly(ε-caprolactone) block copolymers with thermosensitive function. *Macromolecular Research, 15*, 623–632.

Chung, I. (2000). Monte Carlo simulation of free radical telomerization. *Polymer, 41*, 5643–5651.

Chung, J. E., Yokoyama, M., Aoyagi, T., Sakurai, Y., & Okano, T. (1998). Effect of molecular architecture of hydrophobically modified poly(N-isopropylacrylamide) on the formation of thermoresponsive core-shell micellar drug carriers. *Journals of controlled release, 53*, 119–130.

Costioli, M. D., Berdat, D., Freitag, R., Andre, X., & Muller, A. H. E. (2005). Investigation of the telomerization kinetics of N-isopropylacrylamide using 3-mercaptopropionic hydrazide as chain transfer agent. *Macromolecules, 38*, 3630–3637.

Gao, J., & Frisken, B. J. (2003). Cross-linker-free N-isopropylacrylamide gel nanospheres. *Langmuir, 19*, 5212–5216.

Maharjan, P., Woonton, B. W., Bennett, L. E., Smithers, G. W., Desilva, K., & Hearn, M. T. W. (2008). Novel chromatographic separation—The potential of smart polymers. *Innovative Food Science and Emerging Technologies., 9*, 232–242.

Pagonis, K., & Bokias, G. (2007). Temperature and solvent sensitive hydrogels based on N-isopropylacrylamide and N, N-dimethylacrylamide. *Polymer Bulletin, 58*, 289–294.

Zhang, X. Z., & Zhuo, R. X. (2000a). Preparation of fast responsive, thermally sensitive poly (N-isopropylacrylamide) gel. *European Polymer Journal, 36*, 2301–2303.

Zhang, X. Z., & Zhuo, R. X. (2000b). Synthesis of temperature-sensitive poly (N-isopropylacrylamide) hydrogel with improved surface property. *Journal of Colloid and Interface Science, 223*, 311–313.

Zhang, X. Z., Zhang, J. T., Zhuo, R. X., & Chu, C. C. (2002). Synthesis and properties of thermosensitive, crown ether incorporated poly (N-isopropylacrylamide) hydrogel. *Polymer, 43*, 4823–4827.

Preparation and Characterization of Activated Carbon Derived from Waste Rubber Tire via Chemical Activation with $ZnCl_2$: Surface Area and Morphological Studies

C. G. Joseph, G. G. Hoon, Y. L. Sharain-Liew, D. Krishnaiah and M. Massuanna

Abstract Turning waste to wealth is an important aspect in promoting green technology. In this study, activated carbon from waste rubber tire was prepared using chemical activation ($ZnCl_2$) by way of a two-stage activation in self-generated atmosphere method. The preparation parameters examined and compared in this study were activation time, activation temperature, and impregnation ratio. The adsorption of the target pollutant, 2,4-dichlorophenol (DCP), was used to evaluate the efficiency of adsorption capacity of the prepared activated carbon. Results from the experimental work showed that the activated carbon prepared satisfied the Freundlich isotherm and complied with the pseudo-second-order kinetics (not presented here). Other parameters studied, such as the percentage yield of activated carbon, ash, and moisture content and the morphology structure, are presented in this chapter. This study showed that waste rubber tire dehydrated with $ZnCl_2$ with an impregnation ratio of 1:1 and activated at a temperature of 500 °C for 120 min gave the best result (AC5).

Keywords Adsorption · 2,4-Dichlorophenol · Kinetics · Rubber tire

C. G. Joseph (✉) · Y. L. Sharain-Liew
Water Research Unit, School of Science and Technology,
Universiti Malaysia Sabah, 88999 Kota Kinabalu, Sabah, Malaysia
e-mail: collin@ums.edu.my

G. G. Hoon
Faculty of Applied Sciences, Universiti Teknologi MARA Sabah,
88997 Kota Kinabalu, Sabah, Malaysia

D. Krishnaiah · M. Massuanna
Chemical Engineering Programme. School of Engineering and Information Technology,
Universiti Malaysia Sabah, 88999 Kota Kinabalu, Sabah, Malaysia

R. Pogaku et al. (eds.), *Developments in Sustainable Chemical and Bioprocess Technology*, DOI: 10.1007/978-1-4614-6208-8_44,

Introduction

Motor vehicles have become an indispensable and the most popular means of transportation in the world today. In Malaysia, the number of motor vehicles on the road recorded upward trend for the past few years. The Minister of Transport, Datuk Seri Kong Cho Ha said that the total number of registered motor vehicles increased around 11.74 % from the previous year (The Star 2011). Statistics from the Road and Transport Department of the Transport Ministry showed that the total number of registered motor vehicles up to 30 June 2011, was 20,806,554 and approximately 1.4 million rubber tires are dumped every month (Ministry of Transportation 2011). Tires are designed to be extremely resistant to physical, chemical, and biological degradation, making them difficult to be recycled with existing technology. Dumping waste tires at disposal sites poses a high fire hazard risk because they are difficult to be extinguished, besides having a detrimental effect on the atmosphere, soil, and groundwater. In addition, disposal sites for waste rubber tires have become breeding grounds for mosquitoes and rats and have increasingly taken up large amounts of valuable landfill space (USM 2011). A few alternatives for recycling waste tire, such as retreading, reclaiming, incineration, and grinding, have been used to give waste tires a new lease of life (de Marco Rodriguez et al. 2001). Processes such as pyrolysis, gasification, and combustion have been extensively studied for energy recovery and for the production of high commercial value materials such as activated carbon (Betancur et al. 2009; Wojtowizz and Serio 1996; Bajus and Olahová 2011). Oil refineries, coal gasification sites, and petrochemical units generate large amounts of industrial contaminants such as phenols. In addition, phenolic compounds and their derivatives are widely used as intermediate in the industrial synthesis of plastic, pesticides, and dyes (Dąbrowski et al. 2005). Since phenolic compounds are highly toxic, carcinogenic and have poor biodegradability, they are harmful to human beings. Therefore, it is imperative that these should be removed from water sources; 2,4-dichlorophenol (DCP) is a chlorinated aromatic compound with a chlorine atom each at positions 2 and 4 of the phenol ring. It is a common pollutant and a primary reagent in the synthesis of a variety of highly chlorinated phenols and pesticides such as 2,4-dichlorophenoxyacetic acid (Shaarani and Hameed 2010). DCP is a colorless, crystalline solid, slightly soluble in water at neutral pH but very soluble in alcohol, ether, and benzene and has weak acidic properties with pKa of 7.90 (Shaarani and Hameed 2010; Dąbrowski et al. 2005). Removal of phenolic compounds through adsorption is better compared to other techniques in terms of cost and equipment handling. Hence, it is the most frequently used technique (Lu and Sorial 2009; Shaarani and Hameed 2010; Dąbrowski et al. 2005). Adsorption is a process involving the contact of a free aqueous phase with a rigid particulate phase, which has the propensity to selectively remove or store one or more solutes present in the solution. Activated carbon is the most widely used adsorbent for the removal of pollutants from wastewater. This is due to its extended surface area, microporous structure, high adsorption capacity, and high degree of surface reactivity

(Shaarani and Hameed 2010). Many studies (Teng et al. 2000; Nunes et al. 2011; Gupta et al. 2011) have reported the preparation of activated carbon from waste tires using a chemical activation method. Activated carbon prepared from rubber tires has been used widely in adsorption processes such as in the removal of pesticides from wastewater (Gupta et al. 2011), removal of methylene blue (Nunes et al. 2011), adsorption of phenol (Amri et al. 2009), wastewater purification, and gas phase application (Mui et al. 2004). It was reported that a variable such as impregnation ratio effect can cause degradation of the cellulosic material which, in turn, influences the carbonization and activation processes (Olivares-Marín et al. 2006). In this chapter, activated carbon was prepared using a simple two-stage activation method in self-generated method using a muffle furnace (Srinivasakannan and Abu Bakar 2004). The parameters studied were the effects of the activation temperature, activation time, and the ratio of the dry weight of zinc chloride to the dry weight of scrap rubber (impregnation ratio) on the effectiveness of activated carbon prepared.

Methodology

The inner tube of the car tire was used as the precursor in the preparation of activated carbon. This tube was shredded into small particles between 3 and 5 mm in size washed and dried. The shredded tires were placed in a 250 ml conical flask containing $ZnCl_2$ solution (100 ml) in varying concentrations. The impregnation ratios of $ZnCl_2$ solution to the shredded tire were 1:1 and 3:1. The mixed samples were placed in a water bath shaker for a week at constant speed and temperature. After a week, the samples were dried in an oven for 24 h at 110 °C. The dried samples were weighed before semi-carbonizing the sample in the furnace. After carbonizing the samples, the semi-charred samples are activated in the muffle furnace at different activation time and temperature as shown in Table 1. Then, the samples were washed using 0.1 M HCl and rinsed with distilled water to remove dirt, inorganic particles (and oxides), and ash from the samples. The samples were dried in an oven at 110 °C overnight. After that, ash, moisture, and pH analyses

Table 1 Activation temperature

Label	Ratio ($ZnCl_2$:tire)	Semi-carbonization		Activation	
		Temperature (°C)	Time (minutes)	Temperature (°C)	Times (minutes)
AC1	1:1	200	15	500	60
AC2	3:1	200	15	500	60
AC3	1:1	200	15	700	60
AC4	3:1	200	15	700	60
AC5	1:1	200	15	500	120
AC6	3:1	200	15	500	120

were conducted. The product yield and burn-off were determined according to Eqs. 1 and 2:

$$\text{The percentage of yield} = \frac{Y}{W} \times 100\,\% \tag{1}$$

$$\text{Thepercentageofburn} - \text{off} = \frac{W - Y}{W} \times 100\,\% \tag{2}$$

where W is the initial dry weight of the precursor in gram (g) after the impregnation process, and Y is the dry weight of the final product in gram (g).

Surface Area

Surface area is one of the key factors that affects the adsorptive properties of porous materials. In order to determine the surface area of activated carbon, the Brunauer–Emmett–Teller (BET) model is commonly used and is the standard recommended by IUPAC (Mui et al. 2004). Porosity studies are also very important as pores of different sizes determine the overall capacity of the adsorbent in the adsorption process. The structure of activated carbon is composed of pores classified into three groups, namely micropores, mesopores, and macropores (Yalçın and Sevinç 2000). The types of pores present in activated carbon are influenced by the treatment method used to prepare the activated carbon. It should be noted that chemical treatment of rubber tire before activation results in a change of porosity (Mui et al. 2004). BET and surface area analyses were conducted at SIRIM's Advanced Materials Research Centre (AMREC) in Kulim, Kedah.

Scanning Electron Morphology

The scanning electron microscopy (SEM) (LEO 1455 VPSEM) was used to study the surface area of the prepared activated carbon. The SEM was of the variable pressure type. The SEM is a microscope that uses electrons rather than light to form an image. With a large depth of field, a large amount of the sample can be in focus at any one time. It also produces images of high resolution, thus enabling closely spaced features to be examined at high magnification. Preparation of the samples is relatively easy since most SEMs only require samples to be conductive. Therefore, the samples were gold coated after they were mounted onto stubs using carbon-laced double-sided tape. Coating prevents the accumulation of static electric charges on the specimen during electron irradiation and increases signal and surface resolutions.

Results and Discussion

Effect of Impregnation Ratio on Percentage Yield

Table 2 shows the percentage yield of the prepared activated carbon. Generally, the activated carbon prepared at an impregnation ratio of 1:1 (AC1) produced a higher yield (44.88 %) compared to the ones prepared at an impregnation ratio of 3:1. A similar trend was reported by previous researchers; that is, an increase in the value of impregnation ratio resulted in a corresponding increase in weight loss (Teng et al. 2000). The pattern of weight loss by increasing the impregnation ratio from 1 to 3 was due to the large evolution of volatiles during the activation process.

Effect of Activation Temperature on Percentage Yield

The yield of the prepared activated carbon was also influenced by the activation temperature. Table 2 shows that an activation temperature of 500 °C produced a higher yield of activated carbon (AC1 and AC2) compared to the ones produced at an activation temperature of 700 °C (AC3 and AC4). This finding is corroborated by studies by Nunes and co-researchers (Nunes et al. 2011). Therefore, higher activation temperatures translate to higher burn-off of the precursor material.

Effect of Activation Time on Percentage Yield

Different activation time during the preparation of activated carbon from waste tire had a great effect on the product yield of the carbon. Comparisons were made between the samples prepared at an activation temperature of 500 °C with the same impregnation ratio but different activation time. Activation time of 60 min

Table 2 Physiochemical characteristics of the activated carbon prepared

Sample	Temperature (°C)	Activation time (min)	Impregnation ratio ($ZnCl_2$:tire)	Percentage yield (%)	Percentage burned off (%)	Ash content (%)	Moisture content (%)
AC1	500	60	1:1	44.88	55.12	16	1.44
AC2	500	60	3:1	41.94	58.06	17	0.15
AC3	700	60	1:1	21.42	78.58	14	1.07
AC4	700	60	3:1	24.78	75.22	20	1.58
AC5	500	120	1:1	33.10	66.90	19	0.16
AC6	500	120	3:1	25.82	74.18	18	0.66

produced a higher percentage yield of activated carbon, AC1 (44.84 %) and AC2 (41.94 %), while an activation time of 120 min produced a lower percentage yield, AC5 (33.1 %) and AC6 (25.82 %) as shown in Table 2. This indicates that the weight loss of activated carbon increased in proportion to the activation time. Ahmadpour and Do (1996) found that increasing the activation time resulted in a reduced product yield but it increased other properties of the activated carbon such as BET surface area, pore size, and adsorption capacity. Hence, this finding is corroborated by previous studies of Ahmadpour and Do (1996).

Ash Content

Generally, ash is the inorganic content of activated carbon or the amount of impure substance that is left as residue after the complete combustion of carbon. Ash content for the prepared activated carbon is between 14 % and 20 % as shown in Table 2. AC3 prepared at an activation temperature 700 °C had the lowest ash content (14 %), while AC4 also prepared at the same activation temperature had the highest ash content (20 %). No clear trends were observed with regard to how parameters such as impregnation ratio, activation time, and temperature affect the ash content of the prepared activated carbon.

Moisture Content

An important factor that affects the adsorption capacity of activated carbon is its moisture content. By normal convention, a high moisture content of more than 3 % is considered not good for normal applications (Zhou et al. 2001). The moisture content of the activated carbon prepared by us was between 0.15 % and 1.5 %, which is within acceptable limits. No clear trends were observed with regard to how parameters such as the impregnation ratio, activation time, and temperature, affect the moisture content of the prepared activated carbon.

Morphological Study

SEM allows for visualizing samples with excellent spatial resolution and with a large depth of field. This technique reveals detailed graphical information about the morphological structure of the sample that is not apparent when using optical microscopy due to its limited resolution. AC5 was chosen as it gave the best adsorption capacity of 2,4-dichlorophenol in aqueous medium (not presented in this chapter). AC5 was able to remove more than 20 ppm of 2,4-dichlorophenol within 50 min from 200 ml solution. It also satisfied the Freundlich isotherm and complied with the pseudo-second-order kinetics. The SEM micrograph of AC5 shows that the pores are not uniform or evenly dispersed throughout the carbon matrix. The SEM micrograph of AC4 shows that the surface structure of AC4 (prepared at 700 °C with an impregnation ratio of 3:1) is quite smooth and spherical in shape compared to the rugged surface of AC5 with an impregnation ratio of 1:1. This might be influenced by the ratio of activating agent to tires and the activating temperature of the prepared sample. This structure of activated carbon is almost similar to that described by a previous study (Nunes et al. 2011).

Surface Area Analysis

Figure 1 shows the nitrogen adsorption–desorption isotherms of the activated carbon, AC5. The surface area was about 15.42 m^2/g, and the total pore volume was about 739.8 cc/g, while the average pore diameter was about 191.9 Å. The low surface area (which is quite low for activated carbon) is quite normal for activated carbon prepared from waste tires. Other researchers experimenting with pyrolysis of waste tires also obtained similar low surface area in their prepared activated carbon (Cunliffe and William 1999). Active site blocking by the dehydrating agent could also contribute to the low surface area of the carbon. The large average pore diameter and pore volume were responsible for the adsorption process. As AC5 satisfied the Freundlich isotherm, this indicates that this adsorbent was more adept in adsorption at lower concentrations than at higher concentrations. The theoretical explanation for this is that the adsorbed target pollutants formed a non-ideal solid solution on the carbon matrix and the heterogeneity of the active sites on the carbon matrix contributed to the binding of the target pollutant

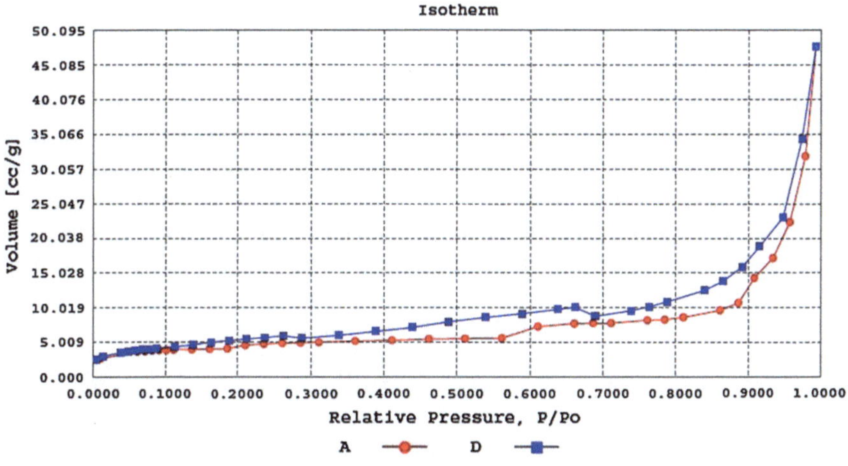

Fig. 1 Nitrogen adsorption–desorption isotherms of AC5

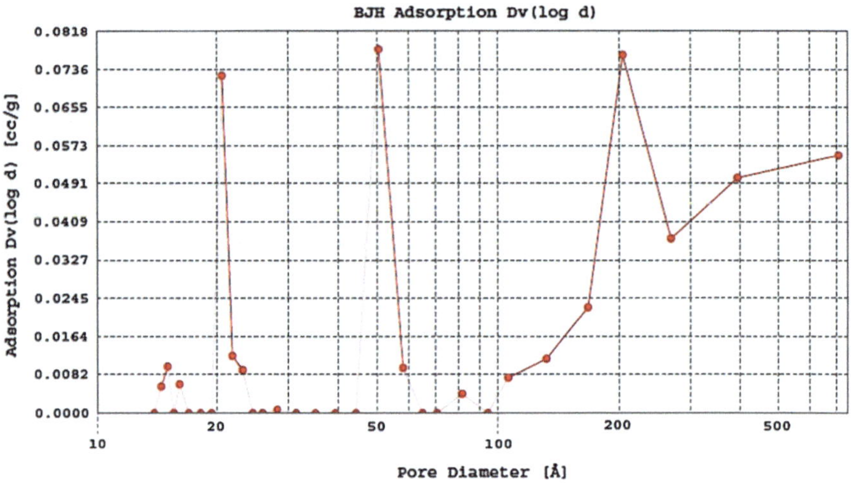

Fig. 2 BJH adsorption pore size distribution graph of AC5

to the surface of the carbon matrix. Therefore, it can be assumed that the pore diameter and pore volume contributed greatly to the adsorption process. The type III isotherm shown in Fig. 1 is indicative of a multilayer adsorption. This finding fits in very well with the compliance of the liquid (target pollutant) adsorption to the Freundlich isotherm and also with the large pore volume and diameter data. The BJH adsorption pore size distribution graph shown in Fig. 2 is indicative that the majority of the pores are mesoporous (>50 Å) or larger. This confirms that the multilayer adsorption was the primary form of adsorption mode for AC5.

Conclusion

Activated carbon was prepared from waste rubber tires using chemical activation ($ZnCl_2$) by way of a two-stage activation in self-generated atmosphere method. The properties of the prepared activated carbon varied with impregnation ratio, activation time, and temperature. These parameters influenced the percentage yield of the activated carbon produced. The weight loss of the activated carbon increased in proportion to the increase in activation temperature, time, and impregnation ratio. SEM morphological analysis showed that changing the preparation parameters produced different carbon surface structures. This finding allows activated carbon to be tailor-made for specific tasks by altering reaction parameters. BET surface area and BJH adsorption pore size distribution graph showed that although the surface area was low, multilayer adsorption by a mesoporous activated carbon can still be effective and efficient in the removal of chlorophenol in wastewater as the strong cohesion force between the adsorbate molecules are formed when water is adsorbed onto hydrophobic activated carbon.

Acknowledgments This research was supported by a Fundamental Research Grant Scheme awarded by the Ministry of Higher Education to Universiti Teknologi MARA Sabah in collaboration with University Malaysia Sabah (Grant No. 600-RMI/ST/FRGS 5/3/Fst (35/2010) and is gratefully acknowledged. The authors would also like to thank **Universiti Malaysia Sabah** for the facilities used in carrying out this research work.

References

Ahmadpour, A., & Do, D. D. (1996). The preparation of active carbons from coal by chemical and physical activation. *Carbon, 34*(4), 471–479.

Amri, N., Zakaria, R., & Mohamad, Z. A. B. (2009). Adsorption of phenol using activated carbon adsorbent from waste tyres. *Pertanika Journal of Science and Technology, 17*(2), 371–380.

Bajus, M., & Olahová, N. (2011). Thermal conversion of scrap tyres. *Petroleum and Coal, 53*(2), 98–105.

Betancur, M., Martínez, J. D., & Murillo, R. (2009). Production of activated carbon by waste tire thermochemical degradation with CO_2. *Journal of Hazardous Materials, 168*(2–3), 882–887.

Cunliffe, A. M., & William, P. T. (1999). Influence of process conditions on the rate of activation of chars derived from pyrolysis of used tires. *Energy and Fuels, 13*(1), 166–175.

Dąbrowski, A., Podkościelny, P., Hubicki, Z., & Barczak, M. (2005). Adsorption of phenolic compounds by activated carbon—a critical review. *Chemosphere, 58*(8), 1049–1070.

de Marco Rodriguez, I., Laresgoiti, M. F., Cabrero, M. A., Torres, A., Chomón, M. J. & Caballero, B. (2001). Pyrolysis of scrap tyres. *Fuel Processing Technology*, *72* (1), 9–22.

Gupta, V. K., Gupta, B., Rastogi, A., Agarwal, S., & Nayak, A. (2011). Pesticides removal from waste water by activated carbon prepared from waste rubber tire. *Water Research, 45*(13), 4047–4055.

Lu, Q., & Sorial, G. A. (2009). A comparative study of multicomponent adsorption of phenolic compounds on GAC and ACFs. *Journal of Hazardous Materials*, 167 (1–3) 15, 89–96.

Ministry of Transportation. (2011) Table 1.1: *New registered motor vehicles by type and state, malaysia, second quarter 2011. Ministry of Transportation.* [http://www.mot.gov.my] (Retrieved 3 Oct 2011).

Mui, E. L. K., Ko, D. C. K., & McKay, G. (2004). Production of active carbons from waste tyres—a review. *Carbon, 42*(14), 2789–2805.

Nunes, M. R., Perez, G. M., Loguercio, L. F., Alves, E. W., Carreilo, N. L. V., Martin, J. L., et al. (2011). Active carbon preparation from treads of tire for removal in waste water. *Journal Brazil Chemical Science, 22*(11), 2027–2035.

Olivares-Marín, M., Fernández-González, C., Macías-García, A., & Gómez-Serrano, V. (2006). Preparation of activated carbon from cherry stones by chemical activation with $ZnCl_2$. *Applied Surface Science, 252*(17), 5967–5971.

Shaarani, F. W., & Hameed, B. H. (2010). Batch adsorption of 2,4-dichlorophenol onto activated carbon derived from agricultural waste. *Desalination, 255*(1–3), 159–164.

Srinivasakannan, C., & Abu Bakar, M. Z. (2004). Production of activated carbon from rubber wood sawdust. *Biomass and Bioenergy, 27*, 89–96.

Teng, H., Lin, Y., & Hsu, L. (2000). Production of activated carbons from pyrolysis of waste tires impregnated with potassium hydroxide. *Journal of the Air and Waste Management Association, 50*, 1940–1946.

The Star. (2011). *Number of vehicles registered on the rise* [http://thestar.com.my/news/story.asp?file=/2011/2/26/nation/8146407&sec=nation] (Accessed 26 Feb 2011).

University of Science Malaysia. (2011) *Penyelidikan USM Hasil Produk Penjerap Minyak Berasas Getah Pertama Dunia.* University of Science Malaysia. [http://www.usm.my/index.php/ms/arkib-berita/8254-penyelidik-usm-hasil-produk-penjerap-minyak-berasas-getah-pertama-dunia.html] (Retrieved 27 May 2011).

Wojtowizz, M. A., & Serio, M. A. (1996). Pyrolysis of scrap tires: Can it be profitable? *ChemTech*, 48–54.

Yalçin, N., & Sevinç, V. (2000). Studies of the surface area and porosity of activated carbon prepared from rice husks. *Carbon, 38*(14), 1943–1945.

Zhou, L., Li, M., Sun, Y., & Zhou, Y. (2001). Effect of moisture in microporous activated carbon on the adsorption of methane. *Carbon, 39*(5), 773–776.

The Effects of Kaolin/PESF Ratios on the Microstructures of Kaolin Hollow Tubes

R. Sarbatly and Z. Kamin

Abstract The effects of kaolin/polyethersulfone (PESf) (k/p) ratios on the microstructures of kaolin hollow tubes have been investigated. Kaolin suspension containing k/p ratios between 1 and 3.5 was used to spin kaolin hollow-tube precursors via a dry-jet wet spinning method at 0-cm air gap. The cross-sectional microstructures of the kaolin hollow-tube precursors were investigated using scanning electron microscopy (SEM). The results show that, at k/p ratios of 1.0, 1.5 and 2.5 (low viscosity) and k/p ratios of 3.0 and 3.5 (high viscosity), the finger like voids dominated the outer regions and inner regions of the cross sections of the kaolin hollow tubes, respectively.

Introduction

A ceramic, hollow-tube membrane works in a similar way as a polymeric hollow-tube membrane in applications such as gas separation, desalination, filtration and membrane reactors. However, a ceramic-based membrane, in advantage, provides higher stability than a polymer-based membrane, in high temperature and pressure as well as in corrosive environments (Clausi et al. 1999; Liu et al. 2005; Kingsbury and Li 2009).

Ceramic, hollow-tube membranes are made of materials such as Al_2O_3 and TiO_2 (Liu et al. 2001; Liu and Li 2003). Kaolin clay is alternatively used as a

R. Sarbatly (✉)
Membrane Technology Research Group, School of Engineering and Information Technology, Universiti Malaysia Sabah, Jalan UMS 88400 Kota Kinabalu, Sabah, Malaysia
e-mail: rslam@ums.edu.my

Z. Kamin
TAS Institute of Oil and Gas, School of Engineering and Information Technology, Universiti Malaysia Sabah, Jalan UMS 88400 Kota Kinabalu, Sabah, Malaysia
e-mail: milakamilia@yahoo.com

R. Pogaku et al. (eds.), *Developments in Sustainable Chemical and Bioprocess Technology*, DOI: 10.1007/978-1-4614-6208-8_45,

membrane material to make both microfiltration and ultrafiltration membranes (Bouzerara et al. 2006) and membrane supports (Mohammadi and Pak 2002, 2003; Mohammadi et al. 2004). However, current methods to fabricate kaolin-based membranes are limited to extrusion, dry pressing and roll pressing processes in which the kaolin-based membranes are bound to only exhibit symmetric microstructures (Mohammadi and Pak 2003; Mohammadi et al. 2004; Bouzerara et al. 2006).

The study of microstructures of hollow-tube membranes is important as these data are used to determine the membrane applications appropriate in either separation or reaction applications. Although studies have been conducted on the microstructures of polymeric hollow-tube membranes (Tan et al. 2001; Shen et al. 2003), such techniques are of limited use because of the large differences between polymeric and kaolin ceramic systems. Therefore, the purpose of this work was to study the effects of k/p ratios on the microstructures of kaolin hollow tube.

Methodology

Materials

Kaolin powder (Sigma-Aldrich) was used as a ceramic material. The kaolin powder was weighted, sieved using a sieve of one mm mesh size, to remove agglomeration and left to dry in desiccators prior to use. The PESf (Sigma-Aldrich) polymer pellets were dried in an oven at 60 °C for 8 h to remove water content. N-methyl-2-pyrrolidone (NMP, Acros Organics, 99 % extra pure) was used as a solvent without treatment. Tap water was used as a coagulation medium for both the internal and external coagulants.

Powder Characterization

The particle-size distribution of the kaolin was determined by a portable, laser-diffraction, particle-size analyser (LSST-Portable, Sequoia).

Preparation of the Spinning Solution and the Measurement of Viscosity

Firstly, NMP in an airtight glass was heated to 60 °C using hot plate (Yellow MAG HS 7 S2, IKA) where PESf was subsequently added and mixed until dissolved using a magnetic stirrer. Next, the polymer solution was left to cool to room temperature. Kaolin was then added in the bottle containing the dissolved polymer. The mixture was mixed using a mechanical stirrer (Ika Labortechnik, Model RW 20n) at a low speed for 30 min. Subsequently, the mixing process of the suspension was continued using a roller shaker (SRT6D, Stuart) at 60 rpm for 48 h to create a homogeneous solution. The composition of k/p at ratio one was determined at 27.0 g kaolin, 27.0 g PESf and 151 g NMP with 27 g kaolin added to the suspension at each ratio increment. The viscosity of kaolin suspension was measured using a Brookfield Programmable HADV-IV+ rheometer at zero s^{-1} shear rate and at 25 °C.

Dry-Jet Wet Spinning

The dry jet wet spinning was done as described in Sarbatly and Kamin (2012) where the ceramic suspensions were degassed and transferred in a stainless steel reservoir, where they were spun in the range of 2.0–2.5 bar using nitrogen. The spinning parameters are shown in Table 1. The nascent tubes emerging from the spinneret were spun directly into a coagulation bath. The tube precursors were then left to soak in a coagulation bath for 24 h to complete the phase-inversion solidification process. The schematic diagram for the spinning process is shown in Fig. 1.

Scanning Electron Microscope

Structures of the spun kaolin hollow tube were observed using a scanning electron microscopy (JEOL JSM-5610LV, Japan) using the procedure described by Tan et al. (2001). In this study, it was observed that the kaolin hollow tubes have exhibited six distinct regions on the cross section of the tube. These regions were classified as shown in Fig. 2.

Table 1 Spinning parameters and condition for spinning kaolin hollow-tube membranes

Spinning parameters	Spinning condition	Spinning parameter	Spinning condition
Solution temperature	27 °C	Spinneret dimension	1.7/0.8 (OD/ID) mm
Internal coagulant and coagulation bath temperature and composition	27 °C and tap water	Internal coagulant injection rate	35 ml min^{-1}
Nitrogen pressure	2.0–2.5 bar	Air gap	0 cm

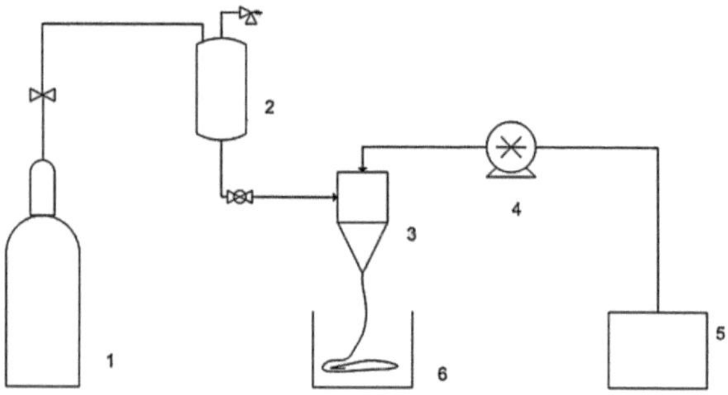

Fig. 1 Schematic diagram of the spinning apparatus: *1* A Nitrogen tank; *2* A pressure vessel; *3* A spinneret; *4* A gear pump; *5* A water reservoir; and *6* A coagulation bath

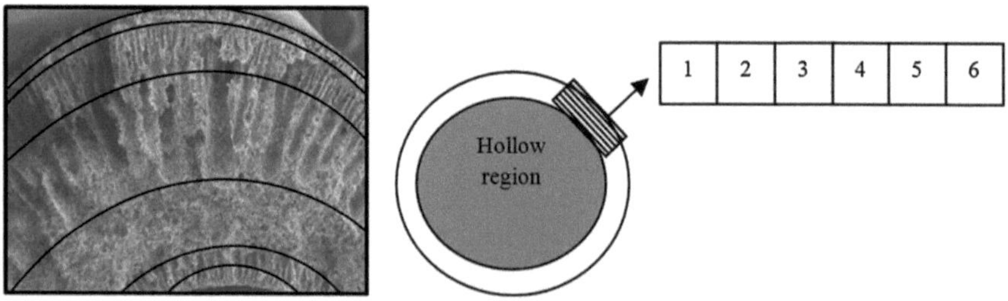

Fig. 2 Classifications of the cross section of a kaolin hollow tube; *1* inner region; *2* near inner region; *3* centre region; *4* and *5* near outer region; and *6* outer region

Results and Discussion

Powder Characterization and Suspension Viscosity Determination

The diameters of kaolin particles were found to be in the range of 1–65 μm. More than 41 % of kaolin particles have diameters that are less than 6 μm (Table 2). The viscosities of the suspensions are tabulated in Table 3, ranging from 1.280×10^3 to 3.310×10^5 cP.

Table 2 A particle-size distribution of clay

Diameter (μm)	<6	6–13	12–29	28–63
Particle-size distribution (%)	41	24	22	13

Table 3 Viscosity of the suspensions

Kaolin/PESf ratio	Viscosities (cP)	Kaolin/PESf ratio	Viscosities (cP)
S1:1.0	1.280×10^3	S4:2.5	3.330×10^4
S2:1.5	1.327×10^3	S5:3.0	4.849×10^4
S3:2.0	7.300×10^3	S6:3.5	3.310×10^5

Effect of Kaolin/PESf Ratios on the Microstructure of Kaolin Hollow Tubes

All kaolin hollow tubes produced in this study have exhibited asymmetric cross section as shown in Fig. 3. The cross sections of the kaolin hollow tubes generally exhibit finger like voids near the outer and inner regions of tube walls, whereas spongelike structures exhibited at the centre region of tube walls. The formations of finger like voids are typically attributed by rapid precipitation by phase inversion and hydrodynamically unstable viscous fingering (Kingsbury and Li 2009), where a viscous suspension is in contact with a low-viscous hydrophilic non-solvent. Meanwhile, the spongelike structures are usually attributed to a slow precipitation (Tan et al. 2001) due to limited contact of suspension with non-solvent.

At low viscosity (k/p : 1.0, 1.5 and 2.5), the finger like voids were seen dominating the outer region (cumulative of regions 4, 5 and 6) of the cross section of the kaolin hollow tubes , which are at 50 % for S1, 58 % for S2 and 51 % for S4

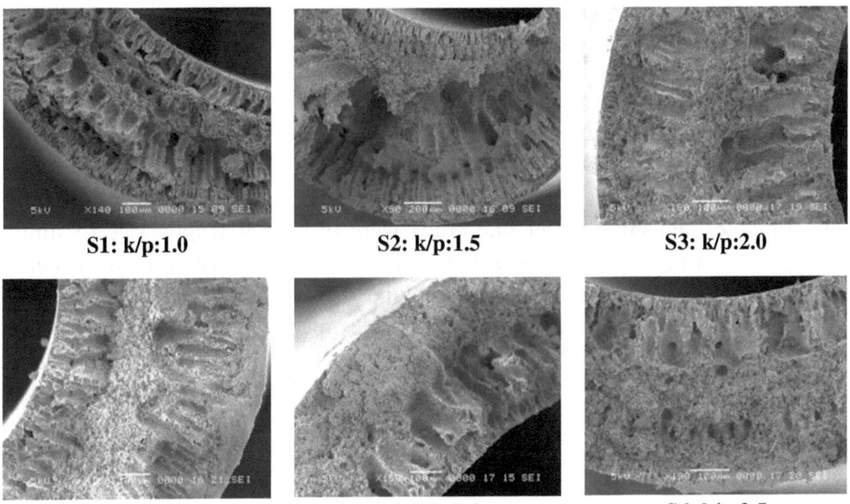

Fig. 3 SEM images of the cross section of kaolin hollow tubes at 1.0–3.5 kaolin/PESf (k/p) ratios, (*S1* Suspension 1; *S2* Suspension 2; *S3* Suspension 3; *S4* Suspension 4; *S5* Suspension 5; and *S6* Suspension 6)

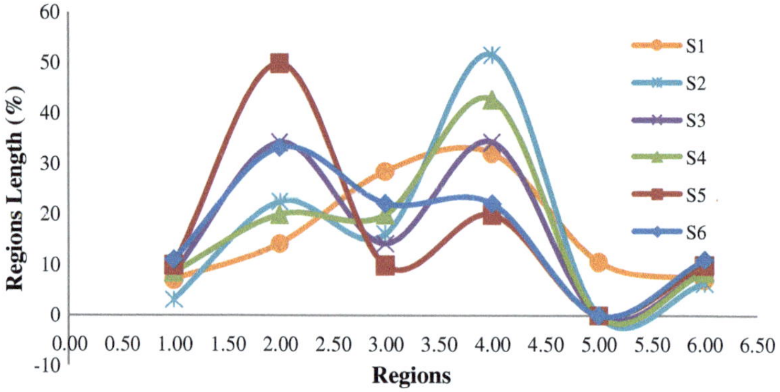

Fig. 4 Region length (%) of the individual regions referred in Fig. 2 for kaolin hollow tube

(Fig. 4). Here, the kaolin particles were not able to undergo molecular orientation to produce graded film layers possibly due to an excessive amount of PESf solution in the suspension. The larger kaolin particles may have been attracted to the tube inner section, thus preventing internal coagulant intrusions into the tube inner skin, resulting delayed de-mixing, as similarly found by Wara et al. (1995) and Wei et al. (2008).

Further, at high viscosity (k/p: 3.0 and 3.5), the finger like voids were seen dominating the inner region (cumulative of regions 1 and 2) of the cross sections of the kaolin hollow tubes , which are at 60 % for S5 and 44 % for S6 (Fig. 4). At these viscosities, the kaolin particles were seen to most likely undergo molecular orientation, thus producing graded film layers. This might be due to a high amount of kaolin particles in the suspension, thus promoting an instantaneous de-mixing. Moreover, the hydrodynamically unstable viscous fingering may have been demonstrated in the inner tube structure, where physical reactions of the larger and smaller kaolin particles would move them equally to the area where water intrusion took place (Wei et al. 2008; Kingsbury and Li 2009).

Meanwhile, for kaolin hollow tubes spun using solution with k/p ratio of 2.0, both the outer and inner finger like voids were in similar lengths due to similar precipitation rates that occurred at both ends. The spongelike structure occupies about 15 % of the cross sections of the tube (Fig. 4). The spongelike structure influences the separation properties and provides mechanical strength to a hollow tube (Kingsbury and Li 2009).

Conclusion

Kaolin hollow tubes spun at 0-cm air gap shows that at low viscosity (k/p ratios: 1.0, 1.5 and 2.5), the finger like voids dominated the outer region of the cross sections of the tube, whereas at high viscosity (k/p ratios: 3.0 and 3.5), the finger like voids dominated the inner region of the tube cross section. Since all of the kaolin hollow-tube precursors have displayed finger like voids at both ends, they are suitable to serve as supports for catalyst particles impregnation; however, they may not be suitable to be used for separation applications.

Acknowledgments The authors would like to thank the Universiti Malaysia Sabah and the Ministry of Science Technology and Innovation, Malaysia, for the usage of facilities and funding this research, respectively.

References

Bouzerara, F., Harabi, A., Achour, S., & Larbot, A. (2006). Porous ceramic supports for membranes prepared from kaolin and doloma mixtures. *Journal of the European Ceramic Society, 26*, 1663–1671.

Clausi, D. T., McKelvey, S. A., & Koros, W. J. (1999). Characterization of substructure resistance in asymmetric gas separation membranes. *Journal of Membrane Science, 160*, 51–64.

Kingsbury, B. F. K., & Li, K. (2009). A morphological study of ceramic hollow fibre membranes. *Journal of Membrane Science, 328*, 134–140.

Liu, S., & Li, K. (2003). Preparation of TiO_2/Al_2O_3 composite hollow fibre membranes. *Journal of Membrane Science, 218*, 269–277.

Liu, Y., & Li, K. (2005). Preparation of $SrCe_{0.95}Yb_{0.05}O_{3-[alpha]}$ hollow fibre membranes: Study on sintering processes. *Journal of Membrane Science, 259*, 47–54.

Liu, S., Tan, X., Li, K., & Hughes, R. (2001). Preparation and characterisation of $SrCe_{0.95}Yb_{0.05}O_{2.975}$ hollow fibre membranes. *Journal of Membrane Science, 193*, 249–260.

Mohammadi, T., & Pak, A. (2002). Making zeolite a membrane from kaolin by electrophoresis. *Microporous and Mesoporous Materials, 56*, 81–88.

Mohammadi, T., & Pak, A. (2003). Effect of calcination temperature of kaolin as a support for zeolite membranes. *Separation and Purification Technology, 30*, 241–249.

Mohammadi, T., Pak, A., Karbassian, M., & Golshan, M. (2004). Effect of operating conditions on microfiltration of an oil-water emulsion by a kaolin membrane. *Desalination, 168*, 201–205.

Sarbatly, R. & Kamin, Z. (2012). Effect of the sintering process on the morphology of kaolin hollow tubes: *12th International Conference on Inorganic Membranes* 9–13 July 2012, University of Twente, Enschede, The Netherlands.

Shen, L.-Q., Xu, Z.-K., Liu, Z.-M., & Xu, Y.-Y. (2003). Ultrafiltration hollow fiber membranes of sulfonated polyetherimide/polyetherimide blends: Preparation, morphologies and anti-fouling properties. *Journal of Membrane Science, 218*, 279–293.

Tan, X., Liu, S., & Li, K. (2001). Preparation and characterization of inorganic hollow fiber membranes. *Journal of Membrane Science, 188*, 87–95.

Wara, N. M., Francis, L. F., & Velamakanni, B. V. (1995). Addition of alumina to cellulose acetate membranes. *Journal of Membrane Science, 104*, 43–49.

Wei, C. C., Chen, O. Y., Liu, Y., & Li, K. (2008). Ceramic asymmetric hollow fibre membranes-one step fabrication process. *Journal of Membrane Science, 320*, 191–197.

On the Alloying Reaction in Synthesis of NiTi Shape-Memory Alloy in Solid State

J. Abdullah and H. H. M. Zaki

Abstract Solid-state synthesis of NiTi alloy has gained much interest recently particularly in its porous form for its promising application in medical implant. In this work, powder synthesis by sintering of elemental Ni and Ti for forming equiatomic NiTi alloy was studied. It was found that the alloying reaction in solid state involves complex interaction between the participating elements, their thermodynamics, and the physical constraint of atomic diffusion. A simple model is proposed to explain the phenomenon. Phase characterization was carried out using X-ray diffraction, scanning electron microscope and energy dispersive X-ray. Differential scanning calorimetry was used to examine the transformation behavior of the NiTi alloy. Multiple phases on the Ni-Ti alloy systems were formed and low transformation heat was observed for the synthesized specimen. This paper identifies possible obstacles for forming single-phase NiTi and proposes further work.

Introduction

Among the many shape-memory alloys scientists have discovered, NiTi alloy has gained much attention for many applications especially in engineering and medical fields. This is due to its excellent mechanical properties (Otsuka and Wayman 1998), good biocompatibility (Jorma 1999; Bansiddhi et al. 2008; Ryan et al. 2006), and superior corrosion resistance (Bansiddhi et al. 2008). The applications so far involve only the bulk or dense NiTi in the form of rods, plates, wires, and thin sheets. An emerging application is its use in medical implants, which particularly requires NiTi in porous form. As implant material, porous NiTi is technically advantageous and superior compared to the traditional dense Ti or steel

J. Abdullah (✉) · H. H. M. Zaki
School of Mechanical Engineering, Universiti Sains Malaysia, Nibong Tebal,
Seberang Perai Selatan 14300 Pulau Pinang, Malaysia
e-mail: mejamal@eng.usm.my

R. Pogaku et al. (eds.), *Developments in Sustainable Chemical and Bioprocess Technology*, DOI: 10.1007/978-1-4614-6208-8_46,

implant currently widely used. Coupled with excellent corrosion-resistant and shape-memory properties of NiTi, its porous structure has been shown to encourage tissue in-growth which naturally supports the implant (Jorma 1999; Michel et al. 2002). This means faster healing process and comfort from ease of mobility of lighter porous implant.

Production of porous NiTi has focused largely on solid-state synthesis where elemental Ni and Ti powders are consolidated and alloyed at below their melting points. However, many earlier studies revealed that this is not a straightforward processing step because it involves solid-state diffusion of the elemental powders which lead to alloying of NiTi. NiTi is one of the three stable phases that exist in the Ni–Ti binary alloy system, thus making it more difficult to control and to ensure single phase of NiTi would be produced. Multiple phases of the Ni–Ti system are produced including Ti_2Ni, NiTi, and $TiNi_3$. The Ti_2Ni and $TiNi_3$ phases are brittle (Yi and Moore 1992) and do not show shape-memory property and thus are undesired products in solid-state synthesis. Up to date, there is limited discussion in the literature on the mechanism of NiTi alloying in solid state. This paper presents the results of an attempt to alloy equiatomic NiTi and discusses the possible mechanism of solid-state alloying process.

Methodology

In this work, high-purity Ni and Ti powders of 74 μm mean particle sizes were used to form near equiatomic mixture. The mixing was carried out by means of tumble milling using ceramic balls for 24 h. The NiTi mixture was then cold pressed at 45 MPa into cylindrical shape of $\phi 12 \times 4$ mm. The compacts were sintered in furnace at different temperatures ranging from 750 to 1,080 °C for 6 h. The sintering was carried out under argon flowing. Sintered specimens were cut using a low-speed diamond cut off wheel and then ground and polished for structural analysis. The microstructure of the samples were examined using a Zeiss 1555 field-emission scanning electron microscope (FESEM) equipped with energy dispersive spectroscopy (EDS) for compositional analysis. X-ray diffraction was performed using a Siemens D5000 diffractometer with Cu-Kα radiation of 1.54 Å wavelength. Phase transformation behavior was studied by means of differential scanning calorimetric (DSC) measurement using a TA Q10 DSC machine at a heating/cooling rate of 10 K/min.

Results and Discussion

Figure 1 shows SEM micrograph of the sample sintered at 850 °C for 6 h prepared from elemental Ti and Ni powders. This temperature is well below the eutectic point at 942 °C, to avoid melting and to assure all solid-state reaction during

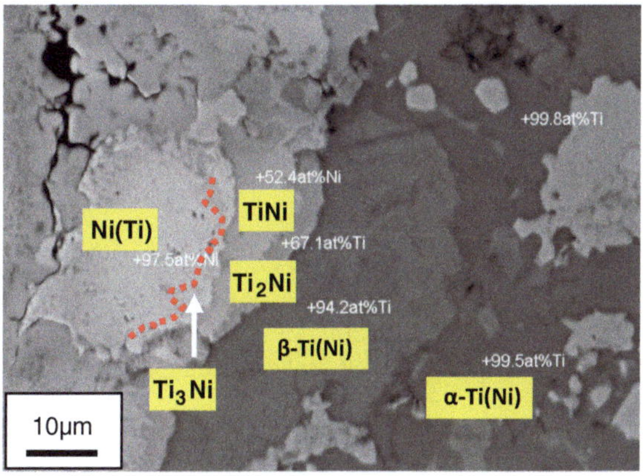

Fig. 1 SEM micrograph of equiatomic sample sintered at 850 °C for 6 h with EDS phase composition

sintering. The phase identifications, as determined by EDS analysis, are marked in the micrographs, including Ti_2Ni, TiNi, $TiNi_3$, and the solid solutions of α-Ti(Ni), β-Ti(Ni), and Ni(Ti). The contrasts of Ni(Ti) and $TiNi_3$ are very similar, thus difficult to differentiate at low magnifications. It is evident that TiNi and Ti_2Ni form in between Ti(Ni) and Ni(Ti).

Figure 2 is the XRD spectra of sample before and after sintering for 6 h at different temperatures ranging from 750 to 1,080 °C. The specimen before sintering showed only elemental Ni and Ti. After sintering at 750 °C, almost no new phase formation took place except Ti_2Ni phase. NiTi in the form of B2-austenite was observed to form in the specimen sintered at 850 °C while Ti_2Ni grew in strength with the increase in the sintering temperature from 750 °C. $TiNi_3$ merged in the specimen sintered at 900 °C; however, this phase was found at 850 °C as shown in Fig. 1. This is due to the limitation of XRD in detecting phases of low-volume fractions. In the specimen sintered at 950 °C, TiNi has become the dominant phase, with Ti_2Ni and $TiNi_3$ being the next two major presences in the microstructure. Some residual Ni is still visible in the XRD spectrum, implying the incompleteness of the primary reactions. In contrast, the Ti has fully disappeared. Increasing the sintering temperature to 1,080 °C led to the disappearance of both Ni and Ti where only three phases were observed, i.e., TiNi, Ti_2Ni and $TiNi_3$ with TiNi being the dominant phase. This indicates that the primary reactions have reached completion at this temperature.

DSC measurement of the transformation behavior of the sintered samples revealed weak transformation events for all samples. The specimen sintered at 750 °C for 6 h shows no sign of martensitic transformation meanwhile the specimen sintered at 850 °C shows very weak A → M transformation on cooling and M → A transformation on heating. The peak temperatures of the transformations are 43 °C for the A → M and 57 °C for the M → A. The transformation enthalpy changes are small, 3.07 J/g for the cooling transformation. This implies low transformation volume. This appears consistent with the SEM observation (Fig. 1), which shows that the NiTi phase is a minority in the matrix of the sample. No transformation was detected for the specimens sintered at 950 °C and 1,080 °C for 6 h.

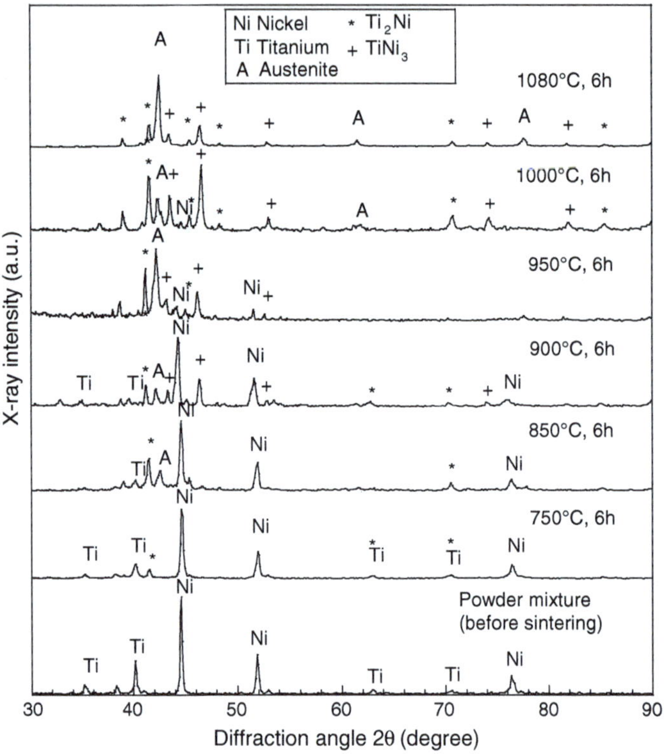

Fig. 2 XRD spectra of samples before and after sintering for 6 h at various temperatures ranging from 750 to 1,080 °C

Fig. 3 Schematic illustration of Ni–Ti diffusion model at different stages of sintering

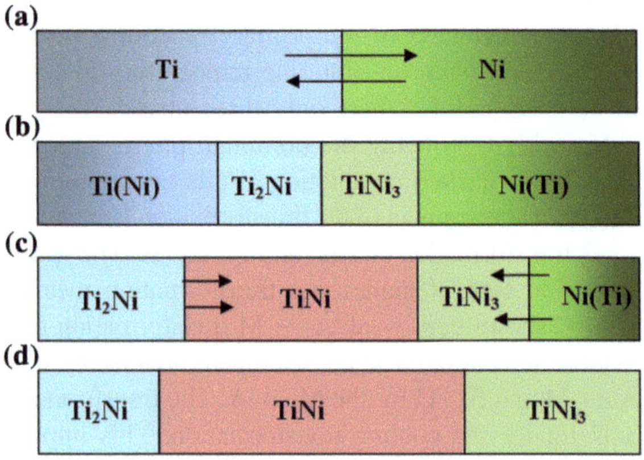

Figure 3 illustrates the expected process stages of phase formation of Ni–Ti. At early stage of sintering, Ni and Ti will diffuse into each other to form Ni(Ti) and Ti(Ni) solid solutions as depicted in (a). Further diffusion of Ni into Ti(Ni) on the Ti side leads to the formation of Ti_2Ni while continuous supply of Ti into Ni(Ti) on the Ni side leads to the formation of $TiNi_3$ as shown in (b). With continued diffusion of Ni and Ti, the Ti(Ni) is noted to disappear first, followed by Ni(Ti). This is attributed to the fact that the diffusion coefficient of Ni into Ti is higher than that of Ti into Ni (Bastin and Rieck 1974; Li et al. 1998). Thus, Ti is converted to Ni-Ti compounds much faster than Ni. In addition, for samples sintered at higher temperatures, Ti(Ni) may disappear due to local melting at near the eutectic at 942 °C. The continuous formation of NiTi expands its boundaries until all the remaining Ni has been consumed. At this stage, NiTi becomes the major phase, which reveals only 3 phases, including Ti_2Ni, NiTi, and $TiNi_3$. The model proposes that solid-state synthesis involves physical migration of atoms across phase boundaries. As such, the diffusion and therefore, the alloying process are influenced by atomic mobility and speed across those boundaries. It is suggested that smaller particle size, hence smaller diffusion distant, accelerates the alloying process. However, physical constraint alone is not sufficient to explain the alloying process and given an indefinite sintering time, the final phase may come into single-phase NiTi, given that this is thermodynamically feasible.

Gibbs free energy change of the primary and secondary reactions at 750 °C can be determined as following and may be classified as primary reactions and secondary reaction. The secondary reactions are reactions involving the remaining pure element and the products of the primary reactions. Among the primary reactions, formation of $TiNi_3$ is the most exothermic and thus most favored, followed by Ti_2Ni. The desired NiTi phase is the least favored to form. This indicates great difficulty to form NiTi from the primary alloying reaction process.

The secondary reactions may take place simultaneously with the primary reactions, upon availability of the reaction products from primary reactions. The secondary reactions are weak event, forming NiTi as the product. This brief thermodynamic analysis indicates the low possibility of single-phase NiTi formation from solid-state alloying of elemental Ni and Ti powders.

Primary reactions:

$$3\text{Ni} + \text{Ti} \rightarrow \text{TiNi}_3\ (\Delta G_{1023\text{K}} = -117.7\ \text{kJ/mol}) \quad (1)$$

$$\text{Ni} + 2\text{Ti} \rightarrow \text{Ti}_2\text{Ni}\ (\Delta G_{1023\text{K}} = -69.7\ \text{kJ/mol}) \quad (2)$$

$$\text{Ni} + \text{Ti} \rightarrow \text{NiTi}\ (\Delta G_{1023\text{K}} = -58.1\ \text{kJ/mol}) \quad (3)$$

Secondary reactions:

$$1/2\text{Ni} + 1/2\text{Ti}_2\text{Ni} \rightarrow \text{NiTi}\ (\Delta G_{1023\text{K}} = -23.3\text{kJ/mol}) \quad (4)$$

$$2/3\text{Ti} + 1/3\text{TiNi}_3 \rightarrow \text{NiTi}\ (\Delta G_{1023\text{K}} = -18.9\text{kJ/mol}) \quad (5)$$

$$2/5\text{Ti}_2\text{Ni} + 1/5\ \text{TiNi}_3 \rightarrow \text{NiTi}\ (\Delta G_{1023\text{K}} = -6.7\text{kJ/mol}) \quad (6)$$

Conclusions

In this study, solid-state synthesis of elemental Ni and Ti powders for alloying of NiTi always results in multiple phases of Ti_3Ni, NiTi, Ti_2Ni and some Ni(Ti), Ti(Ni). The increase in temperature did not produce single-phase NiTi. However, the amount of NiTi is found to increase with increasing sintering temperature. The NiTi phase formed exhibits very small transformation peaks indicating weak shape-memory property. A physical model of inter-diffusion reveals the difficulty for atomic migration across phase boundaries, thus needing consideration of powder particle size used and a longer time allowed for diffusion. Thermodynamic consideration reveals two types of reactions, namely the primary reactions involving the pure Ni and Ti and the secondary reactions involving the remaining pure Ni and Ti and the products of primary reactions. The supply of desired NiTi phase relies heavily on much weaker secondary reactions and a least favored reaction from the primary reactions. These explain the occurrence of multiple phases observed in the solid-state synthesis of NiTi. The major question whether single-phase NiTi possible in solid-state synthesis and thus enable the reliable production of porous NiTi remains unsolved and needed further study.

Acknowledgments Authors would like to thank the Universiti Sains Malaysia and the Ministry of Higher Education via Experimental Research Grant Scheme (ERGS) # 203/PMEKANIK/6730014 for partly supporting this work. The input and experimental support from Professor Yinong Liu of the University of Western Australia is also acknowledged.

References

Bansiddhi, A., Sargeant, T. D., Stupp, S. I., & Dunand, D. C. (2008). Porous NiTi for bone implants: A review. *Acta Biomaterialia, 4*, 773–782.

Bastin, G. F., & Rieck, G. D. (1974). Diffusion in the titanium-nickel system: II. calculations of chemical and intrinsic diffusion coefficients. *Metallurgical Transition, 5*, 1827–1831.

Jorma, R. L. T. (1999). *Biocompatibility evaluation of nickel-titanium shape memory metal alloy*. Oulu: University of Oulu.

Li, B. Y., Rong, L. J., & Li, Y. Y. (1998). Porous NiTi alloy prepared from elemental powder sintering. *Materials Research, 13*, 2847–2851.

Michel, A., Alexandr, C., Michel, A. L., & Charles, H. R. (2002). A new porous titanium-nickel alloy: Part 1. Cytotoxicity and genotoxicity evaluation. *Bio-Medical Materials and Engineering, 12*, 225–237.

Otsuka, K., & Wayman, C. M. (1998). *Shape memory alloys*. Cambridge: Cambridge University Press.

Ryan, G., Pandit, A., & Apatsidis, D. P. (2006). Fabrication methods of porous metals for use in orthopaedic applications. *Biomaterials, 27*, 2651–2670.

Yi, H. C., & Moore, J. J. (1992). Combustion synthesis of TiNi intermetallic compound. *Journal Materials Science, 27*, 5067–5072.

GFRP Composite Material Degradation Under Seawater and Weathering Effect

W. H. Choong, K. B. Yeo, M. T. Fadzlita, Y. Y. Farm and M. Azlan Ismail

Abstract An experimental investigation of the seawater effect on the degradation of GFRP composite material with respect to water absorption was conducted. The GFRP specimens were exposed to seawater and weather environment. Water absorption analysis was conducted to investigate the moisture uptake and degradation of woven GFRP. The woven GFRP percentage weight gain due to moisture uptake were 0, 0.253, 0.513, 0.752, 0.732, 0.752, and 0.997 % of the initial weight, after 1, 2, 3, 5, 8, 48, and 168 immersion hours. Degradation of GFRP was also found to correlate to the corrosion and uptake of seawater at the fibre/matrix interface.

Introduction

Glass fibre-reinforced polymer materials are increasingly utilized in civil and mechanical infrastructure because of their high strength to weight and corrosion resistance. Researchers had shown that properties of GFRP composite systems were compromised in seawater environment (Mourad et al. 2010). One major concern of using GFRP composites in seawater application was attributed to the fibre/matrix interface degradation due to hydrolysis reaction of unsaturated groups within the resin. Seawater degradation could cause swelling and plasticization of the polyester matrix and debonding at the fibre/matrix interface. The swelling and plasticization could reduce the mechanical properties (Aldajah et al. 2008).

Masahiro et al. (2010) investigated the woven GFRP mechanical property by immersion in deionized water at various temperatures. They showed that the woven GFRP strength had a tendency to decrease with hydrothermal ageing. Other

W. H. Choong · K. B. Yeo (✉) · M. T. Fadzlita · Y. Y. Farm · M. Azlan Ismail
Material and Mineral Research Unit, School of Engineering and Information Technology, Universiti Malaysia Sabah, Jalan UMS, 88400 Kota Kinabalu, Sabah, Malaysia
e-mail: nooryeo@yahoo.com

R. Pogaku et al. (eds.), *Developments in Sustainable Chemical and Bioprocess Technology*, DOI: 10.1007/978-1-4614-6208-8_47,

than that, the rate of degradation was also found directly proportional to the water temperature. Mourad et al. (2010) also investigated the seawater and temperature effects on structural properties of glass/epoxy and glass/polyurethane composite materials. The investigation showed that the glass/polyurethane composite degradation mechanism can be accelerated at higher temperature.

Synthesis and Method

The evaluations of seawater and weathering effects on the degradation of GFRP composite material were conducted and elaborated in the following sections.

Material

In-house hand layup fabrication technique of GFRP specimens was conducted. The specimen layup consisting of five layers of E-glass EWR600 cloth was laminated with SHCP 2719 unsaturated polyester resin (400:3 resin to Butanox M-50 mass ratio). The produced five-layer composite laminate of E-glass EWR600 specimen had an average thickness of 3.3 mm.

Specimen Preparation

Experimental techniques for water absorption tests were conducted. In water absorption test, specimens of small rectangular dimensions 10 mm (width) × 25 mm (length) were utilized. The water-uptake specimen was designed such as to allow higher-resolution mass incremental can be measured. These specimen edges were also fine polished for subsequent micrographic analysis purposes.

Testing Method

The effects of seawater and weathering degradation on GFRP were tested. Specimens were exposed to seawater and weathering by immersed into outdoor seawater bath over different periods of time. The seawater bath had salinity content of 33ppt under a temperature range of 25–35 °C (weather dependant).

Water absorption and micrographic tests were also carried out to evaluate the localize effect of seawater and weathering degradation of GFRP material performance. The similar seawater was utilized for the immersion. The seawater mass uptake and the surface micrographic behaviour were systematically recorded before and after the immersion period. The specimen seawater mass gain was measured as weight gain percentage ($\Delta \boldsymbol{m}\,\%$),

$$\Delta \boldsymbol{m}\,\% = \frac{(\boldsymbol{m}_a - \boldsymbol{m}_o)}{\boldsymbol{m}_o} \times 100 \tag{1}$$

where $\boldsymbol{m}_o$ and $\boldsymbol{m}_a$ were specimen weight before (dry) and after immersion. Subsequently, micrographic observation was conducted on the specimen edge surface for examining the degradation effect due to seawater and weathering exposure.

Result and Discussion

Water absorption tests of GFRP composite under seawater bath and weathering effect found a characteristic relation of mass gain and the exposure time period of 1, 2, 3, 4, 5, 6, 7, 8, 48, and 168 h, Fig. 1. In the first 3 h of immersion, specimen mass had linearly increasing and the average mass gained about 0.51 %. This had being attributed to seawater molecular penetration of specimen material, surrounding and particularly at the fibre and matrix interfaces. After 4 h of immersion, the specimen instead showed a mass loss. Sun and Li (2011) had explained this behaviour due to alkali metal oxides leaching from the E-glass in the presence of moisture to form concentrated salt solutions. In this case, aggressive seawater could also cause salinity etching effects to the matrix which attributes the debonding effect at interfaces and voids that caused mass loss. The supplementary diffusion of seawater towards the interface is also increasing the mass.

Sobrinho et al. (2009) found in immersion period between 1 to 8 h, kinetically distinct phenomena occurred with specimen mass increasing and subsequently reduced in a short period. Berketis and Tzetzis (2009) explained the contribution due to occurrence or existing microvoids or porosity and interfaces creating multi-stage diffusion. However, after 168 h of immersion, Fig. 1, the mass average gain

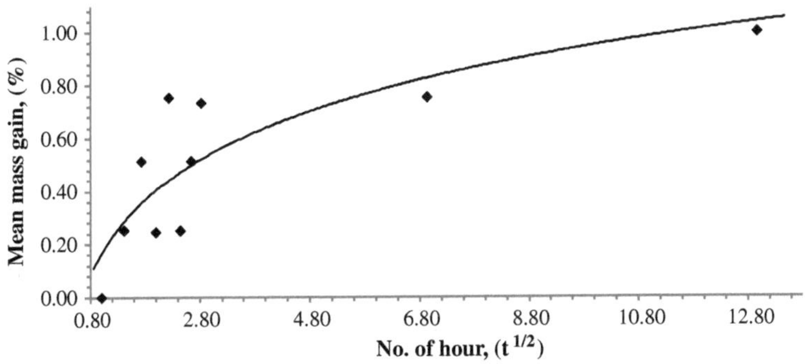

Fig. 1 Specimens weight gain against immersion time period

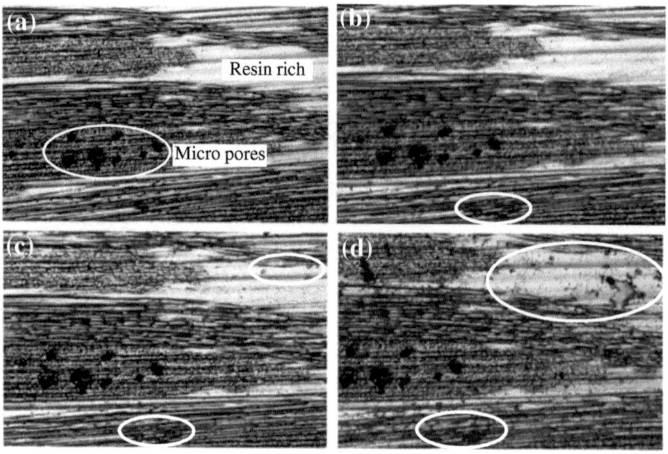

Fig. 2 Dry and seawater-treated specimen micrographs: **a** dry; **b** after 1 h; **c** after 8 h; and **d** after 48 h

up to 1 % indicated a minor physical change such as blister, swelling or crack, and material colour thickening or fading.

The GFRP specimen of microscopic examination before and after the immersion is shown in Fig. 2. In Fig. 2a, the dry specimen examination shows the composite structure formed by two-set fibre orientation (parallel and perpendicular to cutting plane) and resin-rich region. Micropores were formed, and these pores are considered as fabrication defect. This type of fabrication defect is due to the disadvantage of hand layup fabrication method and panel trimming process by circular hand saw machine.

The treated specimen for 1, 8, and 48 h immersion micrographs is shown in Fig. 2b, c, and d. After 1 h, Fig. 2b marked region, indicated new voids formed compared to Fig. 2a. Microscars also developed on resin-rich region, Fig. 2c. As immersion period continues, aggravated micropores or voids are formed due to leaching as shown in Fig. 2d. This qualifies the earlier explanation of increased seawater absorption through micropores and matrix expansion interfaces resulting in increasing specimen mass recorded in Fig. 1.

Conclusions

This research had been carried out to investigate the seawater and weathering effect on degradation of woven E-glass fibre-reinforced polymer with unsaturated polyester resin composite material. The specimens were subjected to seawater and weathering exposure for up to 168 h. The exposure effects were evaluated through micrographic and water absorption analysis. From the water absorption analysis, the specimens had indicated significant moisture absorption against the seawater immersion period. Over a period of 168 h immersion, the GFRP specimen showed an average mass gain of about 1 %, indicating the susceptibility to seawater uptake. Micrographic analysis of specimen surfaces had also indicated fibre break, microcrack or micropores that caused by seawater leaching and etching effects. Gross degradation detrimental to GFRP composite specimen was observed for the 168 h seawater and weathering exposure, suggesting that protective or primer coating would be necessary for safe application in seawater environment.

Acknowledgments The authors would like to express their appreciation to the *Material and Mineral Research Unit, Advanced Composite Research Group, School of Engineering and Information Technology, Universiti Malaysia Sabah*, and the *Ministry of Higher Education of Malaysia* the financial support through the research Grant FRG0249-TK-2/2010.

References

Aldajah, S., Alawsi, G., & Abdul, R. S. (2008). Impact of sea and tap water exposure on the durability of GFRP laminates. *The Journal of Materials and Design, 30*, 1835–1840.

Berketis, K., & Tzetzis, D. (2009). Long-term water immersion ageing characteristics of GFRP composites. *Journal of Materials Science, 44*(13), 3578–3788.

Masahiro, K., Yusuke, Y., Yusuke, T., & Kawada, H. (2010). Study of strength degradation mechanism of woven GFRP in water environment. *Journal of Solid Mechanics and Materials Engineering, 4*, 1574–1584.

Mourad, A. H. I., Abdel-Magrid, B. M., El-Maaddawy, T., & Grami, M. E. (2010). Effect of seawater and warm environment on glass/epoxy and glass/polyurethane composites. *Applied Composite Material, 17*, 557–573.

Sobrinho, L. L., Ferreira, M., & Bastian, F. L. (2009). The effects of water absorption on an ester resin system. *Materials Research, 12*(3), 353–361.

Sun, A. B., & Li, B. Y. (2011). Effects of chemical environment on the durability performances of glass fiber/epoxy composites. *18th International Conference on Composite Materials* (pp. 21–26). August, Korea.

β-Mannanase Production by *Aspergillus flavus* in Solid-State Fermentation of Palm Kernel Cake

Y. P. Wong, H. Y. Saw, J. Janaun, K. Krishnaiah and A. Prabhakar

Abstract *β*-Mannanase production in batch solid-state fermentation (SSF) of palm kernel cake (PKC) was evaluated with flasks and a laterally aerated moving bed (LAMB) bioreactor using *Aspergillus flavus* UMS01. Optimum condition for flask SSF was 110 % moisture content, initial pH 6, 30 °C and particle size 855 μm, yielding 383 U g^{-1} dry PKC after 120 h. Under the same condition and particle size <5 mm, SSF in LAMB produced 276 U g^{-1} dry PKC at an optimal gas flow of 4.4 m s^{-1} in just 96 h. Optimal *β*-mannanase production was consistent with highest fungal growth and mannose production; to support increasing fungal growth, secretion of *β*-mannanase increased to degrade mannans in PKC, producing mannose for microbial consumption. *A. flavus* UMS01 showed promising attributes as a *β*-mannanase producer via SSF of PKC in flasks and LAMB bioreactor.

Introduction

Solid-state fermentation (SSF) is a bioprocess that incorporates cultivation of microorganisms on solid substrates under controlled environment, e.g. moisture level, pH, temperature, and substrate particle size. The substrates act as carbon

Y. P. Wong · J. Janaun (✉) · K. Krishnaiah · A. Prabhakar
Chemical Engineering Program, School of Engineering and Information Technology, Universiti Malaysia Sabah, Jalan UMS, 88400 Kota Kinabalu, Sabah, Malaysia
e-mail: jidon@ums.edu.my

H. Y. Saw (✉)
School of Engineering and Advanced Technology, Massey University, Private Bag, 11 222, Palmerston North 4442, New Zealand
e-mail: sawhorngyuan@gmail.com; H.Y.Saw@massey.ac.nz

R. Pogaku et al. (eds.), *Developments in Sustainable Chemical and Bioprocess Technology*, DOI: 10.1007/978-1-4614-6208-8_48,

sources and/or inert support for microbial growth and generation of primary and secondary metabolites (Krishna 2005; Pandey 2003). SSF has been exploited for various applications, such as productions of enzymes (Krishna 2005; Ong et al. 2004), and substrates nutritional improvement for animal feed production. The choice of microorganisms has included fungi, yeast, and bacteria, while the substrates are generally crops and residues from agricultural sectors and agro-industries (Krishna 2005; Pandey 2003).

The palm oil industry produces palm kernel cake (PKC) as a by-product; PKC is the residue from extraction of oil from palm kernel seeds (Tang 2001). As a cheap commodity and a good source of hemicellulose in the form of mannan (Daud and Jarvis 1992), PKC has found use in SSF for the production of enzymes and animal feed with modified nutritional content. During SSF, the predominant enzyme secreted is β-mannanase; traces of cellulases, β-glucosidase, polygalacturonase, xylanase, amylase, and phytase are also present (Ong et al. 2004). Through the enzymatic activities and hydrolysis of PKC hemicelluloses, reducing sugars are generated to support microbial growth, with mannose as a primary sugar (Dhawan and Kaur 2007). The increase in sugar contents (and thus metabolizable energy) and degradation of mannan, which is indigestible to mono-gastric animals, has been a drive in the development of new feed processing technologies via SSF with a variety of microorganisms and bioreactors.

In a recent work (Wong et al. 2011), PKC was fermented with *Aspergillus flavus* UMS01 to increase the mannose content in the end product. SSF was performed in batch in conical flasks and a laterally aerated moving bed (LAMB) bioreactor. Optimized SSF parameters were determined and fermented PKC had an improvement in mannose level of up to 5.5-fold. However, in the assessment of SSF performance, only fungal biomass and mannose production were measured and reported. β-mannanase, which is an equally important property because it is directly related to fungal growth, mannan degradation, and mannose production, has not yet been analyzed. Therefore this short communication reports an evaluation on β-mannanase production under similar SSF systems and operating parameters. Relationships between β-mannanase secretion, fungal growth, and mannose production were discussed for both conical flask and LAMB systems.

Methodology

Methods for substrate preparation, cultivation and harvest of microorganism for SSF, optimization of SSF in flasks, design of LAMB bioreactor, and SSF in LAMB have been reported in details in previous work (Wong et al. 2011). Optimization of SSF parameters followed a one-factor-at-a-time strategy in the order of moisture content, pH, temperature, and particle size.

Sampling of Fermented PKC

For flask SSF, three flasks (as triplicate) were removed from an incubator every 24 h and subjected to immediate analysis. For LAMB SSF, 2 g of substrate was removed from each of 5 sampling points (height to diameter ratio: 0.9, 2.8, 4.7, 6.6 and 8.5) of the LAMB column every 24 h. Average values were used to represent enzyme production in LAMB, and SSF was done in duplicate.

Enzyme Extraction and β-Mannanase Assay

In the extraction of crude enzyme, 0.2 mM sodium phosphate buffer pH 7 (Fluka, USA) was added to fermented sample in substrate:buffer ratio of 1:10. The mixture was mixed with a vortex machine for 5 min, and centrifuged at 3,250 rpm for 10 min. Resulting supernatant was the crude enzyme used in β-mannanase assay.

For β-mannanase assay, a reaction mixture comprising 2.5 mL of 1 % locust bean gum (Sigma, USA), 0.5 mL of 50 mM sodium phosphate buffer (pH 7.0) and 0.5 mL of crude enzyme was prepared. Distilled water was used to substitute crude enzyme in preparation of blank reaction mixture. The mixture was incubated at 65 °C for 15 min in a water bath, and immersed in boiling water for 2 min to terminate the reaction in the mixture. Upon cooling to room temperature, dinitrosalicylic acid (Sigma, USA) method (Miller 1959) and a Novaspec II spectrophotometer (575 nm) were used in the estimation of reducing sugar (mannose equivalent) liberated in the reaction mixture. Mannose (Sigma, USA) was used as standard. One unit (U) of β-mannanase was defined as the amount of enzyme that released 1 μmol of mannose equivalent per minute under the assay condition.

Results and Discussion

β-Mannanase Production in Flask SSF

In the optimization study, β-mannanase production at every 24 h was monitored for every SSF parameter varied. The results showed that maximum enzyme yields was recorded at 120 h. Optimal moisture content was 110 %; increasing moisture level led to favourable PKC particles swelling and inter-particle space, improving oxygen transfer and fungal growth (Mitchell et al. 1992; Saw et al. 2008; Wong et al. 2011). Lower moisture content caused poor diffusion and accessibility of nutrients for fungal consumption (Pandey 2003; Krishna 2005) and higher moisture could flood inter-particle voids and reduce oxygen transfer (Perez-Guerra et al. 2003; Krishna 2005), both resulted in lower enzyme activity. Optimal pH was 6, where fungal growth was also highest (Wong et al. 2011). It was observed that enzyme production at pH 5 was relatively high, suggesting that β-mannanase was stable under acidic condition and enzyme denaturation could be minimal.

Optimal temperature was 30 °C and associated with highest fungal growth (Wong et al. 2011). A further increase to 40 °C led to lowest enzyme activity, most probably caused by decreasing enzyme stability as temperature increased. At 25 °C, it was observed that enzyme production increased linearly after 48 h (result not shown) and this was consistent with a rather linear increase in fungal growth (Wong et al. 2011). The observation showed that low temperature had delayed fungal growth and hence enzyme production. This also suggested that enzyme activity could possibly reach a maximum in longer fermentation period; no verification was done because experimental works were restricted to 144 h.

Optimal particle size was 855 μm; this size gave better particle swelling, increased inter-particle space, surface area to volume ratio and packing density, and improved gas transfer in SSF (Mitchell et al. 1992; Saw et al. 2008), resulting in highest fungal growth [9]. However, when SSF performances with 655 and 513 μm were compared, the smaller particle size yielded higher enzyme production in spite of lower fungal growth. This phenomenon was unexpected; preceding works have revealed that all three sizes were agglomerates with interstices suitable for microbial intertwinement (Chin 2007; Saw et al. 2011) besides improved swelling with increasing size (Saw et al. 2008). In fact, SSF performance improved in terms of not just fungal growth but also reducing sugar production with the bigger particles (Wong et al. 2011). Hence any plausible reasons were unknown.

The optimum condition for β-mannanase production was 110 % moisture content, initial pH 6, 30 °C and particle size 855 μm, consistent with the condition for optimized fungal growth and mannose production (Wong et al. 2011). Figure 1a shows relationships between enzyme activity and fungal growth, and Fig. 1b shows links between β-mannanase and mannose production. Fungal growth lag phase was from 0–12 h, exponential phase from 12–96 h, and stationary phase from 96–144 h (Wong et al. 2011). At 24 h, significant quantity of β-mannanase was produced because sugars in PKC was limiting; β-mannanase was generated to degrade PKC mannan into mannose to support increasing fungal growth. Between 24 and 72 h, enzyme yield remained constant at 192–227 U g^{-1} dry PKC while fungal biomass and reducing sugar continued to increase.

This suggested that the mannose produced from PKC degradation was sufficient to support fungal growth; extra enzyme production was not needed. From 72 to 96 h, due to continuous fungal biomass, enzyme secretion increased to produce more mannose for growth support. In the stationery phase, the rise and drop in enzyme activity corresponded well; enzyme and reducing sugar productions were inter-dependent.

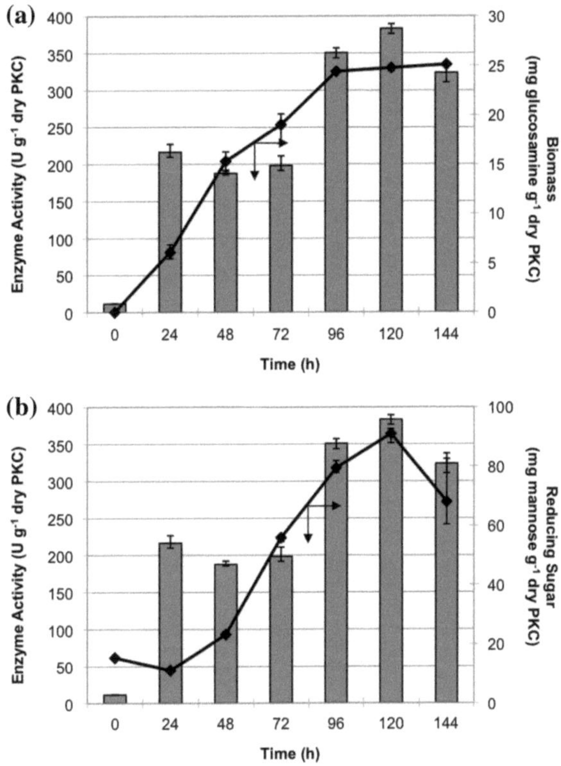

Fig. 1 Relationships between β-mannanase production and (**a**) fungal growth and (**b**) mannose production in flasks SSF

β-Mannanase Production in LAMB Bioreactor

Maximum enzyme yields at 96 h for three different superficial velocities in LAMB SSF are listed in Table 1. Particle size smaller than 5 mm was used because 855 μm particles contributed to only 11 % in the particle size distribution of PKC (Saw et al. 2011); incorporating the whole range of sizes would be more practical in large-scale SSF. Optimal enzyme activity was 276 U g^{-1} dry PKC at 4.4 m s^{-1} gas flow and 96 h. This value was an average calculated with samples from five sampling points. Wong (2008) has reported that differences in enzyme yield as affected by sampling locations were in the range of 3–10 %; hence the average value was representative of the enzyme production in LAMB. Enzyme yields at higher superficial velocities were lower as a predominant result from the drying up of PKC bed. With increasing velocity, bed moisture content dropped from 110 % to below 60 %, while bed moisture level was maintained at about 60 % at 4.4 m s^{-1} gas flow (Wong et al. 2011). Comparing enzyme activities in flask (383 U g^{-1} dry PKC) and LAMB, production in LAMB was lower by 28 % and this was very likely caused by particle size effect. In the unscreened PKC, smaller particles could have occupied void spaces between particles and decreased the void fraction (Mitchell et al. 1992), causing adverse effect on enzyme production.

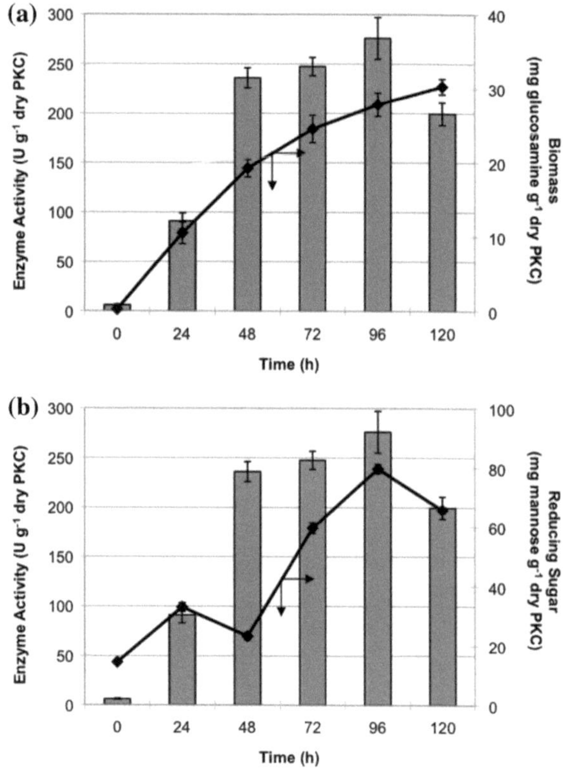

Fig. 2 Relationships between β-mannanase production and (**a**) fungal growth and (**b**) mannose production in LAMB SSF

This problem could be reduced if a more consistent particle size was used; this would require mechanical assistance on size reduction of larger particles and screening off smaller ones.

Relationships between enzyme production and fungal biomass, and connections between enzyme production and mannose yield in LAMB SSF are shown in Fig. 2a, b respectively. As fungal biomass continuously increased, reducing sugars in PKC were consumed to support growth and became limiting at 48 h, therefore β-mannanase secretion increased significantly from 24 to 48 h to convert mannans in PKC to mannose. With increased production of enzyme over the period of 48–96 h, mannose production improved, peaking at 96 h. Subsequently, both enzyme and mannose levels dropped as fungal growth entered into stationary phase. β-Mannanase, mannose, and fungal biomass productions were consistent with one another.

Comparison on β-Mannanase Production with Other Works on SSF of PKC

Comparison on β-mannanase production between this study and other works on SSF of PKC; published works are limited and focused on *Aspergillus niger* in flasks SSF (Ong et al. 2004) and SSF in a packed bed with axial flow of humidified air (Abdeshahian et al. 2010). *A. niger* produced considerably higher enzyme at lower moisture level. It also seemed to grow well on particles up to 2 mm; this could be attributed to relatively higher swelling for particles in the range of 1.5–2.7 mm (Saw et al. 2008). As for enzyme production, higher yield could also be a result of longer period for crude enzyme extraction (24 h). The duration in this study was only 5 min. Both strains performed well in acidic condition (pH 5–6) and ambient temperature (30–33 °C), and were robust towards forced aeration regardless of flow direction. An advantage with *A. flavus* was that it required shorter fermentation time for optimal production, especially in the LAMB bioreactor. However, potential mycotoxin production with this strain has not yet been addressed.

Conclusions

In SSF of PKC with *A. flavus* UMS01 to produce PKC with improved mannose content for poultry feed production, β-mannanase secretion was directly related to fungal growth and mannose production; optimal enzyme yield was achieved with highest fungal biomass and optimal mannose production under similar optimized parameters. This study also showed that *A. flavus* UMS01 was as a good β-mannanase producer with PKC as the SSF substrate, and the LAMB was a suitable bioreactor to host industrial-scaled β-mannanase productions.

Acknowledgments Financial support came from Universiti Malaysia Sabah Fundamental Research Grant B-0101-13/PRU037 and PKC was provided by IOI Edible Oils Sdn. Bhd.. All experimental works were performed in Universiti Malaysia Sabah.

References

Abdeshahian, P., Samat, N., Hamid, A. A., & Yusoff, W. M. W. (2010). Utilization of palm kernel cake for production of β-mannanase by *Aspergillus niger* FTCC 5003 in solid substrate fermentation using an aerated column bioreactor. *Journal of Industrial Microbiology and Biotechnolology, 37*, 103–109.

Chin, H. C. (2007). Solid-state fermentation of palm kernel cake by *Bacillus subtilis* ATCC3366. *M.Sc. thesis*. Universiti Malaysia Sabah, Sabah, Malaysia.

Daud, M. J., & Jarvis, M. C. (1992). Mannans of oil palm kernels. *Phytochemistry, 31*(2), 463–464.

Dhawan, S., & Kaur, J. (2007). Microbial mannanases: An overview of production and applications. *Critical Reviews in Biotechnology, 27*, 197–216.

Krishna, C. (2005). Solid-state fermentation systems—an overview. *Critical Reviews in Biotechnology, 25*, 1–30.

Miller, G. L. (1959). Use of dinitrosalicylic acid reagent for determination of reducing sugar. *Analytical Chemistry, 31*, 426–428.

Mitchell, D. A., Targonski, Z., Rogalski, J., & Leonowicz, A. (1992). Substrate for processes. In H. W. Doelle, D. A. Mitchell, & C. E. Rolz (Eds.), *Solid substrate cultivation* (pp. 29–52). England: Elsevier Science.

Ong, L. G. A., Abd-Aziz, S., Noraini, S., Karim, M. I. A., & Hassan, M. A. (2004). Enzyme production and profile by *Aspergillus niger* during solid substrate fermentation using palm kernel cake as substrate. *Applied Biochemistry and Biotechnology, 118*, 73–79.

Pandey, A. (2003). Solid-state fermentation. *Biochemical Engineering Journal, 13*, 81–84.

Perez-Guerra, N., Torrado-Agrasar, A., Lopez-Marcias, C., & Pastrana, L. (2003). Main characteristics and applications of solid substrate fermentation. *Electronic Journal of Environmental, Agricultural and Food Chemistry, 2*(3), 343–350.

Saw, H. Y., Janaun, J., & Subbarao, D. (2008). Hydration properties of palm kernel cake. *Journal of Food Engineering, 89*(2), 227–231.

Saw, H. Y., Janaun, J., Kumaresan, S., & Chu, C. M. (2011). Characterization of the physical properties of palm kernel cake. *International Journal of Food Properties*. doi:10.1080/10942912.2010.492543.

Tang, T. S. (2001). Quality and characteristics of Malaysian palm kernel cakes/expellers. *Palm Oil Developments, 34*, 1–3.

Wong, Y. P. (2008). Solid-state fermentation of palm kernel cake by *Aspergillus flavus* UMS01 in radial flow packed bed bioreactor. *M.Sc. thesis*. Universiti Malaysia Sabah, Sabah, Malaysia.

Wong, Y. P., Saw, H. Y., Janaun, J., Krishnaiah, K., & Prabhakar, A. (2011). Solid-state fermentation of palm kernel cake with *aspergillus flavus* in laterally aerated moving bed bioreactor. *Applied Biochemistry and Biotechnology, 164*(2), 170–182.

Author Biography

Prof. Dr. Ravindra Pogaku is a Professor of Chemical and Bioprocess Engineering at University Malaysia Sabah. He is now the Director for Oil and Gas Engineering Studies and Training Unit at the School of Engineering and Information Technology, University Malaysia Sabah (UMS). Pogaku Ravindra's work has appeared in more than 200 publications and 4 patents in ISI/SCI/Scopus journals. He has published eight books, six book chapters and edited two scientific books. Prof. Pogaku is also on the Editorial Board of several international journals and has reviewed over 500 journal manuscripts for reputed international journals. He has delivered various keynotes, plenary lectures, and chaired the prestigious conferences, seminars, workshops, technical events both nationally and internationally. He has been honored with senior membership of American Institute of Chemical Engineers for his credentials in the field of chemical and bioprocess engineering. He was a Visiting Scientist at Cornell University. He was UNESCO consultant. Prof. Ravindra is bestowed with the distinguished Chemical Engineer Award from the Indian Institute of Chemical Engineers, India. He has been recognized with Excellence Award consecutively for the last three years from the University Malaysia Sabah. Prof. Pogaku Ravindra is awarded with gold and silver medals for his research contribution for bioprocess industry at Geneva and Malaysia.

Prof. Dr. Ravindra Pogaku
Professor of Chemical and Bioprocess Engineering
UNESCO energy consultant; Editor in Chief BMBR
Director Oil and Gas Engineering Studies and Training Unit
Phone: 088-320533/320000 (ext 3661)
Fax: 088-320539/320348
Email: ravindra@ums.edu.my or dr_ravindra@hotmail.com

R. Pogaku et al. (eds.), *Developments in Sustainable Chemical and Bioprocess Technology*, DOI: 10.1007/978-1-4614-6208-8,

Prof. Dr. Awang Bono

Awang Bono (BSc., PhD.) is a Separation Science and Technology Professor at Universiti Malaysia Sabah (UMS). Prof. Bono completed his PhD work from Chemical Engineering Department, University of Surrey, United Kingdom in 1986 and has been involving in teaching, research as well as industry for more than 20 years. To date, he has over 150 scientific publications. Currently, Prof. Bono is serving as a Deputy Dean at School of Engineering and Information Technology, UMS.

Associate Professor Dr. Chu Chi Ming

Christopher Chi-Ming Chu is the Head of Chemical Engineering Programme at the School of Engineering, Universiti Malaysia Sabah. He received his B.Sc. and Ph.D. from the University of Birmingham, United Kingdom. His main research interest is in thermal systems and currently works on chimney performance enhancement using wire mesh and passive cooling systems. He is the Head of Thermal and Environment Group in the Material and Mineral Research Unit. He has published and reviewed international journal papers and conferences.

Index

R. Pogaku et al. (eds.), *Developments in Sustainable Chemical and Bioprocess Technology*, DOI: 10.1007/978-1-4614-6208-8,

B

D

J

K

T

MIX
Papier aus verantwortungsvollen Quellen
Paper from responsible sources
FSC® C105338

If you have any concerns about our products,
you can contact us on
ProductSafety@springernature.com

In case Publisher is established outside the EU,
the EU authorized representative is:
Springer Nature Customer Service Center GmbH
Europaplatz 3, 69115 Heidelberg, Germany

Printed by Libri Plureos GmbH
in Hamburg, Germany